Social Predation

Social Predation: How Group Living Benefits Predators and Prey

Guy Beauchamp

AMSTERDAM • BOSTON • HEIDELBERG • LONDON
NEW YORK • OXFORD • PARIS • SAN DIEGO
SAN FRANCISCO • SINGAPORE • SYDNEY • TOKYO

Academic Press is an imprint of Elsevier

Academic Press is an imprint of Elsevier
32 Jamestown Road, London NW1 7BY, UK
225 Wyman Street, Waltham, MA 02451, USA
525 B Street, Suite 1800, San Diego, CA 92101-4495, USA

First edition 2014

British Library Cataloguing-in-Publication Data
A catalogue record for this book is available from the British Library

Library of Congress Cataloging-in-Publication Data
A catalog record for this book is available from the Library of Congress

ISBN: 978-0-12-407228-2

For information on all Academic Press publications
visit our website at elsevierdirect.com

Typeset by TNQ Books and Journals
www.tnq.co.in

Printed and bound by CPI Group (UK) Ltd, Croydon, CR0 4YY
14 15 16 17 18   10 9 8 7 6 5 4 3 2 1

Contents

Part B
Prey 65

Preface

In the Darwinian struggle to survive, predation represents one of the most dramatic cases in point. An inattentive zebra may pay with its life while a nonchalant lion risks losing a meal. Most of us in the Western world are very unlikely to face predation in our lifetime, except for the few unfortunate surfers or campers attacked by sharks or grizzly bears. Nevertheless, as judged from the reaction to these few cases each year in the popular media, predation still holds a visceral appeal that probably harks back to our own evolutionary past as both predator and prey. To this day, we are fascinated by the ploys and counter-ploys of predators and their prey. Striking camouflage by prey and sophisticated hunting tactics by predators in groups are very familiar examples of predator–prey relationships.

Predation is a cornerstone of ecological research and has emerged as a key factor in population regulation. Students of animal behaviour have also been interested in predation focusing not only on the tactics used by predators to catch prey but also on the non-lethal effects of predation for prey, such as the allocation of time to vigilance and habitat choice.

Research on predator–prey relationships was originally developed with solitary species in mind. This is understandable given the complexities involved in predicting the behaviour of even a single animal. For instance, the classic model of Lotka and Volterra predicted population changes over time for a solitary predator and its main solitary prey. One of the better known examples of predator–prey relationships involving solitary species comes from my country, Canada. The Hudson's Bay Company historically tallied the number of pelts captured by trappers over time in Canada. Tallies made it possible to see cyclical changes in population size for lynx and snowshoe hare, the main prey of lynx, as predicted by simple models that apply to such solitary species. As a further example, optimal foraging theory, which is concerned with the adaptive value of foraging tactics, was very influential in behavioural ecology when it emerged in the late 1960s. Originally, optimal models of prey choice or habitat selection were all concerned with solitary foragers. Social foraging theory emerged later from the need to apply similar concepts to species foraging in groups.

It has become increasingly clear that what happens when predators and prey forage in groups cannot be easily deduced from what we know about solitary foragers. The following examples illustrate some of the unique problems faced by social animals. Hunting in groups may allow predators to catch larger prey, and thus substantially alter their ecological niche, but at the cost of having to share each meal. These costs and benefits may not simply increase linearly with group size but vary in a complex fashion. For prey in groups, consider the simple fact that if a predator can capture only one member of a group during an attack, the impetus for a prey animal may be to outrun its companions rather than the predator. Such new perspectives required new modelling approaches, because the best course of action for a predator or prey may also depend on the behaviour of group members. The advent of game theory in animal behaviour research in the early 1970s proved a catalyst for theoretical developments on the adaptive value of group living for both predators and prey. Advantages and disadvantages of group living had been documented for decades, but the framework provided by these new models allowed a resurgence of interest in group living that persists to this day.

I propose the term 'social predation' to capture the complexities of finding prey and avoiding predation in groups. The purpose of this book is to explore the ways group living can benefit predators and prey as well as the potential disadvantages that may accrue. Books can fall anywhere on a continuum from philosophical to encyclopaedic. Rather than documenting all costs and benefits related to group living, I aim to provide a firm theoretical basis for each theme and then explore relevant assumptions and predictions. Technical details related to particular models can be examined, but assumptions and predictions are essential to empirical testing. I include empirical findings from the widest possible range of species.

The first part of the book focuses on predators, here defined as those individuals that consume other species, or at least some of their parts, for feeding. Prey may include live animals, such as a zebra for a lion, recently deceased animals like a carcass for a vulture, or parts of a species like seeds for an herbivore. I exclude species that specialize in decomposing matter and those that seek hosts to lay their eggs, a searching behaviour that is performed solitarily. Topics covered here include how group living influences food finding and how the presence of companions affects the amount of resources obtained by each group member.

The second part of the book deals with prey and describes ways that group living may reduce predation risk through factors such as vigilance, risk dilution, and confusion. Vigilance, in particular, has been studied extensively over the last 40 years, and I will explore in detail the relationship between vigilance and group size.

The third part of the book is concerned with issues of general concern to predators and their prey. In light of the costs and benefits associated with group living, predators and their prey may seek to forage in groups that maximize fitness. I will show that the expected group size depends on who controls entry in the group. In addition, animals may also pay attention to the composition of their groups. I will explore these issues in single- as well as mixed-species groups.

It has long been recognized that predators and prey may be locked in an arms race with adaptation by one resulting in selection pressure to counter-adapt by the other. Many of the models discussed in the two preceding sections have simplified this issue by assuming a fixed strategy for the predator or for the prey. More complex models allow for co-evolution between predators and prey, and their insights are presented in this section.

Species vary extensively in the expression of sociality. Comparative analyses, using information about the evolutionary relationships between species, can shed light on factors that have promoted the evolution of social predation. I will review these analyses in a wide range of species. This part of the book along with the previous parts forms a whole that explores social predation from different angles, but with the same view of increasing our understanding of this fascinating topic.

It is a pleasure to thank my collaborators over the last 15 years who have kept me in touch with the academic world: Peter Alexander, Peter Bednekoff, Marc Bekoff, Dan Blumstein, Esteban Fernández-Juricic, Luc-Alain Giraldeau, Eben Goodale, Philipp Heeb, Andrew Jackson, Roger Jovani, Chunlin Li, Zhongqiu Li, Raymond McNeil, Olivier Pays, Graeme Ruxton, Étienne Sirot, and Hari Sridhar. For their useful comments on some chapters, I thank Esteban Fernández-Juricic, Eben Goodale, and Graeme Ruxton. The weaknesses that remain are entirely my own. I am grateful for the wonderful front cover illustration done by Gabriela Sincich. The staff at Academic Press, Kristi Gomez and Pat Gonzalez, have been most helpful in producing this book.

Eunice and Heather Cail have provided a home away from home during my field trips to study semipalmated sandpipers in New Brunswick, for which I am most thankful. Many naturalists flock each year to watch sandpipers in the Bay of Fundy. It is always a pleasure to swap tips and stories with them: David Christie, Dick and Irma Dekker, Diana Hamilton, Peter Hicklin, and Colin McKinnon.

The most common word in this book, not surprisingly for a treatise on living in groups, is 'companion.' It is thus quite fitting to acknowledge my companion in life, Susan Lemprière, whose way with words and understanding of biology vastly improved this book. She allowed me to take time away each year for field work and to write this book, for which I am most grateful.

Guy Beauchamp

Predators

Finding and Exploiting Food in Groups

1.1. INTRODUCTION

Parasitoids lay their eggs inside a host species and their growing larvae feed off the body of this host until they are ready to emerge. This peculiar type of development occurs frequently in insects, particularly in wasps (Hawkins, 1994). By laying eggs directly in the food larder, parasitoid mothers have solved the problem of finding food for their developing larvae. For most species, however, the search for resources consumes considerable time and energy. For example, the wandering albatross, a large seabird of the southern oceans, may cover up to 15,000 km in a single foraging trip before returning to the nest (Jouventin and Weimerskirch, 1990). One solution to the problem of food procurements has been the evolution of group foraging: the pooling of individual efforts to find and exploit resources. The multiple, independent evolution of group foraging in many species of animals, which I cover in Chapter 9, implies that in the evolutionary past group foraging provided enough benefits to offset the obvious cost of sharing resources with other group members.

Social Predation. http://dx.doi.org/10.1016/B978-0-12-407228-2.00001-9

Group foraging takes different forms across the animal world: from loose associations between as few as two individuals, to millions in the swarms of marine invertebrates (Ritz, 1994). In addition to this tremendous variation in the number of individuals involved, group foraging also encompasses a wide range of interaction between group members. In the simplest cases, a few individuals may search rather independently and only gather to share the large prey or food patches discovered by any group member. Cooperative hunting, at the other extreme, represents the most spectacular expression of group foraging, involving elaborate tactics and often specialized roles to gather resources. For example, pods of orcas flush seals from ice floes by using cooperative wave-washing behaviour (Pitman and Durban, 2012). Groups of Harris's hawks, a raptor species from southern North America, attack rabbits by swooping down from different directions (Bednarz, 1988). Some of the hawks flush the prey, while others wait in ambush to capture the fleeing animals. In other species, individuals may even specialize by adopting the same role over many attacks, as witnessed in groups of female lions (Stander, 1992). Similarly, in bottle-nose dolphins foraging off the coast of Florida, some individuals specialize in herding fish prey, while others keep the prey from escaping by acting as a barrier (Gazda et al., 2005). Perhaps less spectacular, but still illustrating the various ways in which group foraging can benefit individuals, is the finding that aggregations of insect larvae on host plants generate more food *per capita* by overcoming plant defences (Fordyce and Agrawal, 2001). In all these examples, group foraging increases the ability to find or capture prey, which is a defining feature of group foraging.

Although it makes intuitive sense for individuals to gather in groups to detect predators more easily or to defend themselves (see Chapters 3 and 4), it is not immediately obvious why individuals would gather in groups for the purposes of foraging. Indeed, each predator could simply stake out a food territory and exclude conspecifics. However, many types of resources exploited by foragers are unevenly distributed in both space and time. This patchiness implies that in a territorial system, the mean amount of resources may be insufficient for many individuals. Even when territoriality breaks down, individuals could still forgo foraging in groups and simply compete with one another for finding and exploiting resources. Group foraging, therefore, would evolve when finding and exploiting resources in groups provide more net benefits than foraging alone or defending resources.

Resource distribution in both space and time has long been known to influence grouping patterns in animals (Crook, 1965; Crook and Gartlan, 1966; Jarman, 1974; MacDonald, 1983). For instance, in birds, solitary foraging occurs mainly in species that forage on insect prey that are too small to share or escape when disturbed by the presence of companions. Gregarious foraging, on the other hand, occurs with food types unpredictably distributed in both space and time, such as seeds and fruits. Comparing different species exploiting different types of resources has been a powerful method to determine the ecological factors that facilitated the evolution of group foraging. I shall return to this approach in Chapter 9. An alternative approach, which I follow here, consists in comparing the success of solitary and group-living members of the same species living in the same environment. This approach identifies the current costs and benefits of group foraging in a species, and thus suggests what selection pressures may have favoured group foraging in the evolutionary past.

What actually constitutes group foraging still remains controversial. An aggregation of foragers in both space and time certainly represents the minimum criteria for defining group foraging, but exactly where a group starts and ends remains difficult to define, and may vary between species due to differences in sensory abilities (Fernández-Juricic and Kowalski, 2011). However, co-occurrence in time and space may not be sufficient to define group foraging, because an aggregation may form without providing any benefits to foragers. Indeed, animals may be found together at the same location because of independent attraction to the same resources (Wilson and Richards, 2000). In addition, resource patchiness may force foragers to remain together because staying in a rich patch with others may represent the best option when alternatives are scarce (Fretwell and Lucas, 1970). Chance aggregation or limited foraging opportunities can thus lead to group foraging, but without any forces to keep foragers together. Most definitions of group foraging imply the action of forces that keep foragers together over some time period (Pitcher and Parrish, 1993; Wilson, 1975).

The purpose of this chapter is to examine the various ways in which predators benefit from foraging in groups. These benefits involve finding and exploiting resources and constitute the forces that keep foragers together. But although group foraging provides opportunities to increase foraging efficiency, it also involves unique costs that may influence the evolution of group foraging in animals. I will also review these costs in this chapter. My discussion will be restricted to foraging groups; the evolution of sociality for reproductive purposes has been covered elsewhere (Bourke and Franks, 1995; Frank, 1998).

1.2. BENEFITS OF GROUP FORAGING

1.1.2.1. Detecting Resources

A solitary forager must find resources alone. By contrast, individuals in groups can rely on one another to locate resources. If finding resources is equated with getting a specific number after rolling a die, it is easy to see that the odds of getting this particular number are much higher when many individuals, as opposed to just one, roll their die at the same time. Not only will the average time between successes be reduced, but runs of bad luck also become less likely, thus reducing variance in success as well.

It has long been suspected that group foraging increases the efficiency with which resources are detected. Tristram, back in 1859, noted that griffon vultures search for carcasses over a very wide area but maintain contact with one another while foraging. When one individual locates a carcass, others are soon alerted and congregate more rapidly than would be expected on the basis of individual detection (Tristram, 1859). Similarly, other early researchers noted that although one individual may search for food fruitlessly over a long period of time, prey are less likely to escape detection when many individuals are searching at the same time and alert one another after a food discovery (Miller, 1922; Pycraft, 1910).

Experimental investigation provides evidence that foraging in groups can reduce the time needed to locate resources. For example, goldfish can locate hidden clumps of food more rapidly as group size increases (Pitcher et al., 1982) (Fig. 1.1). In social insect larvae aggregating on plants, a similar mechanism is at work, but here feeding in groups increases the chances that at least one individual will manage to overcome the plant defences, attracting other companions to the exposed resources (Ghent, 1960). Notice that to benefit from group foraging, predators must be able to determine when companions have detected or obtained resources so that they can join their discoveries. Pitcher et al. (1982) suggested that specific postures associated with feeding act as clues of food discovery in goldfish. Without the ability to detect food discoveries made by others, individuals would not be expected to find food faster in groups. The use of cues provided by companions to locate food has been referred to as local enhancement (Thorpe, 1956). Local enhancement has been documented in a very broad range of taxa and appears to be a universal mechanism (Galef and Giraldeau, 2001).

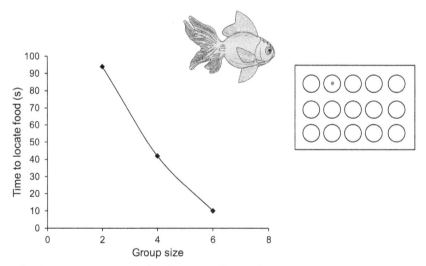

FIGURE 1.1   Food detection and predator group size: goldfish locate hidden clumps of food more rapidly when foraging in larger groups. Food was placed in one of many cups located at the bottom of an aquarium (inset). *Adapted from Pitcher et al. (1982).* (For colour version of this figure, the reader is referred to the online version of this book.)

Predators can also find resources more quickly if individuals increase their rate of searching when in groups. Searching at a faster rate may take place because competition intensity increases in larger groups (Shaw et al., 1995) or because individuals in groups allocate more time to foraging (Caraco, 1979a) (see Chapter 4). Increased foraging activity in groups is sometimes called social facilitation (Zajonc, 1965). Social facilitation and local enhancement constitute the two broad categories of factors that enhance detection of resources in groups.

Local enhancement works by establishing a network of foraging individuals, each investing time to detect resources. Transfer of information takes place on the foraging grounds as individuals in the group search for sharable resources while maintaining contact with one another. It has been proposed that exchange of information about food patch location can also occur away from the foraging grounds at central gathering locations that all foragers visit regularly, such as breeding colonies or communal resting areas. The information-centre hypothesis (ICH) proposes that such gatherings allow unsuccessful foragers the opportunity to follow knowledgeable companions to distant resources (Ward and Zahavi, 1973). An even more elaborate transfer of information occurs in social insects, such as bees, where the bearing and distance to distant resources can be communicated to group members at the colony (Seeley, 1985; von Frisch, 1967). In other insects (Hölldobler and Wilson, 1990), and surprisingly, even in one mammal, the naked mole-rat (Judd and Sherman, 1996), individuals that discover resources leave trails behind that are followed by naïve companions to locate new food sources. In birds and mammals, where the ICH has been applied most often, transfer of information is thought to operate by following knowledgeable companions from a central gathering area to a distant food source.

The ICH has met with many challenges over the years and remains on the margins of active animal behaviour research (Box 1.1). Although the scope of information transfer at central locations may not be as broad as first thought, the hypothesis does provide a mechanism for locating resources away from the foraging grounds if naïve individuals are able to identify knowledgeable companions. More recent theories emphasize information transfer among group members when individuals are on the move, rather than at a central location, without the need for informed or uninformed individuals to recognize each other (Couzin et al., 2005).

1.2.1. Acquiring Resources

In the previous section, I showed that groups can detect resources more quickly than solitary individuals. Once resources are detected, predators in groups can also acquire resources more efficiently by subduing prey more easily or by making prey more readily available through flushing or herding.

1.2.1.1. Subduing Prey

The presence of companions in a group multiplies the weaponry and force available to capture prey. Group foraging can thus greatly increase the range

BOX 1.1 The Information-Centre Hypothesis

The ability to exchange information about distant resources represents a major benefit of gathering at central locations, such as breeding colonies or communal roosts, according to the ICH. In particular, non-knowledgeable foragers returning to a central gathering location can benefit by following more successful companions to distant, sharable patches of food, thereby increasing their foraging efficiency. Earlier adaptive hypotheses for the evolution of central gatherings had focused mostly on antipredator benefits (Eiserer, 1984; Wittenberger and Hunt, 1985). The ICH proposed, instead, a major role for increased foraging efficiency. While the idea that information transfer about distant foods can take place at a central gathering was already firmly established in social insects, the ICH met with considerable resistance when applied to vertebrate species, such as birds and mammals, and remains controversial to this day (Barta and Giraldeau, 2001; Bijleveld et al., 2010; Danchin and Richner, 2001; Richner and Heeb, 1995).

Mock et al. (1988) formalized the necessary conditions under which the ICH can operate. Transfer of information is only necessary if food patches are difficult to locate and only possible if patches are large enough to accommodate several individuals. Patches must also last long enough at the same location to allow individuals to return after a visit to the central gathering area. Knowledgeable individuals must provide direct or indirect cues of success, which would allow non-knowledgeable individuals to recognize and follow them to the distant patch (Mock et al., 1988).

One obvious difficulty for the ICH is why knowledgeable individuals should return to a central location. Indeed, knowledgeable individuals that return to a central gathering area attract competitors to the patches they discover, and thus suffer a potential cost. This is not an issue in a breeding colony because all individuals, knowledgeable or not, must return to take care of their young. Why knowledgeable individuals should return to a central location is more problematic for aggregations that may be joined on an opportunistic basis, such as communal roosts. Such central locations should provide additional benefits to knowledgeable individuals, such as a reduction in predation risk, to compensate for the cost of sharing resources with others (Weatherhead, 1983). The cost of sharing resources may be negligible if patches are large and if individuals actually benefit from the presence of companions when exploiting a patch (Richner and Heeb, 1997). In the original formulation of the ICH, knowledgeable individuals return to the central gathering area in exchange for the opportunity to follow a more successful companion in the future, a reciprocal altruism arrangement.

One empirical difficulty with the ICH has always been to rule out alternative mechanisms for the build-up of foragers at a patch following a return to the central location. Local enhancement on the foraging grounds can lead to an increase in the number of foragers at a patch without the need to invoke information transfer at the central gathering area. Obviously, one must also rule out the possibility that foragers at a distant patch can be detected from the central location (Andersson et al., 1981).

BOX 1.1 The Information-Centre Hypothesis—cont'd

Two recent studies examined the movements of individually marked foragers in response to differential foraging success, providing the most compelling support for the ICH. In one study, ignorant hooded crows were more likely to visit a new patch when a previously successful companion visited the same patch again, implying that discovery of a new patch can be facilitated by the presence of more knowledgeable companions at the roost (Sonerud et al., 2001). Such following by non-knowledgeable companions has also been noted in another study involving ravens searching for carcasses (Marzluff et al., 1996). However, in both cases, the reciprocal altruism argument that forms the basis of the ICH was not assessed, and it is possible that recruitment occurred at the roost to increase foraging benefits at the food patch. We are still awaiting the last word on the ICH.

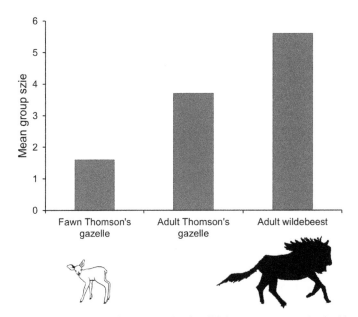

FIGURE 1.2   Prey size and predator group size in wild dogs: mean group size in this species increases when hunting larger prey. *Adapted from Fanshawe and FitzGibbon (1993).*

of prey available to capture. Consider the case of female lions hunting in the African savannah. Although a lone female lion can easily capture a warthog, groups are more successful at hunting larger quarry like zebra or buffalo (Scheel and Packer, 1991). Similarly, wild dogs can bring down larger prey when hunting in bigger groups (Creel and Creel, 1995; Fanshawe and FitzGibbon, 1993) (Fig. 1.2). This effect has also been reported for other mammalian species (Caro, 1994; Lührs et al., 2013; Murie, 1944). Although images of mammals hunting in the savannah come to mind when thinking about the ability to subdue larger prey, the same idea applies to other types of predator-prey systems. For instance, a small number of bark beetles may inflict little damage on a large pine tree, but large numbers of these beetles can overcome a healthy tree (Berryman et al., 1985). Similarly, larger nets in social spiders can capture larger prey (Yip et al., 2008); larger mucous traps produced by group-living triclads, an aquatic invertebrate predator, capture a larger range of prey (Cash et al., 1993); and a large number of caterpillars can more easily overcome plant defences (Fordyce and Agrawal, 2001).

Subduing larger prey often carries a risk for the predators. Examples of prey-inflicted injuries or even deaths are plentiful in the literature (Mukherjee and Heithaus, 2013). Therefore, hunting in groups may not only allow individuals to capture larger prey but also to reduce the *per capita* risk of injury when tackling larger, dangerous prey. Whether it makes economic sense, in general, to pay the costs of capturing larger prey and sharing returns with several companions will be considered more fully in Chapter 7.

1.2.1.2. Flushing Prey

The presence of several predators in the same group has long been thought to temporarily increase food availability. Prey flushing was one of the earliest mechanisms proposed to enhance food availability in groups (Belt, 1874; Dewar, 1905; Hingston, 1920; Swynnerton, 1915). In roving mixed-species flocks of birds, for instance, individuals at the forefront disturb insects while searching for food, which are then caught by those behind. This so-called beater effect has also been documented in mammals (Struhsaker, 1981), fish (Arnegard and Carlson, 2005), and spiders (Uetz, 1989). By reducing the probability that any prey escapes, prey flushing thus increases the pool of prey available to all group members, and will decrease the variance in food intake rate at the individual level by spreading resources across the whole group.

A group of predators often adopt a formation whose aim appears to increase the amount of ground covered while foraging. For example, foraging grey herons align with each other when walking across a field (Ritchie, 1932). Similar formations have been documented in other species foraging on mobile prey (Källander, 2008). While such formations are compatible with the prey flushing hypothesis, foraging formations are difficult to induce experimentally, and their adaptive value remains speculative.

Direct evidence for benefits related to prey flushing is relatively scant. In a laboratory experiment with gulls exploiting prey fish, foraging success increased with the number of gulls in the group as fish escaping from one attacker were more likely to be captured by another (Götmark et al., 1986) (Fig. 1.3). Similarly, in colonial spiders, insect prey bounce from one net to another before being captured, increasing individual food intake rate and decreasing variance in foraging success by spreading prey across the colony (Rypstra, 1989). Such studies thus provide evidence that prey flushing may be a benefit of group foraging.

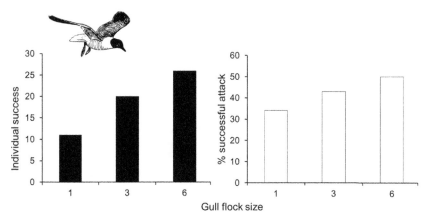

FIGURE 1.3 Prey evasion and predator group size: flocks of black-headed gulls obtain more food and attack prey more successfully when in larger groups. Fewer fish prey could escape when attacked by several gulls. *Adapted from Götmark et al. (1986).*

1.2.1.3. Herding Prey

Predators in groups can corral prey into smaller areas, thus increasing the density of prey and reducing the cost of exploitation. Prey herding resembles prey flushing by reducing the ability of prey to escape, but generally herding involves active restriction of prey movement. The most complex forms of prey herding involve many individuals encircling prey, but prey herding may also be invoked for groups as small as two when one individual drives the prey towards the other, as seen in many avian raptor species (Ellis et al., 1993). Prey herding has been documented in a large range of species, and this strategy may represent an integral part of foraging on dispersed resources in birds (Anderson, 1991; Ryan et al., 2012; van Eerden and Voslamber, 1995), fish (Schmitt and Strand, 1982), amphibians (Bazazi et al., 2012), and mammals (Benoit-Bird and Au, 2009; Similä and Ugarte, 1993; Vaughn et al., 2010).

Spinner dolphins foraging on small, dispersed prey provide a remarkable example of prey herding (Benoit-Bird and Au, 2009). Using sonar equipment, the authors identified several stages in the herding of prey (Fig. 1.4). At first, dolphins swim in a line perpendicular to the shore. Spacing then decreases as the dolphins swim towards shore and push the prey in front of them. The dolphins then form a circle about 28 m to 40 m in diameter, closing the circle from offshore. While the prey fish are encircled, pairs of dolphins from opposite sides of the group dart inside to feed. The dolphins maintain their formation so that fish are confined to a tight vertical cylinder during the duration of the feeding bout. Prey density through herding can increase 200-fold from the initial to the final stage of group formation.

Prey herding represents a collective effort that increases the density of prey through an active mechanism. Whether prey herding can be considered cooperative foraging remains a contentious issue (Bailey et al., 2012). To invoke

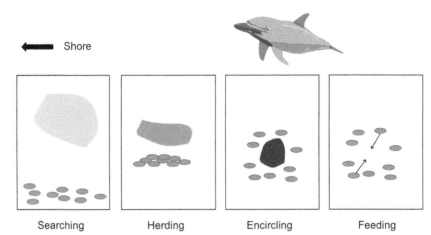

| Searching | Herding | Encircling | Feeding |

FIGURE 1.4   Cooperative foraging in spinner dolphins: the different stages of fish herding by groups of spinner dolphins illustrate the cooperative nature of foraging in this species. In the first stage, searching dolphins swim towards shore in a loose group and congregate to push prey in front of them (herding). Prey fish are then encircled, allowing pairs of dolphins from opposite sides of the group to take turns feeding inside. Dolphins are shown as green dots and prey fish as the grey shaded area (a darker area indicates a higher density). *Adapted from Benoit-Bird and Au (2009).* (For interpretation of the references to colour in this figure legend, the reader is referred to the online version of this book.)

cooperative foraging, the individuals involved in the collective effort must adopt some role and also restrain their tendencies to feed on their own and disrupt the process of prey herding (Wilson, 1975). The key aspect of cooperative foraging must be a potential cost to the individual that allows all individuals in the group to obtain more resources subsequently through the collective effort. For an individual dolphin, for example, the time lost while herding prey and the sporadic access to food while prey are encircled is largely compensated by the increase in prey density, which leads to a large increase in feeding success rate.

Finding that prey herding increases foraging success does not necessarily imply cooperative foraging. Increased success may occur through passive collective effort. For instance, two bears moving from opposite directions along a small creek capture salmon at a higher rate than when they forage alone because each bear forces the prey into a smaller area from which it is more difficult to escape (Stringham, 2012). Despite the increase in feeding success, bears rarely adopt this type of foraging, and typically flee from one another instead of coming closer. To consider prey herding as a foraging tactic, care must be taken to rule out the possibility of such accidental benefits.

Because collective foraging in the field cannot be induced, the adaptive value of prey herding must be based on a comparison of the costs and benefits experienced by solitary foragers and those in group formation, which provides at least an indirect way to assess the function of prey herding. Although the benefits of corralling prey are obvious, the costs have not always been carefully examined. In addition to the time needed to herd prey, which could be spent instead searching for prey, prey herding also brings many foragers together in a small space, potentially increasing foraging interference and the risk of collision. A high density of prey may also create confusion in predators (see Chapter 3), perhaps reducing attack success rate. That herding may not always be profitable has been suggested in pelicans encircling prey fish (Saino et al., 1995). In tight formations, the rate of head dipping underwater to capture prey decreased, paradoxically, as group size increased, presumably because individual birds needed to wait longer to synchronize their activities in larger flocks so as to avoid hitting one another. Documenting such costs will help us better understand under which circumstances prey herding can evolve.

1.2.2. Exploiting Resources

Time allocated to foraging represents a major determinant of foraging efficiency. Individuals that allocate more time to foraging can expect to encounter and acquire more resources, and thus increase their fitness. This is true whether animals forage alone or in groups. Animals must allocate a finite amount of time to a number of fitness-enhancing activities, thus creating a trade-off between the time spent foraging and, for example, resting and detecting predators (Caraco, 1979b; Marshall et al., 2012). By contrast to solitary foraging, group foraging allows individuals to reduce their investment in antipredator defences through many mechanisms that will be reviewed in the second part of this book, such as risk dilution and collective detection. In particular, time allocated to antipredator vigilance can decrease considerably with group size (Elgar, 1989), which means that in large groups individuals can, in principle, allocate more time to foraging and thus increase their foraging success. Greater protection in larger groups could then translate directly into greater foraging success even if predators in groups do not locate or acquire resources more efficiently. This prediction is most relevant to predators that face predation threats themselves.

However, this hypothesis has been challenged by recent empirical research with birds. In a review of changes in mean food intake rate with increasing group size, a sizable number of studies (23%) have failed to report an increase in foraging success despite a significant decrease in antipredator vigilance (Beauchamp, 1998). One reason for this lack of association may be related to increased competition in larger groups. An increase in free time caused by a reduction in antipredator vigilance may be used not for foraging but rather for monitoring threatening group members (Favreau et al., 2010), scrounging food discoveries made by companions (Beauchamp, 2007a; Coolen, 2002), or alleviating the cost of increased competition (Sansom et al., 2008). The end result would be no changes in food intake rate as group size increases. Such effects also appear to occur in other taxa. In a coral reef fish, the increase in time spent foraging in larger groups lead to lower, not higher growth rates as individuals invested more time to competing with one another (White and Warner, 2007).

Time-budgeting advantages associated with group foraging are by no means universal, and must be examined on a case by case basis.

1.2.3. Defending Resources

A large group size provides an asset when contesting disputed resources. In many species, attackers in groups can overcome smaller groups defending resources, providing extra resources to share with all. In coral reef fish, for instance, territorial species can prevent non-schooling, non-territorial species from accessing their feeding territory, but they cannot defend against schooling species because they are too numerous (Robertson et al., 1976). Lions in groups can obtain more food by hunting together but, and perhaps more importantly, they can more easily thwart scavenging attempts by hyaenas (Cooper, 1991). Small packs of wild dogs are less able to defend their kills against other species, such as lions and hyaenas, which may explain, in part, their recent decline in Africa (Carbone et al., 1997) (Fig. 1.5). The ability of large groups to defend resources against other species has also been reported in wolves (Vucetich et al., 2004), social spiders (Cangialosi, 1990), and birds (Ridley and Raihani, 2007). Such increased losses incurred by smaller groups due to food stealing, a form of inverse density dependence, may render populations more vulnerable in the

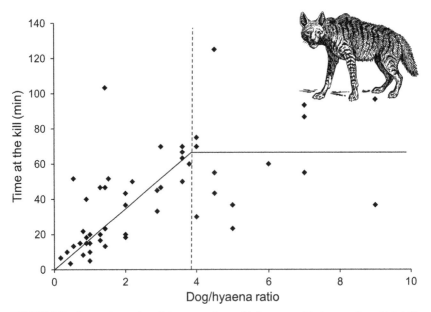

FIGURE 1.5 Scavenging and predator group size: wild dogs are evicted sooner from their kills when the ratio of wild dogs to hyaenas decreases. Wild dogs would remain approximately 65 min at a kill when greatly outnumbering hyaenas. *Adapted from Carbone et al. (1997).*

long term (Courchamp et al., 1999), and may be one factor that favours the formation of large groups in predators.

Overcoming defences also works between individuals of the same species. Indeed, groups of immature ravens were more likely to overcome an adult pair defending a carcass, and feed at a higher rate, than did immature birds in smaller groups or alone (Marzluff and Heinrich, 1992). Such contests between groups of the same species are believed to be an important factor in the evolution of group living in primates (Isbell, 1991; Janson and Goldsmith, 1995; Wrangham, 1980). In some sense, this benefit of foraging in groups for predators resembles the pooling of resources to capture larger prey, which I described earlier. Here, the defended food patch represents the equivalent of a large prey that would remain inaccessible to solitary foragers.

A large group size may also be an asset in territorial encounters between neighbouring groups not directly linked to any particular contested resource. A larger group size may translate into a larger territory, providing a larger pool of resources to divide among group members. In many territorial encounters between rival groups of the same species, the likelihood of winning increases as the difference in group sizes increases (Creel and Creel, 1995; Crofoot et al., 2008; Kruuk, 1972; Packer et al., 1990; Radford, 2003).

1.2.4. Managing Resources

Group foragers have been thought to manage resources more efficiently (Miller, 1922). Miller first proposed that group foraging was the most efficient way of exploiting resources as it would allow foragers to avoid searching in areas that had already been partly or totally exploited. This hypothesis has reappeared sporadically in the literature, mostly for birds and primates foraging on resources characterized by standing crops (Cody, 1971; Schoener, 1971; Terborgh, 1983).

The relative lack of interest in this so-called efficient resource management (ERM) hypothesis stems, perhaps, from the perception that managing resources relies on group selection. Earlier interpretations implied that groups must maintain cohesion to reap the benefits of efficient management, but failed to specify why individuals should stay in the group rather than search on their own. Nonetheless, group foraging provides several advantages, as this book attests, and foraging alone may not represent a viable option for many species. In addition, individuals foraging alone may compete with larger groups searching for the same resources, and thus fare worst (Zemel and Lubin, 1995). Therefore, managing resources in groups may result from individual selection after all.

Recent theoretical analyses have sought to explore these issues more thoroughly for foragers exploiting resources that renew at a slow rate and are thus worth avoiding shortly after exploitation (Beauchamp, 2005). The ERM hypothesis has usually been applied to resources characterized by standing crops, such as fruiting trees or grasses, which eventually produce a new crop or re-grow following exploitation. However, it may also apply to mobile prey that hide after near encounters with predators, and only become available again after a time delay.

In such a system, group foraging is thought to allow individuals to avoid visiting areas exploited earlier, and thus waste time searching in patches with largely depleted resources. In our model, Ruxton and I (2005) considered solitary and group foragers exploiting the same patchy resources scattered across the habitat. By keeping together, instead of foraging singly in many areas at the same time, group foragers are more likely to encounter unexploited patches of food, but must spend more time travelling between patches because group foragers deplete these patches more quickly. By contrast, solitary foragers that encounter a full patch can obtain all the food therein, but run a higher risk of visiting areas already visited previously by others. The costs and benefits of resource exploitation thus differ for solitary and group foragers. We sought to determine the best strategy in response to variation in the rate of resource renewal in food patches. When the rate of food renewal in patches was quite low, food intake rate over a fixed time horizon was higher when all individuals foraged in the same group rather than separately (Fig. 1.6), providing support for the ERM hypothesis. We also considered whether a defector, namely an individual that forages on its own rather than in the group, could enjoy a higher food intake rate than the remaining group foragers exploiting the same resources at the same time. This step is essential to determine whether group foraging represents an evolutionary stable strategy, a strategy that cannot be invaded by rare alternative strategies. Under almost all of the conditions we evaluated, an individual foraging alone enjoyed a higher food intake rate than the remaining group foragers (Beauchamp and Ruxton, 2005). The solitary foraging strategy eventually dominates the population over several generations (Fig. 1.6). At least under the conditions explored in our model, ERM is unlikely to have been a force promoting the evolution

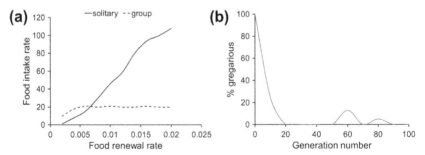

FIGURE 1.6   Harvesting resources in groups: foragers exploiting resources in a group obtain more food than solitary foragers only when the rate of renewal of resources is low (a). However, in a habitat where renewal rate favours group foragers, a population that consists of 100% of group foragers can be readily invaded by solitary foragers after a few generations (b), suggesting that group foraging may not be an evolutionarily stable strategy in this system. *Adapted from Beauchamp and Ruxton (2005).* (For colour version of this figure, the reader is referred to the online version of this book.)

of group foraging. However, if predators must remain together, say, to better defend their resources or to locate resources more efficiently, ERM may represent a benefit of group foraging.

The ERM hypothesis has been applied to species that forage almost exclusively in groups or within a territory (Cody, 1971; Cords, 1990a; Janson, 1998; Prins et al., 1980; Rowcliffe et al., 1995; Watts, 1998), and that would not face challenges from solitary foragers. To properly assess the ERM hypothesis in such species, we would have to manipulate the number of groups exploiting resources in the same habitat, while maintaining population size constant. I am not aware of any empirical studies of this nature.

1.2.5. Decision-Making

Animals in foraging groups must make choices about when to start foraging, where to travel, and how long to stay in the current patch. There is, of course, the fascinating issue of how a consensus in the group is reached and group cohesion maintained (Conradt and Roper, 2005). More relevant to my discussion of the evolution of group living is the hypothesis that group living allows for more accurate decision-making. In the context of migration, for example, pooling information about compass direction from all available individuals can increase the ability of the group to reach the final destination (Simons, 2004). Similarly, when sampling a foraging patch, individuals in a group may possess different information about patch quality. Pooling such information can lead to increased foraging efficiency by enabling foragers to leave patches at a more optimal time (Valone, 1989).

Such pooling at the group level raises many questions concerning proximal mechanisms. For instance, how do individuals value their own information versus that gleaned from others, and is it possible to gather the relevant information from many companions in a timely fashion? Recent theories have explored how groups can achieve more accurate outcomes by gathering information from a restricted set of companions (Berdahl et al., 2013). Quorum-sensing, in particular, appears to be a useful principle for guiding decision-making in groups (Sumpter, 2010). With quorum-sensing, the probability that an individual adopts the same choice as others in the group increases in a non-linear fashion with the number of individuals that have already made that choice. Obviously, such following behaviour can result in many individuals making the wrong choice. However, in a large group, it is more likely that at least some individuals possess the right information, and these individuals will act as leaders guiding the rest of the group, through the quorum-sensing mechanism, in the right direction (Couzin et al., 2005; Sumpter et al., 2008; Ward et al., 2011).

Quorum-sensing would appear to be very relevant to understanding one of the main benefits of social foraging that I discussed earlier—local enhancement. In a network of searching predators, an individual at some point must choose whether to continue searching or join a patch discovered by others.

The choice to join a patch may be based on the number of foragers already present, and when this number exceeds a threshold, the individual becomes more likely to join the patch, which, in all likelihood, is probably quite profitable given the number of individuals already exploiting it. Whether quorum-sensing applies to local enhancement in the field remains to be established.

1.3. COSTS OF GROUP FORAGING

Just as the benefits of foraging in groups for predators are varied, so are the costs. These costs come in many forms, although all of these costs increase with group size. The main costs are related to competition for resources.

1.3.1. Competition for Resources

1.3.1.1. Exploitative Competition

When a group exploits a patch of limited size, each food item obtained by a forager reduces the number or amount available to other foragers. Even without any direct interaction between foragers, such as aggression, each individual in a group obtains a smaller share of the resources available in the patch than would a solitary forager. This is called exploitative competition (Sutherland, 1996). Sharing resources with companions perhaps represents the most widespread cost of group foraging. This cost is, at best, inversely related to group size if resources are divided equally but may, under some circumstances, be greater for some individuals in groups due to unequal competition (see below). To be profitable, group foraging must always provide enough benefits to compensate for the n-fold decrease in food intake that results from dividing resources from a patch among n foragers.

The cost of decreased food intake per individual does not occur in two special situations. In the first case, patches are large enough to satiate all foragers before depletion. When exploiting large patches, food intake rate will not decrease with group size, and may actually increase if the number of meals available from the patch is larger than group size (Clark and Mangel, 1986). This advantage applies when foragers abandon a patch after their meal and resume search after a resting time proportional to the degree of satiation. An example of this scenario is provided by a group of predators that abandon a carcass to scavengers after completing their meal (Carbone et al., 1997).

In the second case, the lifetime of a patch is limited by events unrelated to patch exploitation. For instance, a swarm of insects may disperse after a change in the weather and become unavailable to predators (Brown, 1988). The advantage to a solitary forager lies in the ability to obtain exclusive access to all resources in a patch, which assumes that there is no limit to the time needed to deplete the patch. However, if patches are ephemeral and suddenly disappear before the meal is completed, a solitary forager will obtain a smaller proportion of the resources available (Fig. 1.7). Because a larger group exploits resources in each patch more quickly, the imposition of a time limit on patch exploitation

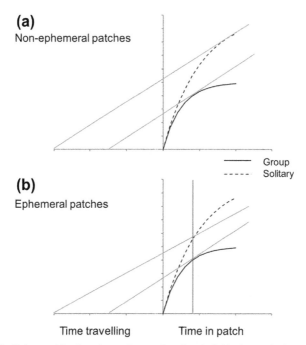

(a)
Non-ephemeral patches

Group
Solitary

(b)
Ephemeral patches

Time travelling   Time in patch

FIGURE 1.7   Ephemeral food patches and group foraging: individuals can obtain more food from ephemeral patches when foraging in groups than alone. The *x*-axis illustrates total time spent foraging, which is partitioned into travelling time to a patch to the left of the *y*-axis and exploitation time within the patch to the right. Individuals should leave a patch when the current rate of food intake drops below the average expected for the habitat, which is given by the slope of the green line. When patches are not ephemeral (a), the mean rate of food intake is similar whether individuals forage alone or in groups. Individuals in groups locate patches more quickly but must divide the food among more companions, resulting in no net advantages. However, when patches last a limited amount of time, indicated by the red line, individuals may be forced to leave before the optimal time, resulting in a lower food intake rate for solitary individuals (b). (For interpretation of the references to colour in this figure legend, the reader is referred to the online version of this book.)

will have a much lower effect on their food intake rate. In fact, patch ephemerality will have no effect on the food obtained from a patch if a group depletes patches before these patches are expected to disappear. Models indicate that when patches are ephemeral, the rate of food intake will actually increase with group size, thereby favouring group foraging (Clark and Mangel, 1986; Pulliam and Millikan, 1982).

For patches that are not too large or ephemeral, exploiting resources in groups will generally lead to faster food depletion. Faster patch exploitation can be costly since it generally forces foragers to look for other patches sooner in order to fulfil their daily food requirements. Time spent travelling, which is at the expense of foraging, is thus expected to represent a larger proportion of the time budget in group foragers. Such an increase in travel costs with increasing group size has been documented in many species, including primate groups that travel together within their territory and exploit patchy resources such as fruiting trees (Wrangham et al., 1993). Similarly, wolves travel much longer each day when in larger packs (Vucetich et al., 2004) (Fig. 1.8). Forming smaller groups may become advantageous when the increase in travel costs fails to be recouped by an increase in energy gained (Chapman and Chapman, 2000). Food sharing can thus lead to not just direct losses, but also indirect costs due to an increase in time spent searching for resources.

Exploitative competition is also found in relatively stationary or sessile predators that forage on drifting resources. Those that occupy more frontal positions can obtain more food, thereby reducing the amount of resources available to those behind (Kent et al., 2006; Wilson, 1974). This has been called shadow interference, although decreased foraging success may not be caused by direct interactions between foragers. For more mobile predators, however, individuals may compete to obtain these more favourable positions, a form of contest competition (see below).

Another form of exploitative competition takes place when prey are able to respond to the close proximity of predators and become temporarily immune to predation. This process is the exact opposite of prey flushing, which actually increases prey availability for group members. For example, mudflat amphipods retreat to their burrows when they encounter a shorebird, a process that creates an area of low prey availability in the vicinity of any foraging group member

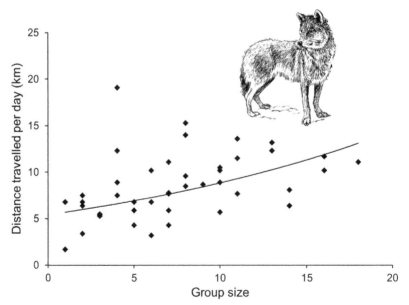

FIGURE 1.8   Travel time and predator group size: wolves travel longer each day when foraging in larger groups. *Adapted from Vucetich et al. (2004).*

(Minderman et al., 2006; Selman and Goss-Custard, 1988). In semipalmated sandpipers, a small shorebird species, which feed on this type of prey, individuals compensate for the decreased food availability caused by increased flock size by foraging more quickly (Beauchamp, 2012b). Adjustments in search paths to temporarily avoid depleted food areas may also allow individuals to reduce the costs of foraging in larger groups (Stillman et al., 2000).

1.3.1.2. Scramble and Contest Competition

The typical conceptualization of group foraging entails equal sharing of resources, but individuals in groups may compete with one another in order to obtain a larger share of resources and thus increase their relative fitness. Such competition may take place through differences in the speed at which resources are acquired (scramble competition) or involve direct interactions between foragers that produce a short-term reduction in food intake rate (contest competition) (Sutherland, 1996). As the intensity of competition typically increases with group size, any individual trait that reduces relative competitive ability in the group will lead to a decrease in relative food intake for at least some individuals. The smaller share of resources may not be sufficient to compensate for the costs associated with group foraging. Competition is also costly as it takes time and energy away from foraging, as illustrated below.

The direct costs of scramble competition are related to an increase in the speed of resource exploitation. Models of group foraging predict that foraging intensity, which can be related to any aspect of the biology of a species that translates into relatively greater foraging gain per unit time, should increase with group size (Bednekoff and Lima, 2004; Lima et al., 1999; Shaw et al., 1995). For example, when competing for limited resources, fish will increase their swimming speed to obtain a larger share of the limited resources (Fig. 1.9) (Shaw et al., 1995). The increase in speed represents a metabolic cost that is expected to increase with group size, which in the end decreases the net energetic value of each contested item. I documented a similar effect in semipalmated sandpipers competing for dwindling mudflat invertebrates: as the density of birds increased, which translates into an increase in competition intensity, individuals ran more quickly when searching for food, incurring a higher energy cost and perhaps less efficient prey selection (Beauchamp, 2012b). In coots contesting resources in a scramble, a decrease in antipredator vigilance allows individuals to allocate more time to foraging and thereby increase their relative share of resources (Randler, 2005a). However, the proportionately lower vigilance may leave foragers more vulnerable to attacks by predators. This potential cost of scramble competition remains to be documented.

As group size increases, predators may resort to more direct means to obtain a greater share of resources. Literature reviews indicate that aggression levels typically increase with group size in birds (Beauchamp, 1998), primates (Isbell, 1991), and fish (Grant, 1993). Although the primary purpose of aggression may

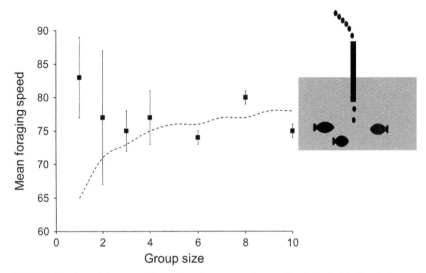

FIGURE 1.9   Scramble competition and predator group size: predicted (dashed line) and observed changes in foraging speed in larger groups of fish competing for limited resources. The inset illustrates the experimental aquarium and the food delivery system. Error bars show one standard error. *Adapted from Treganza et al. (1995).* (For colour version of this figure, the reader is referred to the online version of this book.)

be the establishment of dominance hierarchies rather than accessing food, an increase in aggression levels usually affects foraging because it reduces the time available to forage and entails unequal access to resources (Caraco, 1979b). However, aggression is costly, and depending on the type of resources contested and the number of foragers competing for resources, aggression levels may actually decrease when groups become too large (Dubois et al., 2003; Pagel and Dawkins, 1997).

Unequal access to resources in groups can translate into fitness losses, especially for the less successful competitors. For instance, an increase in group size has been shown to reduce growth in fish (Booth, 1995) and frogs (Griffiths and Foster, 1998), and survival in spiders (Ulbrich and Henschel, 1999), birds (Brouwer et al., 2006), and mammals (Clutton-Brock et al., 1999a). Such frequency-dependent negative effects of group size on fitness components probably set upper limits to the size of a group. These issues will be explored more fully in Chapter 7.

Aggression is one obvious form of contest competition. However, contest competition may be costly even in the absence of overt aggression. In some cases, the steps needed to avoid more dominant companions can be just as costly. Monitoring other group members may take precious time away from foraging (Cresswell, 1997; Favreau et al., 2010; Treves, 1999). Physically avoiding other group members may also decrease time spent foraging and force individuals to areas with fewer resources (Stillman et al., 1997; Whitehouse and Lubin, 1999).

A decreased feeding rate in the absence of overt aggression and resource depletion has been referred to as cryptic interference (Bijleveld et al., 2012). In a
shorebird species, for example, food intake rate decreased by as much as 93%
when group size increased from 2 to 8, despite no evidence for resource depletion or overt aggression or food stealing (Bijleveld et al., 2012).

1.3.1.3. Kleptoparasitism and Scrounging

In kleptoparasitism, a predator loses some or all its prey to competitors from
either the same or different species (Brockmann and Barnard, 1979). Kleptoparasites obtain food without actually expending time and energy searching and
capturing prey. Kleptoparasitism is a form of contest competition as it results in
the loss of resources through direct interactions. Here, I will consider the relationship between group size and kleptoparasitism when it involves members of
the same or different species.

Loss of prey to competitors from other species may be substantial, as
described earlier for wild dogs. One solution for reducing such losses is foraging
in groups, as explained earlier. In particular, groups may be able to defend their
resources more easily against intruding species. Even when defence is ineffec-
tive, losses to kleptoparasites can be divided among all group members, reducing
individual burden, as documented in colonial spiders (McCrate and Uetz, 2010).

Another solution consists in tackling smaller prey that can be eaten before
competitors arrive (Lamprecht, 1978). However, foragers in groups must often
tackle larger prey in order for individuals to obtain a net benefit (Kruuk, 1972;
Packer et al., 1990). If competitors can arrive at a large prey before individuals
are satiated (or steal food while individuals are sated), then food usurpation
represents a cost of hunting in groups. This cost may be partly offset by the
greater effectiveness of larger groups in repelling food stealers, as noted earlier,
but larger prey may also attract a larger number of competitors, resulting in proportionately greater loses in larger groups. In their analysis of food stealing by
ravens, Vucetich et al. (2004) assumed that losses to ravens were independent of
wolf group size, but this may not be the case. The relationship between losses to
scavengers and predator group size deserves more attention.

Many instances of food usurpation are intraspecific, although they need not
always take place in the context of group foraging. For instance, water crickets
forage alone but conspecifics will share any large prey captured (Erlandsson,
1988). Communal food sharing, in this case, represents a breakdown in food
defence due to the presence of several competitors and is not the outcome of
group foraging. Similarly, mussel stealing occurs frequently among oystercatchers, a large shorebird, feeding in the same patch. However, the presence of
several oystercatchers in the same patch reflects the limited number of suitable
food patches rather than any attraction among individuals (Goss-Custard, 1996).

Kleptoparasitism is also known to occur in group-foraging species. Communal exploitation of resources can take place if prey items are large enough or
if a patch discovered by a forager contains several items that can be shared. In a

group foraging context, all individuals are expected to search independently for resources, and all obtain a share when resources are discovered by one group member. Kleptoparasitism, in this context, is costly because some individuals exploit patches or food items discovered by others without investing time and energy searching for the resources. Models of group foraging predict that the proportion of individuals relying on others to find resources (that is, scrounging) should increase with group size (Giraldeau and Caraco, 2000). Frequency-dependent food usurpation implies that the ability to locate resources by the group will not increase as rapidly as expected with group size. The more moderate increase in food finding ability in a group with scroungers may not be sufficient to compensate for the n-fold decrease in food intake, leading to a decrease in food intake with group size. Such issues are explored more fully in the following chapter.

1.3.2. Overlap in Search Areas

From a perceptual point of view, the more eyes (or the equivalent in another sensory modality (Dechmann et al., 2009)) available to scan the habitat for food, the more likely food will be located rapidly. This process, however, is not expected to provide ever-increasing benefits as group size increases because of overlap in search areas between foragers (Mangel, 1990; Ruxton, 1995). Although avoiding already searched areas may be difficult for a solitary forager, it is probably unavoidable when many foragers search the same general area at the same time. In his model of group foraging, for instance, Ruxton (1995) showed that in a group where individuals move randomly when searching for food, the time to locate a food patch decreases with group size but at a slower rate than expected if all individuals searched without interfering with one another.

In Ruxton's model, individuals move randomly in search of food patches. In the field, foragers may follow, instead, specific search paths aimed at maximizing prey encounter rate. However, in the presence of companions, individuals may stray from their preferred paths (Cresswell, 1997; Vahl et al., 2005). This may force individuals to adopt suboptimal paths, which may partly explain why areas that have already been searched are revisited more often in larger groups. Comparing foraging paths for solitary and group foragers would help us determine to what extent overlap in search areas is caused by the disruption of search paths. In some cases, it has been suggested that adopting specific spatial formation while searching for resources reduces overlap in search areas and increases foraging efficiency (Dermody et al., 2011; Haney et al., 1992), but it is not known if this strategy is common in other taxa.

Empirical evidence for a decreased rate of food finding in larger groups comes from experimental studies in birds (Hake and Ekman, 1988; Krebs et al., 1972) and fish (Pitcher et al., 1982). The difficulty here is that the decrease in the rate of food finding may also arise because fewer individuals search for resources at any one time in larger groups, as argued earlier in the case of scrounging. Data on the occurrence of multiple visits to the same patches as a function of group size are needed to rule out the effect of scrounging.

Overlap in search areas has also been invoked as a factor causing exploitative competition in primate groups. For species that do not deplete their food patches or that feed on dispersed resources, such as insects, overlap in search areas will reduce encounter rate with resources, forcing individuals to increase the area that must be searched (Gillespie and Chapman, 2001; van Schaik and van Hooff, 1983). This 'pushing forward' will also increase travel costs, as was the case when groups depleted resources more quickly.

1.3.3. Increased Detection

Just as prey in larger groups may be more vulnerable to predators because of their increased conspicuousness (Chapter 3), a larger group of predators may be more easily detected, allowing prey to escape sooner. Indirect support for this prediction comes from the observation that predators that rely on surprise, such as tigers and falcons, typically forage alone (Gittleman, 1989). More direct evidence can be gleaned from the reaction of prey to variation in the group size of their predators. Most studies, however, report the reaction of prey to a single predator, usually a single human approaching their position (Stankowich and Blumstein, 2005). In studies using approaches by two humans, instead of one, flushing took place at a significantly greater distance in one species of bird (Geist et al., 2005) and a lizard (Cooper et al., 2007). However, it is not clear whether animals fled sooner because they detected intruders in pairs sooner or because the presence of more intruders was perceived as riskier. Data on detection distance, that is, the moment when prey become aware of the presence of intruders, are needed to distinguish between the two effects.

1.3.4. Increased Predation Risk

Many species of predators face little predation risk themselves due to their sheer size and weaponry. For species like wolves and orcas, the main costs of foraging in groups are related to food competition. For smaller and less protected predator species, an additional cost of foraging in groups is the conspicuousness or vulnerability of their group to their own predators. For example, semipalmated sandpipers prey on amphipods and are themselves prey to falcons, which show a tendency to attack larger groups more often (Beauchamp, 2008a; Sprague et al., 2008).

A large body of literature suggests that attack rate varies as a function of group size (Cresswell, 1994; FitzGibbon, 1990; Lindstrom, 1989). Much of the data comes from studies of insect larvae and coral reef fish (see Table 3.1 in Chapter 3, for references). However, most of these studies are observational, which raises the possibility that variation in predation rate with group size reflects the action of uncontrolled risk factors, like habitat type or location, rather than group size *per se*. In addition, groups of different sizes may vary in conspicuousness (see Chapter 3), making it difficult to determine whether some groups are attacked more often because it is more profitable to do so or simply because they are encountered or detected more easily.

More persuasive evidence comes from studies where groups of different sizes are simultaneously available to predators. Using this cafeteria-style approach, fish predators showed a preference for the larger of two groups of prey presented simultaneously (Krause and Godin, 1995). In a field study, natural predators preferred to attack a shoal of guppies rather than a single individual when available at the same time (Botham et al., 2005) (Fig. 1.10). A similar field study in birds revealed the opposite pattern: sparrowhawks showed a higher probability of capture when hunting redshanks, a small shorebird, in smaller groups and closer to shore. When faced with a choice between groups of different vulnerabilities at the same time, sparrowhawks attacked the more vulnerable in 66% of the cases (Cresswell and Quinn, 2004). In view of the limited and conflicting nature of the results thus far, more work is needed to assess the relationship between group size and attack rate.

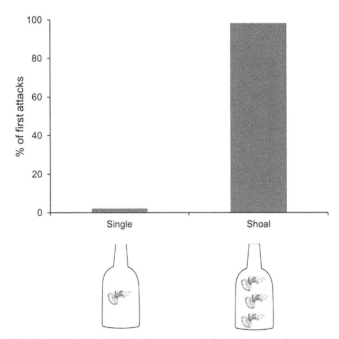

FIGURE 1.10   Group size and attack preference: natural fish predators prefer to attack guppies in shoals of 10 rather than solitary guppies. The guppies were held in bottles, which were presented simultaneously to predators. *Adapted from Botham et al. (2005).*

1.4. CONCLUDING REMARKS

Predators can benefit in many ways from the presence of companions. In particular, predators in groups may detect prey faster, acquire larger prey, spend more time foraging, defend their prey or displace other groups more easily, harvest resources more efficiently, or make more accurate choices during foraging. Nevertheless, foraging in groups may also be costly because individuals may compete more intensely for resources, either directly or indirectly, produce resources at a lower rate than expected, or attract more predators. The costs of exploiting food in groups may be avoided in large patches or patches that last a limited amount of time.

In many species, there may be several costs and benefits at play, and the relative value of group foraging will typically vary as a function of group size and habitat characteristics, such as patch type and richness. The end result will be that a given species tends to prefer a certain group size. These issues are explored more fully in Chapter 7 in the context of optimal group size.

The relationship between vulnerability to predation and group size clearly represents an area that deserves more attention. Many species of predators are themselves subject to predation, and the net value of living in groups will depend not only on their ability to find and exploit resources, but also on their ability to escape predation. Krause and Ruxton (2002) noted in their monograph on group living that vulnerability to predation in relation to group size was one of the least well understood costs of grouping. I found very little additional information since the publication of their book, and the evidence reviewed here is clearly contradictory. It is certainly not difficult to imagine that prey may respond to the choices of their predators, and vice versa, creating an arms race whose outcome varies from species to species, perhaps explaining why different patterns arise in different species. I explore co-evolution between predators and their prey more fully in Chapter 9.

Further studies on overlap in search areas are also likely to reveal novel, interesting effects. This potential cost of foraging in groups may be partly compensated for by changes in search paths. However, at the moment, it is difficult to determine whether the less than linear increase in patch finding rate in larger groups results from overlapping search areas or more simply from a decrease in search effort caused either by a greater investment in scrounging or by the avoidance of threatening companions.

Predators in groups must move together and maintain cohesion while facing a vast array of choices, such as where to go and how long to remain in the current patch. Different individuals in the group may possess information of varying quality about the habitat and may face different challenges, even within the same group. For example, some individuals may be more susceptible to foraging interference or more vulnerable to predation. With the recent surge of interest in collective animal behaviour and decision-making, it may be possible in the near future to determine how groups can reach a consensus about such collective decisions and maintain cohesion despite the various needs and challenges of different group members.

Producing and Scrounging

2.1. INTRODUCTION

In the land tenure system of medieval Europe, herders grazed their cows on a shared piece of land known as a commons. An unscrupulous herder that brought additional cows to the commons benefitted by producing more animals but reduced the amount of resources left for everyone else. The temptation to bring more animals to the commons should be irresistible when considering the potential individual benefits. Unfortunately, widespread adoption of this strategy meant that in the end the commons became irreparably damaged to the benefit of no one. The tragedy of the commons—the depletion of shared limited resources—arises from the short-term pursuit of individual interests (Hardin, 1968).

The tragedy of the commons illustrates two key concepts in the study of group foraging. First, the best course of action for a group member often depends on the behaviour adopted by everyone else. In the tragedy of the commons, the benefits from bringing more cows to the commons, a defection from the point of view of the public good, depend on how frequently the same behaviour is adopted by the other herders. Indeed, the payoffs will be highest when few herders follow the same course of action, which means that payoffs from defection tend to be frequency dependent. High payoffs when defection is rare drive the initial wave of selfish behaviour. Second, individuals tend to adopt a course of action that benefits them rather than the group. In the tragedy of the commons, a restriction on the number of cows that each herder can bring to the commons would preserve the land longer and benefit all herders the most over the long-term. But without any external incentive, the temptation to bring more cows will be strong and in the end everyone will fare worst.

Animals also exploit resources conjointly, and it is no surprise that selfish, exploitative strategies have also evolved in their groups. These exploitative strategies extend the theme, which I developed in the previous chapter, that foraging in groups while often beneficial can also be costly. Food usurpation among species represents a case in point. For example, in the plains of east Africa, packs of wild dogs lose many of their prey to spotted hyaenas, a factor that may have contributed to their recent decline (Gorman et al., 1998). The same dynamics also occurs among individuals of the same species: for instance, carcasses discovered by one vulture are eventually shared with other vultures attracted to the food bonanza.

As food usurpation clearly illustrates, foraging animals can obtain resources on their own or exploit the food discoveries of others, which is known as scrounging. This is especially true for group-foraging species where finding resources is harder to hide. Although a species may become totally dependent on another to obtain resources, as the occurrence of parasitic species exemplifies, the same situation appears unlikely within a species because most individuals must at one time or another find resources on their own. Nevertheless, the pattern of allocation of time between finding resources and exploiting resources produced by others may vary between individuals, with some finding more than others.

This chapter deals with the dynamics of scrounging resources within species. Reflecting the tragedy of the commons, I will show that the payoffs from scrounging are frequency dependent. In addition, when everyone selects the best course of action for themselves, everyone in the group fares worst. Theoretical studies of scrounging rely heavily on game theory, which I shall introduce briefly. Since the early 1980s, when the first model of scrounging was developed, a cottage industry of models has sprung up in the literature. I will describe the basic scrounging model, and then examine variations upon the basic theme. I will then review the empirical evidence for the assumptions and predictions derived from these models.

A glance at the literature on scrounging reveals that most of the empirical evidence comes from captive birds. Models also tend to be formulated with birds in mind, probably reflecting the greater familiarity of modellers with avian species. Nevertheless, scrounging is not restricted to bird species, and has been documented in many other animal groups including insects

(Slaa et al., 2003), fish (Hamilton and Dill, 2003; Ryer and Olla, 1992), mammals (di Bitetti and Janson, 2001; King et al., 2009), and even humans (Blurton-Jones, 1984). Although studies have been mostly conducted in the laboratory, there is no reason why scrounging cannot be examined in the field. Therefore, this chapter should be of general interest.

2.2. DEFINITION

Many terms have been used to describe food usurpation. The use of terms such as 'robbing' (Kushlan, 1978), 'stealing' (Goss-Custard et al., 1982), 'piracy' (Burger and Gochfeld, 1981), or 'theft' (LeBaron and Heppner, 1985) smacks of anthropocentrism when applied to animals. Even in the human literature, 'tolerated theft' (Blurton-Jones, 1984) has been replaced by 'food transfer,' a more neutral concept (Gurven, 2004). The term 'kleptoparasitism' avoids the connotation of intent (Brockmann and Barnard, 1979). As the name implies, the loss of resources following an interaction should be costly to the victims. It is not clear to me that this is always the case. Tolerance of food theft in humans certainly suggests that victims may eventually be repaid in the form of recipro-cal exchanges (Gurven, 2004). In animal species, one can imagine that the loss of food may be compensated by other benefits of foraging with scroungers, such as a reduction in predation risk. Even supposedly straight cases of kleptopara-sitism among different species have been reinterpreted in the light of potential gains by alleged victims (Radford et al., 2011). The burden of proof should definitely lay with researchers when using terms like kleptoparasitism.

The term 'scrounging' seems more neutral and captures the essence of obtaining food from others without paying the cost of actually producing it (Barnard, 1984). Scrounging may take different forms. Aggressive scrounging involves the use of force or threats to obtain resources produced by others like when a male lion displaces females from a kill. Scramble scrounging implies that scroungers are unable to displace food finders and instead must scramble to obtain a share of the resources. As an example, when a mackerel holds a fish in its mouth, companions can only snatch morsels from the whole fish (Allen, 1920a). Convergence of several individuals at a food source discovered by others has been variously labelled in the past as 'local enhancement' (Thorpe, 1956) or 'area copying' (Krebs et al., 1972). The means of achieving scroung-ing matter less than the outcome: the loss of resources obtained at a cost which scroungers avoid entirely.

2.3. THE BASIC PRODUCING AND SCROUNGING MODEL

The ability to obtain resources produced by others was introduced early in group-foraging models (Clark and Mangel, 1984; Clark and Mangel, 1986; Ranta et al., 1993; Ruxton et al., 1995). In these models, all group members are assumed to search independently for resources and to converge rapidly on any patches discovered by a group member. Such models are known as information-sharing models. Information-sharing models assume that all individuals search for resources without fail when they are not exploiting a patch. Essentially, these models embody the idea that individuals can search for food and pay attention to others in the group simultaneously. Therefore, the occurrence of scrounging has no effect on the rate at which food patches are discovered at the group level.

A different view proposes that individuals must invest time to find resources on their own but also to monitor their surroundings to locate patches discovered by others. Crucially, these two activities are assumed to be mutually exclusive (Barnard and Sibly, 1981). The two alternative ways to obtain resources are labelled 'producer' (searching for resources) and 'scrounger' (exploiting resources produced by others). These labels can also be attached to individuals, but it should be clear that an individual can only use one tactic at a time to obtain food, producer or scrounger. Incompatibility between producing and scrounging departs strongly from the simultaneous assumption of information-sharing models.

The producer-scrounger (PS) paradigm implies that not all individuals participate in food finding at any one time. Therefore, in stark contrast to the prediction from information-sharing models, the rate at which food patches are discovered must decrease when the use of scrounger increases in the group. The fewer patches discovered at the group level will also be exploited by a larger number of foragers, a double whammy that induces negative frequency-dependent payoffs for tactic use. To summarize, payoffs in the information-sharing paradigm vary as a function of the total number of foragers in the group because all foragers in the group search for and share resources. In the PS paradigm, payoffs depend not only on the total number of foragers but also on the proportion of individuals using the scrounger tactic. PS models aim to predict the stable pattern of producing and scrounging in a group as a function of a host of ecological variables, including patch richness and group size.

Game theory provides a natural framework to analyse problems involving frequency-dependent payoffs, namely, when the best course of action for an individual depends on the actions taken by everyone else (Box 2.1). Game theory also features prominently in Chapter 4, which deals with antipredator vigilance.

An alternative option, N-person symmetric game represents the simplest possible way to model producing and scrounging. In this game, a forager can only use one option at a time, producer or scrounger, and the payoffs from the chosen option depend on the actions taken by the remaining $N-1$ companions that participate in the game. Symmetry means that once a stable solution has been achieved, all individuals obtain the same payoffs when using the same tactic. Later, I shall examine cases where different individuals obtain different payoffs when using the same tactic. For the purposes of presentation, I rely

BOX 2.1 Game Theory in a Nutshell

Game theory is a set of mathematical models aimed at studying interactions between rational decision-makers. Developed by economists in the mid-1940s, the theory was soon applied to animal behaviour research (Maynard Smith, 1974) to study conflict and cooperation among individuals. Game theory has been used to study a vast array of problems in animal behaviour, ranging from social foraging to communication and learning (Dugatkin and Reeve, 1998). One of the better-known applications of game theory to animal behaviour deals with the evolution of fighting tactics (Maynard-Smith, 1976). Properties of this game are explored below to get general insights into game-theoretical arguments.

When competing for food, animals may use aggression to win resources or use displays instead to avoid direct interactions. The model considers two tactics, labelled hawk and dove. These labels apply to the tactics and do not refer to the different species. Individuals using the hawk tactic fight to obtain resources. A hawkish individual may win outright but also risk losing with a costly injury. With the dove tactic, individuals use a low-cost display to intimidate opponents and never suffer injuries related to a fight. In order to find which strategy natural selection will favour, payoffs must be calculated for each possible type of contest. Ideally, payoffs should be expressed in terms of the expected number of descendants produced over many generations, an evolutionary fitness measure. Practically, a proxy for fitness may be used, such as the amount of food obtained after the contest, assuming that a positive relationship exists between the proxy and fitness. In the following game, payoffs are expressed in imaginary fitness units proportional to the expected number of descendants.

In a contest between two animals, four possible types of encounters are possible: hawk or dove meets an opponent that can play either hawk or dove. Average payoffs for each type of encounter are calculated assuming that a winner gets 50 units, a loser gets 0 units, a fighting injury costs 100 units, and a display costs 10 units. When a hawk meets a hawk, each has the same probability of winning the contest but only the loser pays the injury cost. Similarly, when a dove meets a dove, each individual is equally likely to win but each pays the cost of display. A hawk wins every contest against a dove. The following payoff matrix ensues:

Attacker	Opponent	
	Hawk	Dove
Hawk	0.5*50+0.5*–100	50
Dove	0	0.5*(50 – 10)+0.5* – 10

In a population where everyone follows the dove tactic, one hawk would obtain a higher expected payoff (50) than one dove (15). The hawk tactic would thus spread quickly in the population, suggesting that playing dove all the time is not a stable strategy. Consider now a population that consists entirely of hawk players. One dove in this population would obtain a higher expected payoff (0) than one hawk (–25) and spread quickly in the population at the expense of the other. Playing hawk all the time is not a stable strategy either. The payoff matrix suggests that a mixture of tactics may be stable. A mixed strategy implies that at any one time

(Continued)

BOX 2.1 Game Theory in a Nutshell—cont'd

individuals play one tactic with probability p and the alternative with probability $1 - p$. Individuals select one of the two alternatives according to these probabilities before the contest. This is similar to the choice of tactics in the schoolyard rock-paper-scissor game: the choice is made before the actual play. The solution identifies the value p where the expected payoffs from the two tactics are equal. If this were not the case, selection would tweak the probability of playing one tactic to even out the payoffs. If p represents the probability of playing hawk during one contest, expected payoffs from playing hawk or dove are given by:

$$\text{Hawk} = -25p + 50(1 - p)$$

$$\text{Dove} = 0p + 15(1 - p)$$

Solving for p after setting the two expected payoffs equal provides the solution to the game. In this case, playing hawk with probability 7/12 is a stable solution. Any player that uses hawk with a different probability would have lower fitness.

This particular solution to the game is referred to as a mixed evolutionary stable strategy (mixed ESS). The term 'mixed' indicates that the solution involves a mixture of two tactics, while the term 'stable' indicates that no unilateral deviation from the solution can provide higher fitness to an individual. In evolutionary terms, the ESS represents a strategy that when adopted by all members of a population cannot be invaded by any rare alternative strategy. In contrast to games in economic settings, where players know the payoffs and try to predict the consequences of their choices, the ESS thinking views payoffs in fitness terms and strategies are assumed to be biologically inherited. The stable solution is achieved over several generations and alternative strategies can arise through mutations. This is why the ESS is expressed in terms of resistance to invasion by a rare mutant strategy.

Some general points deserve mention. Expected payoffs at the ESS are often much lower than those that would arise if individuals tried to maximize their fitness. In a pure ESS, with exclusive use of one tactic at equilibrium, expected payoffs from playing dove all the time (15) are much higher than those from playing hawk all the time (-25). The optimal solution, the one that maximizes fitness, should thus be to play dove all the time, but alas this strategy is not stable as illustrated above. In the end, individuals adopting the mixed ESS experience a lower fitness (6.25) than the optimal value (15). As pointed out earlier with the tragedy of the commons, the pursuit of individual interests reduces fitness for all.

The second point concerns how a mixed strategy should be expressed in a population. Two possible outcomes are at least theoretically possible: all individuals play hawk with probability p, as I assumed above, or a proportion p of all individuals play hawk. In the producer-scrounger game and in the vigilance game, which is explored in Chapter 4, specialized use of one tactic by some individuals is considered rare and all individuals allocate some time to each tactic.

The value of game theory lies in providing a framework to analyse interactions among individuals when the best course of action depends on the choices made by everyone else. Such models are most often used heuristically to uncover broad principles but when payoffs can be documented empirically, nothing prevents testing quantitative predictions from game theory models.

on a model where the payoffs are expressed in terms of mean food intake rate (Vickery et al., 1991).

Vickery et al. consider N foragers exploiting patches of food each containing a fixed number F of indivisible items. The finder of a patch can obtain a items $(0 < a \leq F)$ before the first scrounger arrives. The proportion of the patch that can be exploited exclusively by its finder is referred to as the producer's advantage. The remaining food items are shared equally among all individuals exploiting the patch thereafter. The finder of a patch cannot exclude other scroungers and scroungers cannot evict a finder. For simplicity, all players in the game use the two tactics and all obtain the same payoffs at equilibrium when using the same tactic. When searching for patches of food, scrounger is used with probability p and producer with probability $1 - p$. Finding a patch occurs at rate λ, and foraging takes place over a period of time T. Since finding a patch is thought to occur at a low rate, it can be safely assumed that patches are discovered sequentially by the pool of producers.

When one individual locates a patch, the expected number of foragers joining the patch should be pN. All scroungers arrive instantaneously at the patch and the $F - a$ available food items are divided equally among the $1 + pN$ foragers. The expected food intake rate when using the producer tactic is given by $I_{producer}$:

$$I_{producer} = \lambda T \left(a + \frac{F - a}{1 + pN} \right) \tag{2.1}$$

Notice that the time needed to exploit a patch is not included in the calculation and is therefore assumed to be negligible. This assumption has been relaxed in subsequent models. Payoffs for the scrounger tactic, $I_{scrounger}$, depend on the expected number of patches discovered by food finders, which will increase with the probability of playing producer:

$$I_{scrounger} = \lambda N (1 - p)T \left(\frac{F - a}{1 + pN} \right) \tag{2.2}$$

The expected food intake rate when playing producer decreases as the probability of playing scrounger increases in the population (Fig. 2.1).

In a population where all individuals search for food without fail ($p = 0$), a mutant individual that waits instead for patches to be found by others will be able to exploit food patches discovered by each food finder, thus reducing the expected time needed to find a patch. This scrounger will only fail to feed in a given time interval if all producers fail to locate a patch, a much lower probability than failing to find a patch for any single producer independently. Now consider the other extreme. In a population where scrounging is very common ($p = 1$), an individual that searches for food patches independently will get the same share of food from a patch as all the others but also benefit from the producer's advantage, thus getting a higher food intake rate. The stable solution should therefore take the form of a mixed ESS with intermediate probabilities

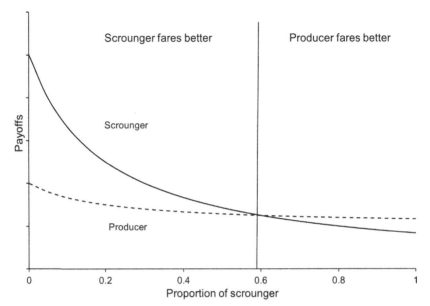

FIGURE 2.1   The producer-scrounger game: payoffs for the producer and the scrounger tactics decrease as a function of the proportion of individuals using scrounger in the population. The ESS occurs at the intersection of the two payoff curves. Payoffs for scrounger exceed payoffs for producer when scrounger is rare in the population but not when it is common.

of playing scrounger. At the mixed ESS, payoffs from the two tactics should be the same and the two payoff curves will intersect.

The ESS level of scrounger in the population can be found by solving for p when the two expected payoff functions are set equal to one another. At equilibrium, the stable solution takes the following form:

$$\hat{p} = 1 - \frac{a}{F} - \frac{1}{N} \tag{2.3}$$

It can be shown mathematically that any unilateral deviation from the stable policy will lead to a reduction in expected food intake rate. The stable policy indicates that the probability of playing scrounger should increase with group size and decrease when the producer's advantage increases. Notice that the corporate rate of patch finding λT is not a part of the stable policy. In other words, the probability of using scrounger is independent of the rate of patch finding. This prediction has been challenged in more recent models (see below).

The prediction regarding the effect of the producer's advantage can be couched in ecological terms related to food patch characteristics. For a fixed producer's advantage, a reduction in the size of a patch will reduce the use of scrounger in the population, which of course drops to zero when a food finder can monopolize the whole patch. For a patch of a fixed size, the producer's advantage will decrease as the time needed to handle food items increases. A conifer cone, for instance, may provide many seeds but extracting and husking one seed may take a long time, which should allow for a greater use of scrounger.

The model presented above considers patches with indivisible items. However, scrounging may also occur on single, divisible items. The mackerel example discussed earlier represents a case of a single item that can be divided into several morsels. In this case, the producer's advantage represents the portion of fish secured in the mouth of the food finder. Large fruits can also be considered single, divisible items (di Bitetti and Janson, 2001). When food items are divisible, the use of scrounger in the population should decrease when items get smaller.

Patches may contain a very large number of food items in which case the producer's advantage will cause a negligible dent in food availability. Consider, for instance, a large school of fish or a swarm of flying insects: any prey removed by the food finder has little effect on the overall number of prey items and the ratio a/f must be effectively close to zero. The model predicts an increase in the use of scrounger when exploiting very large food patches.

Thus far, I have presented the producer's advantage as a space construct, namely, the physical fraction of a patch or of a food item monopolized by the food finder. The producer's advantage may also be extended to the temporal dimension. Some patches, such as insect swarms, may be ephemeral in duration as prey disperse or hide due to forager activity. In such a case, the producer's advantage may be thought of as the proportion of time the food finder can exploit the patch alone. As another example, the number of prey items available to shorebirds on a mudflat can be very large but foraging causes prey animals to bury deeper, effectively creating a desert zone around each bird passing through the patch (Minderman et al., 2006). The temporal producer's advantage when entering a patch must be close to 1 here, forcing individuals to spread out rather than converge at the same location.

2.4. NEW THEORETICAL DEVELOPMENTS

Several new models of producing and scrounging have seen the light of day in recent years. These new theoretical developments relax some assumptions made in the original model, which in the end increases the range of predictions. I now examine some of the most relevant developments here.

2.4.1. Risk-Sensitive Scrounging

The basic PS model assumes that selection favours foragers that maximize their expected food intake rate. Food intake rate is considered a proxy for fitness, with the expectation that the higher the food intake rate, the higher the fitness. Expected food intake rate represents the mean payoffs that could be achieved over several instances of the game. It should be clear that the mean payoffs will not be achieved in every single game because of inherent randomness in patch encounter. A run of bad luck, for instance, can reduce the rate of encounter with food patches and thus reduce food intake rate. Therefore, following one tactic or another should lead to a distribution of payoffs, rather than a single value, for which the mean should be the most likely outcome.

Variance in payoffs may be an important consideration when calculating the ESS because variance can influence survival just as much as mean payoffs (Caraco, 1981). As an example, consider two foraging tactics that provide the same mean payoffs but with different variances. The best option may be to maximize the probability of exceeding some minimum requirement for survival rather than maximizing the mean rate of food intake. In other words, under some circumstances, fitness may be greater when choosing a tactic with more variable outcomes even if the mean payoffs are lower.

Variability in outcomes makes intuitive sense in the PS game. As mentioned earlier, a scrounger benefits from the pooled searching abilities of many producers and should therefore encounter foraging opportunities with less uncertainty than any single food finder. Caraco and Giraldeau (1991) presented a model of producing and scrounging where the goal was not to maximize expected food intake rate, as in the basic model, but rather to minimize the probability of an energy shortfall. Food items obtained through producing or scrounging provide energy to meet some requirements for survival, and a shortfall occurs when the accumulated food intake falls below a critical physiological limit. With such a currency, both the mean and variance associated with producing and scrounging now become relevant (Caraco and Giraldeau, 1991).

The model considers a group of N foragers divided into P producers and S scroungers. A producer must accumulate enough food to exceed the minimum requirement R (in units of food intake). The scroungers must exceed the same limit but also pay a scrounging cost. An example of a scrounging cost may be the time needed to travel to food patches discovered by others during which no food can be obtained. A producer shares a patch with all scroungers, but the producer's advantage was not modelled the same way as in the basic model. Here, each item from the first to the last is contested by all $S + 1$ foragers at the patch. The probability that the producer obtains a food item is θ and each scrounger has a probability $(1 - \theta)/S$ of obtaining the same item. All scroungers are considered equal competitors when contesting resources.

The producer's share of a patch can vary in one of two possible ways. In the so-called producer-priority rule, the producer maintains the same probability of obtaining a food item regardless of the number of scroungers present. In biological terms, this may correspond to a situation where the producer initially secures a good position in a patch and retains a competitive advantage throughout patch exploitation. For instance, the first bird to arrive at a conifer cone may hold a better grip on the cone and obtain more food than the scroungers until the cone is depleted. The second scenario, referred to as the scramble-competition rule, assumes that each item is contested with the same effectiveness by all foragers present, including the producer. Here, the producer has no advantage over the scroungers.

Because patch finding and contests over every single food item are random processes, the number of food items collected over a period of time will be a distribution of values with a mean and a variance that may differ for producers and scroungers. From the distribution of the expected number of food items obtained by a producer or a scrounger over the duration of a foraging episode, it is then possible to calculate the probability of an energy shortfall for each as a function of the proportion of scroungers in the group. Just as in the basic model, the intersection of the two payoff curves provides the stable solution to the game.

Results from the model vary according to the way the producer's share of a patch is calculated (Fig. 2.2). Under the producer-priority rule, the probability of an energy shortfall for a producer is independent of the number of scroungers in the group as long as there is one, which follows from the assumption of a constant advantage of the producer over scroungers when contesting resources. In contrast, under the scramble-competition rule, the probability of an energy shortfall for a producer increases with the number of scroungers in the group. The probability of a shortfall always increases for scroungers as the proportion of scroungers increases in the group.

Different stable solutions emerge in this game. One solution involves a group composed entirely of producers. This is a stable solution as long as the probability of an energy shortfall for a producer is less than that for a single individual adopting the scrounger tactic. More typically, the stable solution entails a mixture of producers and scroungers. At equilibrium, the probability of an energy shortfall will be the same for producers and scroungers. For this

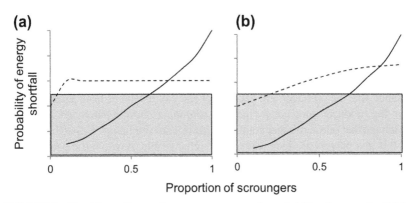

Proportion of scroungers

FIGURE 2.2   The risk-sensitive producer-scrounger game: the probability of energy shortfall for producers (dashed line) and scroungers (solid line) changes in groups with different proportions of scroungers. The solution to the game lies at the intersection of the two curves. The probability of energy shortfall varies depending on the way resources are shared in a patch ((a): producer-priority rule, and (b): scramble-competition rule). The shaded box illustrates the area where physiological requirements are met. *Adapted from Caraco and Giraldeau (1991).*

solution, the equilibrium proportion of scroungers is expected to be low when (1) the cost of scrounging is high, (2) groups are small, (3) patch finding rate is low, and (4) the producer has a relatively strong advantage when competing for resources. The equilibrium value will also vary as a function of physiological requirements in a rather complex fashion.

The effect of group size in the risk-sensitive model parallels the finding from the basic model. It also makes sense that a high cost of scrounging or a strong competitive advantage for producers should reduce the prevalence of scrounging. In the basic model, scrounging was independent of patch finding rate, but patch finding rate emerges as a determinant of scrounging in the risk-sensitive version. As patches are found more rapidly by a producer, opportunities to scrounge become more readily available and a higher prevalence of scrounging can emerge.

The fact that physiological requirements influence the use of scrounger arises as no surprise in a model based on minimizing the probability of an energy shortfall. The authors made the following predictions: when producers have a competitive edge over scroungers, the stable proportion of scroungers should decrease with an increase in physiological requirements. By contrast, when producers enjoy a smaller advantage over scroungers, an increase in physiological requirements should lead to more scrounging. Therefore, altering physiological requirements may lead to more or less scrounging depending on the relative competitive ability of producers and scroungers when exploiting patches. The challenge for testing a risk-sensitive model of scrounging must be to obtain sufficient information about relative competitive ability, physiological requirements, and the cost of scrounging.

2.4.2. Dominance

The PS models considered thus far are fashioned for symmetric players, players that achieve the same payoffs when adopting the same tactic at equilibrium. A symmetric game need not imply that all players will get the same rewards all the time. The producer-priority rule in the risk-sensitive game assumes, for instance, that the producer will get a larger share of resources from a patch. However, the producer does not exclude the scroungers from a patch and enjoys an advantage that one scrounger could also get if it arrived first at the patch. In these models, the players are identical and interchangeable. Asymmetric games, with unequal players, depart from symmetric games because at equilibrium different players can get different payoffs. The need to develop asymmetric games stems from the observation that different individuals may have different phenotypic traits, which may influence the payoffs they get from producing and scrounging.

In many species, heavier or more aggressive individuals can displace others from their food discoveries. As an example, heavier pigs in a pair can displace lighter companions to monopolize a food patch (Held et al., 2000). Unequal encounters between individuals have also been documented in several avian species (Liker and Barta, 2002; Rohwer and Ewald, 1981; Stahl et al., 2001). Other traits may favour the use of scrounger for some individuals. A recent study found that shy barnacle geese scrounge more presumably because they lack the exploratory tendencies necessary to locate new food patches on their own (Kurvers et al., 2010). These findings certainly suggest that payoffs from producing or scrounging may vary among individuals.

Phenotype-limited game theory models are perfectly suited for PS games with asymmetric players (Parker, 1984). In a phenotype-limited game, where phenotypes vary on a continuous scale and two tactics are possible, the solution is for each individual to select the tactic that yields the highest payoffs for its phenotype. A threshold phenotype exists such that all individuals whose pheno-type lies above the threshold choose one tactic and all those whose phenotype lies below the threshold use the other. Only at the threshold will individuals use the two tactics (Gross, 1996). At equilibrium, individuals with different pheno-types may experience different payoffs.

A phenotype-limited model of producing and scrounging was developed with the aim to predict the relative use of scrounger when individuals differ in their ability to compete for resources (Barta and Giraldeau, 1998). The model considers a group of foragers exploiting patches containing a fixed number of indivisible items. The group consists of a number of producers and scroungers. Now, all individuals are not equal and can be ranked in terms of their ability to obtain shares of each patch: a more dominant group member, with a higher rank, can obtain a larger share of resources from a patch than a less dominant companion. Notice that competitive ability is not related to the effectiveness of patch finding when not exploiting a patch. As usual, the finder of a patch can exploit the patch alone before the scroungers arrive, and thus benefit from the producer's advantage. The remaining food items are shared between the producer and all scroungers and the amount of food obtained by any forager is proportional to its relative competitive ability.

To find the ESS in this game, the authors assigned a tactic to each forager, calculated the payoffs for each group member, and searched for a combination of tactic use among all foragers that would be stable against unilateral changes in tactic use. A combination was deemed stable when no forager could achieve a higher payoff when adopting the alternative tactic. Results indicate that at the ESS, the proportion of scrounger in the group should decrease as the producer's advantage increases (Fig. 2.3). This is qualitatively the same prediction made in the basic model. However, introducing unequal competition always reduces the use of scrounger, for a given producer's advantage, especially when unevenness among foragers increases. Predictions from the two models converge when the producer's advantage becomes very large: in both cases the use of scrounger is expected to be very low.

The model also predicts a relationship between dominance rank and scrounger use. When disparity in competitive ability is relatively small, all indi-viduals regardless of rank are predicted to use the same mixture of producer

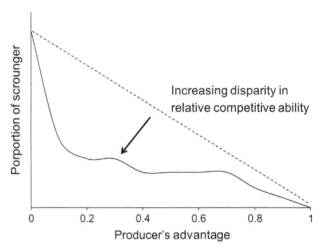

FIGURE 2.3 Dominance in the producer-scrounger game: increased disparity in competitive ability among foragers decreases the stable equilibrium frequency of scrounger, especially when the producer's advantage takes on intermediate values. *Adapted from Barta and Giraldeau (1998).*

and scrounger. As disparity increases, higher ranking individuals use scrounger more often than low-ranking companions. One extreme solution involves a threshold dominance rank: all individuals with a higher dominance rank play scrounger all the time and all individuals with a lower dominance score play producer all the time. The step function that was expected from phenotype-limited game theory models only takes place in small groups composed of very unequal competitors.

At equilibrium, more extreme specialization in tactic use is associated with divergent payoffs: payoffs are smaller the lower down the hierarchy. Lesser reliance on scrounger, when disparity in competitive ability increases, should increase mean payoffs across the group. However, the results indicate that it is mostly the more dominant group members that benefit from this reduction in scrounging. Due to reduced scrounging, dominant individuals get access to more patches and face fewer competitors in each patch. Accordingly, their relative share of each patch increases at the expense of producers.

The model assumes a group of fixed size. Solutions to the asymmetric game may be very different if individuals have the option to leave the group. For instance, if the expected payoffs from staying in the group are smaller than those achieved when foraging alone, subordinate individuals may simply leave. Including such considerations changes the PS game to a mixture of PS and group-membership games. Models of phenotypic assortment in response to differences in competitive ability already exist (Ranta et al., 1996), but more work is needed to couple the two types of games to predict both group structure and scrounger use in a population with more fluid group membership.

2.4.3. Spatial Games

Foragers in a group are spread over two or three dimensions. Yet models of producing and scrounging have typically ignored the spatial structure of groups. Consider, for instance, the assumption that all scroungers arrive at a patch instantaneously and simultaneously, which is untenable in a 2- or 3-D group where individuals must spend a nontrivial amount of time travelling to a patch. Position in the group, whether at the edge or the centre, has also been shown to have a large impact on foraging success and predation risk in many animal species (Krause, 1994a). It is thus reasonable to ask how a group with producers and scroungers should be arranged in space.

Barta et al. (1997) explored the influence of scroungers on the geometry of a group by creating an explicitly spatial version of the PS game. In the model, patches contain a fixed number of indivisible items and are spread out randomly over a large, two-dimensional area. At any time step, all foragers can scan their surroundings for possible threats or choose to forage. Scanning for threats takes time away from foraging and lasts longer when foragers are further apart. This new feature in a scrounging model reflects the fact that individuals typically face predation threats and can warn one another more easily when closer (see Chapter 3). When foraging, individuals can look around to locate other group members with probability p or look for food with probability $1 - p$. When looking for food, a producer moves in a random direction within a fixed radius to locate patches of food. A scrounger looks around to locate scrounging opportunities. Such opportunities are less likely to be detected when an exploited patch is farther away. If detected, the scrounger moves instantaneously to the patch and shares the available food items with everyone there. If no opportunities are detected, the scrounger wastes one unit of time.

In addition to the stable number of producers and scroungers in the group, the model was used to predict other behavioural parameters, including step length and the range of choices for movement direction. Combining these different parameters determines inter-individual distances and the relative position of producers and scroungers in space. Results indicate that in a population where scrounging takes place, scroungers tend to occur at the centre of the group and producers at the edges (Barta et al., 1997). A group containing producers and scroungers also spreads out less than a group composed of producers only (Fig. 2.4).

A wider spread allows producers to avoid locating patches already exploited by others. A smaller spread in a group with scroungers implies that selection favours tighter groups in which scrounging opportunities are less likely to be missed. The occurrence of producers at the edges of the group also reduces interference among food finders.

The model fixed the role of each individual, but what would happen if foragers could use a mixture of the two tactics? Even if individuals were exactly similar, the prospect of getting higher payoffs from scrounging when in the middle of the group should favour a conditional strategy such as: scrounge when in the middle of the group or produce otherwise. Spatial variation in payoffs can thus create an asymmetric game with otherwise perfectly symmetrical foragers. This is similar to the finding that dominant group members scrounge more than subordinates in the previous phenotype-limited model, except that in the spatial game the asymmetry is related to the location of an individual rather than being a consequence of a phenotypic attribute such as size. In contrast to the dominance game, producers and scroungers experience the same payoffs at equilibrium.

How scroungers travel to patches exploited by others constitutes a rather restrictive assumption in the model. Despite the spatially explicit nature of the model, scroungers arrive almost instantaneously at any exploited patch they detect. While distant patches are less likely to be detected, scroungers still pay no cost to reach any detected patch. What this means is that scroungers tend to arrive early in patch exploitation and never face a depleted patch. In addition, scroungers at different distances from an exploited patch still all experience the same share of resources since they all arrive at the patch at the same time. In a setting with non-negligible distances between foragers, any attempt to exploit the food discoveries of others must cost time and energy. Time and energy constraints can be ignored if distances between patches are relatively small. This is probably the case in the various laboratory experiments described in the following section, in which patches are spread over a small area. However, a different set of assumptions is needed for situations where distance matters.

To this end, I explored what happens to producing and scrounging when travelling is costly (Beauchamp, 2008b). In the model, foragers are spatially

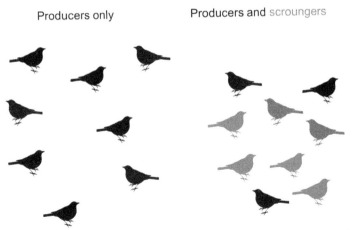

Producers only   Producers and scroungers

FIGURE 2.4 Spatial organization of producers and scroungers: groups composed entirely of producers are more spread out than groups composed of a mixture of producers and scroungers. In mixed groups, scroungers tend to occupy more central positions. *Adapted from Barta et al. (1997).* (For colour version of this figure, the reader is referred to the online version of this book.)

dispersed and can only move one step at a time when searching for food or when attempting to join patches discovered by others. Hence, a small patch may become depleted before a distant scrounger arrives. All foragers are assumed to be equal and instead of fixing roles for each player, all individuals can use a mixture of the two tactics. The goal of the model was to determine the stable pattern of scrounging when individuals attempt to maximize their food intake rate in a spatially explicit habitat.

In contrast to the basic model, this spatially explicit model predicts that scrounging should decrease when food patches are encountered more frequently (Fig. 2.5). Decreased reliance on scrounger when patches can be discovered rapidly makes sense in a spatially explicit setting. When the habitat contains many food patches, a scrounger could encounter unexploited patches on the way to join another patch, a rather high opportunity cost. The risk-sensitive scrounging model, which I described earlier, predicts a decrease in the use of scrounger when the encounter rate with patches increases but only if the size of a patch is small relative to group size (Caraco and Giraldeau, 1991). In the spatial model, the negative relationship held for small and large patches and for

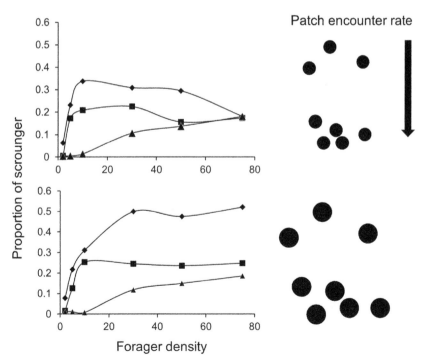

FIGURE 2.5   Spatially explicit producer-scrounger game: stable equilibrium frequency of scrounger use in a spatially explicit population decreases when patch encounter rate increases and when patches (filled circles) are smaller, especially when forager density takes on intermediate values. *Adapted from Beauchamp (2008).*

small and large groups. The two models use a different fitness currency and also differ in the modelling details of the scrounging costs. It is, therefore, not simple to determine why their predictions are different. At the very least, I uncovered a relationship between the rate of encounter with food patches and scrounging in a rate-maximizing, spatially explicit model.

The spatial model predicts an increase in the use of scrounger as group size increases. However, when the rate of encounter with food patches is low, scrounger use can level off or actually decrease in larger groups (Fig. 2.5). Since scrounger use is already high when patches are discovered more slowly, fewer patches will be exploited by a larger number of foragers. More intense patch exploitation probably means that late-arriving scroungers will get a very small share of the resources, especially when exploiting small patches. Indeed, the negative effect of group size on scrounger use was only documented for smaller food patches (Fig. 2.5).

I also explored another consequence of modelling scrounging in a spatially explicit world. Attraction to feeding conspecifics temporarily inevitably increases the density of foragers in certain parts of the habitat. This spatial concentration of foragers may make scrounger use more likely for two reasons: (1) the travel time to nearby food discoveries will tend to be small and (2) foraging interference between producers all searching in a restricted area will reduce their payoffs. To address this issue, I documented the stable pattern of scrounging when individuals resume their food searches in the vicinity of patches they have just depleted—the baseline condition—or when scroungers are reallocated randomly in the habitat after depleting a patch, which breaks down the spatial concentration of foragers. As expected, scrounging decreased in the habitat with random reallocation of foragers. Scrounging thus appears to be a self-organizing phenomenon (Camazine et al., 2001) in which the use of scrounger creates conditions that enable the use of more scrounging.

A more recent spatial model also suggests that foragers influence the spatio-temporal distribution of resources, which in turn influences the spatial distribution of food searchers and those that seek food searchers to locate food (Tania et al., 2012). Such an interaction between foragers and their food supply makes it possible to extend the PS paradigm to field situations where patches deplete both in time and in space. The above models of scrounging highlight novel predictions that emerge when considering scrounging in a spatially explicit world.

2.4.4. Learning the Stable Pattern of Scrounging

The models explored thus far identify the stable proportion of scrounger and producer in a group as a function of several ecological variables like patch size and group size. How a solution is actually achieved in a population has received much less attention. The ESS literature typically envisions a gradual elimination over evolutionary times of genetically determined phenotypes that perform relatively poorly. In games that involve behavioural options, learning represents a different means of achieving a solution. Individuals may learn how others have responded in the past to different courses of action and adopt the best course of action based on these prior expectations. By contrast, individuals may tend to select courses of action that have been productive in the past and ignore less successful options based entirely on their own experience. The first scenario has rarely been considered in animal behaviour games but features quite prominently in games with rational agents (Fudenberg and Levine, 1998). The second scenario involves the use of learning rules, rules that weigh failures and successes in the past and provide a means to select the best course of action at any given time. Can such learning, based on successes and failures, allow individuals to reach the stable equilibrium frequencies predicted by ESS models?

Learning rules have been applied to other behavioural games in the past, such as the spatial distribution of foragers competing for food (Houston and Sumida, 1987). The use of learning rules in PS games has been explored theoretically and empirically (Beauchamp, 2000b; Hamblin and Giraldeau, 2009; Morand-Ferron and Giraldeau, 2010). At this point in the discussion, it is reasonable to question the validity of the ESS concept in a game where the payoffs are allegedly learned. Giraldeau and Dubois (2008) proposed to use the term 'behaviourally stable solution' (BSS) when natural selection acting on the frequency of genetically determined phenotype alternatives cannot be invoked. In a BSS, the stable equilibrium frequency of a tactic can be reached through other mechanisms, such as learning, in the lifetime of an individual (Giraldeau and Dubois, 2008). Obviously, natural selection may be involved indirectly in tweaking the learning process, but the main feature of a BSS is the gradual achievement of a stable solution during the lifetime of an individual.

Learning rules in general have two distinct parts: (1) a mechanism to update the value of behavioural options as information accumulates and (2) a mechanism for choosing an alternative based on the value of each option. Updating the value of an option involves weighing past successes and failures using a relevant currency like food intake rate. Several mechanisms can be used to select among behavioural options, including maximizing, where the option with the highest payoffs is selected; or matching, where the probability of choosing an option is based on the value of this option relative to that of the others.

The relative payoff sum rule represents perhaps the most discussed learning rule in animal behaviour research (Harley, 1981). For this rule, the variable S refers to the value of an option I, say, in terms of food intake rate. The rule determines the value of that option at time t based on its value at the previous time period:

$$S_i(t) = xS_i(t-1) + (1-x)r_i + P_i(t) \qquad (2.4)$$

where x $(1 > x > 0)$ denotes a memory factor, r_i $(r > 0)$ is the residual value associated with option i, and finally P_i represents the current payoffs from choosing option i at time t. With this rule, current payoffs (P) are never devalued. When an option fails to provide any payoffs over several time steps, the value of that option decreases geometrically to the residual value. This residual value can be thought of as the intrinsic value of an option prior to any weighing of evidence. The memory factor puts a weight on the past value of an option: a higher value devalues recent payoffs. The relative payoff sum rule uses matching for selecting among alternative options. The probability of choosing option i at time $t + 1$ is based on the relative value of tactic i with respect to an alternative j:

$$Pr(t+1) = \frac{S_i(t)}{S_i(t) + S_j(t)} \qquad (2.5)$$

The linear operator rule is also frequently considered in the context of foraging (Beauchamp and Fernández-Juricic, 2005; Bernstein et al., 1988; Hamblin and Giraldeau, 2009). The linear operator rule embodies a geometrically moving average of past payoffs. The updating mechanism for the value of option S has the following form:

$$S_i(t) = xS_i(t-1) + (1-x)P_i(t) \qquad (2.6)$$

Here, in contrast to the relative payoff sum rule, the value of an option can gradually decrease to zero. The choice mechanism between options can be based on maximizing or matching.

To examine whether the stable equilibrium frequency predicted by ESS models can be reached in a population using a learning rule, it is necessary to predict the equilibrium frequency and then to simulate changes in the use of producer and scrounger through time as each forager experiences successes and failures associated with the use of each tactic. I thus set out to compare the predicted use of scrounger with the simulated end-point achieved in a population using various learning rules (Beauchamp, 2000b).

Results from my simulations indicate that the linear operator rule, with strong discounting of recent payoffs ($x > 0.99$) and a matching choice mechanism, can provide a very good estimate of the stable equilibrium frequency predicted for the game by ESS models (Fig. 2.6). The relative payoff sum rule, with strong discounting, also provides good estimates. Maximizing in general performs more poorly and leads to groups composed of specialist producers and scroungers. Putting more emphasis on recent past, which arises with smaller values of the memory factor, always leads to more scrounging than predicted (Fig. 2.6).

This first step made it possible to evaluate how a given learning rule fares with respect to the predicted stable equilibrium frequencies. It is clear that blind choices based entirely on individual failures and successes in the past can allow a group to reach the predicted equilibrium. In addition, the results also suggest that building features of learning rules matter: for instance, matching performs better than maximizing as a choice mechanism, and higher values of the memory factor provide a better fit.

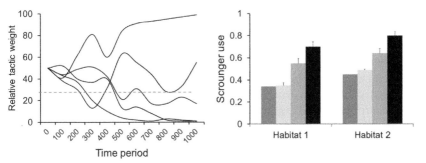

Time period

FIGURE 2.6 Learning in the producer-scrounger game: the relative value of producing and scrounging in a group changes as learning takes place over time. The panel on the left shows divergence among five foragers in the relative weight of the two tactics: when the relative weight is larger than 50, the scrounger tactic is more likely to be selected at any time period. The red dashed line denotes the stable equilibrium frequency in the population. The panel on the right illustrates the stable equilibrium frequency of scrounger in two habitats (red bars) and the frequencies predicted by the linear operator rule with a matching choice mechanism. The memory factor for the linear operator was 0.99 (pale grey bars), 0.95 (medium grey bars), or 0.85 (black bars). The number of foragers in habitat 1 is five and 10 in habitat 2. Bars show one standard error based on 100 simulation results. (For interpretation of the references to colour in this figure legend, the reader is referred to the online version of this book.)

Another matter altogether is whether a population that uses a learning rule with those specific features could be invaded by an individual using different rules or the same rule with different parameter values. It is therefore important to pit learning rules against one another to determine whether such strategies are stable. In the above simulations, I showed that when all members of a population use the linear operator rule with strong discounting, no other rules can invade when rare, but the reverse was not true, and several rules actually provided very similar payoffs when rare. I only performed simulated invasions of this type using a restricted range of parameter values, making it difficult to establish whether one rule can be definitely considered stable in the PS game.

A different approach was used recently to determine whether such rules can be stable in the face of invasion by rare alternative options (Hamblin and Giraldeau, 2009). In their model, characteristics of the habitat are fixed and the best combination of parameter values for a given rule was determined after many learning trials in the population. The key innovation is that different individuals in the population can use different rules simultaneously, each with different parameter values. For instance, at any one time, a number of individuals can use the linear operator rule, each with different memory factor values, while others use the relative payoff sum rule, each again with different memory factor and residual values. After pitting these rules against one another over many learning trials, one rule may emerge the winner. At equilibrium, any individual that would use an alternative rule or the same rule with different parameter values would have lower payoffs.

Results indicate that the relative payoff sum rule can become predominant in the population when contesting against other rules, including the linear operator. This is especially the case in groups larger than 10 individuals, which may explain why my earlier simulations in smaller groups often failed to find a clear winner in pairwise contests. A weakness in the linear operator rule, as described above, resides in the fact that once the value of an alternative gets to zero, this alternative effectively disappears from the choices available to a forager. This is never the case with the relative payoff sum rule since an option that provides little benefit still has a residual value, which allows individuals to sample this option from time to time. As suggested by Hamblin and Giraldeau (2009), this key difference may provide a means to test which rule may be followed by foragers in the real world. For example, producing or scrounging will never be abandoned completely, even when the tactic fails consistently over time, if individuals use the relative payoff sum rule rather than the linear operator rule.

In the spatial model described above including food searchers and exploiters, Tania et al. (2012) considered the possibility that individuals can switch between tactics depending on their perception of the relative success of each tactic. For instance, if foragers using the searching tactic do better, exploiters are more likely to switch to this tactic in the next time frame. This approach differs quite considerably from the use of learning rules since decisions to switch are not based on individual successes and failures (the past experiences of one forager) but rather on the perception of differences in the average food intake rate achieved by those using each tactic within a given radius of interaction. Interestingly, with this approach, the proportion of foragers actively searching can fluctuate cyclically through time as opposed to reaching a fixed, stable value. This approach assumes that individuals are able to pool information about food intake rate from neighbours and to distinguish between the two types of foragers. Whether this can be achieved in the field remains an empirical issue.

2.5. EMPIRICAL EVIDENCE

2.5.1. Testing the Assumptions

2.5.1.1. The Incompatibility Assumption

Incompatible search modes characterize the PS game. Indeed, an individual passes opportunities to find resources on its own when devoting any time to scrounging. Information-sharing models, by contrast, assume that the two tactics can be used concurrently with no trade-off. Despite the key importance of this assumption, there is surprisingly little direct evidence to support it. One possible way to investigate the incompatibility assumption is to find search modes uniquely associated to producing or scrounging. The nutmeg mannikin, a small granivorous species of bird originally from south-east Asia, has become the champion of laboratory research on producing and scrounging, and much of the empirical research described below focuses on this and other related species. These birds forage in the laboratory for seeds hidden in food patches distributed over a relatively small area, reproducing the conditions captured in the basic PS model described in the preceding section. Nutmeg mannikins hop on the ground with their head up or down. It turns out that the more birds hop with their head down, the more food they find on their own. Conversely, the more they hop with their head up, the more patches they scrounge (Coolen et al., 2001; Wu and Giraldeau, 2005). In a further experiment that prevented scrounging altogether, hopping with the head up disappeared from the flocks (Coolen et al., 2001).

These findings suggest that producing and scrounging tend to be associated with specific postures that are incompatible with one another. The results are not clear-cut, as the strength of the association between postures and tactics varied quite extensively from flock to flock. In addition, it is not clear to which extent hopping with the head up precludes finding new food patches or whether hopping with the head down interferes with scrounging. For instance, even individuals that obtained most of their food from scrounging still hopped head down quite frequently. In the end, it is not possible to tell which posture was used for each food discovery. The real answer to the incompatibility issue can only come from experiments designed to test the ability of individuals to detect scrounging opportunities when producing or vice versa.

Incompatibility between concurrent activities has also been an issue in the vigilance literature. It was long thought that vigilance against predators, which requires scanning the surroundings for potential threats, could not occur when animals were foraging head down. However, recent findings show that vigilance, albeit of a lower quality, can be carried out when looking down (Lima and Bednekoff, 1999). Clearly, it is important to keep in mind that the field of view of an animal may allow detection of relevant stimuli over a broad range of postures (Fernández-Juricic, 2012). Many species can apparently search for food and monitor companions simultaneously with little loss of information. For such species, the incompatibility assumption is shaky at best.

Scrounger use in general will depend on the degree of incompatibility between the two search modes. Vickery et al. (1991) considered a different alternative to producer and scrounger, called opportunist, which can carry out producing and scrounging to varying degrees at the same time. When there is no trade-off between producing and scrounging, it is clear that one should expect a population of pure opportunists to arise (Fernández-Juricic et al., 2004a). When producing reduces the ability to scrounge, or vice versa, the stable equilibrium may include a mixture of opportunists and scroungers but rarely pure producers. Determining the degree of incompatibility between producing and scrounging thus has important ramifications for the evolution of scrounging. The potential occurrence of opportunists also makes it more difficult to assign scrounging to a tactic in a group since both opportunists and scroungers can join a patch discovered by others.

The incompatibility assumption was also tested in the field to determine whether individuals reduce their own foraging effort to detect profitable scrounging opportunities (Smith et al., 2002). In food patches provided with suet chunks, individual European blackbirds that spent more time scrounging food from a companion did not experience a decrease in the returns from their own personal foraging effort, suggesting that detecting scrounging opportunities did not detract from their ability to find food alone. In conclusion, the empirical evidence for the incompatibility assumption is rather weak at the moment. This is troublesome not only for PS models but also for information-sharing models, which assume no incompatibility between search modes.

2.5.1.2. Negative Frequency Dependence

Greater use of scrounger in a group should reduce payoffs for both producer and scrounger tactics. For equilibrium to arise, the impact of scrounging on payoffs should be stronger for one of the two tactics allowing the two payoff curves to intersect. What is the empirical evidence to support negative frequency dependence? In an experimental study with nutmeg mannikins, the cost of producing was increased by adding a weight to lids that partially covered each food patch (Giraldeau et al., 1994). Additional work to produce a patch should increase reliance on the scrounger tactic, making it possible to document payoffs associated with each tactic when the prevalence of scrounging increases. As expected, the addition of weights induced more scrounging and decreased payoffs, especially those associated with the scrounger tactic (Fig. 2.7).

Further support for the assumption of negative frequency dependence comes from an experiment carried out in the field in Barbados with carib grackles. These common birds are readily attracted to food sources provided

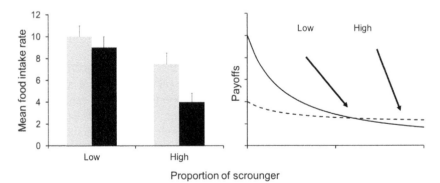

FIGURE 2.7 Empirical payoffs in the producer-scrounger game: negative frequency dependence of payoffs in flocks of nutmeg mannikins foraging under conditions that favour low or high scrounger use. The left panel illustrates payoffs for producing (grey bars) and scrounging (black bars). The panel on the right illustrates the theoretical payoff curves predicted from the basic model of producing and scrounging. Error bars show one standard deviation. *Adapted from Giraldeau et al. (1994).*

by people. Grackles often dip bits of dry food in puddles to soften the morsel before swallowing it. These dunked food items, like the clumped suet chunks used in the blackbird study, are attractive to other birds and attempts at stealing are quite frequent (Morand-Ferron et al., 2004). The cost of producing a dunked item was manipulated by increasing the distance a bird had to travel to reach a puddle. In another part of the study, the cost of scrounging was manipulated by changing the shape but not the area of the puddle so as to increase the distance between a producer and a scrounger. Scrounging, as expected, occurred more frequently when the cost of producing increased or when the cost of scrounging decreased (Morand-Ferron et al., 2007). Crucially, the payoffs decreased with an increase in the number of scroungers, suggesting again negative frequency dependence.

Other studies have also documented similar adjustments in scrounging patterns when the cost of producing is altered (Barrette and Giraldeau, 2006; Bugnyar and Kotrschal, 2002).

Researchers also documented frequency-dependent payoffs by establishing a relationship between payoffs and the degree of attraction of particular group members to scroungers. In Mexican jays feeding on food patches within their natural territories, scroungers prefer to follow individuals with a high propensity to find food patches. As expected, individuals that attract more scroungers experience lower payoffs (McCormack et al., 2007).

Although the above studies typically show a decrease in payoffs when scrounging increases, it remains difficult to document tactic payoffs over a large range of scrounger frequencies. For instance, in situations where the producer tactic provides a clear advantage over scrounger, it may be difficult to coax individuals to use scrounger to document their payoffs. Ideally, tactic payoffs should be documented in groups with different ratios of producers to scroungers. In theory, different ratios could be achieved by assigning different individuals to each tactic. However, in practice, it is difficult to entice individuals to use only one tactic because most foragers tend to adopt a mixture of the two tactics.

A clever experiment circumvented this issue by forcing individuals to adopt one tactic or the other, but not both (Mottley and Giraldeau, 2000). In the experiment, a wall confined individual nutmeg mannikins to one of the two sides of an experimental cage. Food hoppers were only available to individuals on the producer side of the apparatus (Fig. 2.8). Individuals on the scrounger side could only feed when a producer released food on the opposite side. However, the food that was produced fell in a container accessible from both sides of the apparatus. Varying the number of individuals on both sides of the cage allowed the experimenters to create various producer-to-scrounger ratios. Also of note, the producer of a patch always obtained more food from a patch than any of the scroungers on the other side, simulating the producer's advantage. Payoffs, in terms of food intake rate, clearly decreased for both producers and scroungers when the proportion of scroungers increased in the group (Fig. 2.8). In a further

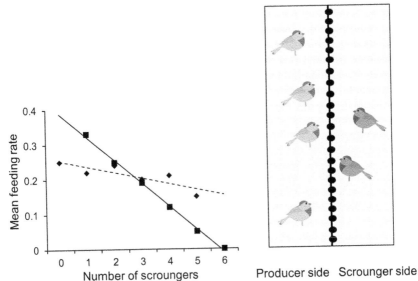

FIGURE 2.8 Convergence on the stable equilibrium frequency of scrounging: negative frequency dependence of payoffs for the producer (dashed line) and the scrounger (solid line) tactic in one flock of six nutmeg mannikins as a function of the number of scroungers in the group. The panel on the right illustrates the apparatus used to confine birds to one of two sides of the cage. The producer side allowed birds to release seeds in one of 22 possible patches (filled circles in the middle of the cage). Birds on the scrounger side could only obtain food when a producer released seeds. *Adapted from Mottley and Giraldeau (2000)*. (For colour version of this figure, the reader is referred to the online version of this book.)

experiment where birds could move freely between the two sides of the apparatus, the experimenters determined that birds converge on the predicted stable equilibrium frequency of scrounger, and that the two tactics provide equal payoffs at equilibrium.

Results from the above experiments provide strong evidence for negative frequency-dependent payoffs in the PS game, at least for small species of birds foraging in controlled environments. Evidence suggests that birds can also converge on the stable equilibrium frequency of scrounger predicted by rate-maximizing PS models and that the payoffs for the two tactics are equal at equilibrium.

2.5.2. Testing the Predictions

2.5.2.1. Effect of Group Size

PS models predict an increase in the use of scrounger in larger groups except in very large groups exploiting patches that deplete rapidly. So far, two laboratory experiments have explored this question with small groups of birds. Notice that information-sharing models also predict an increase in joining frequency in larger groups: in a group of N foragers, the joining frequency should be $N - 1/N$, which is an increasing function of group size. As group size increases, the corporate rate of patch finding decreases in the PS paradigm because individuals that rely on scrounger cannot discover any new food patches. In the information-sharing paradigm, by contrast, the corporate rate of patch finding should never decrease when group size increases as joining and searching for food patches can be performed simultaneously.

In a laboratory experiment with nutmeg mannikins foraging in groups of various sizes, scrounging was more prevalent in the larger groups. In addition, the rate of patch finding at the level of the group decreased in the larger groups as predicted from the PS model (Coolen, 2002). I also documented a positive relationship between scrounging and group size in zebra finches, a closely related species with similar foraging habits (Beauchamp, 2007a). Similarly, the proportion of goldfish searching for food also decreased in larger groups although the PS dynamic was not investigated (Stenberg and Persson, 2005). In one field study with chacma baboons involving a limited number of groups, scrounging occurred to the same extent in the small and in the large group (King et al., 2009).

2.5.2.2. Effect of the Producer's Advantage

An increase in the producer's advantage should lead to a decrease in scrounger use. A simple way to vary the producer's advantage is to increase the total number of food items in a patch: as more food items are available, the producer's advantage decreases and scrounger use should thus increase accordingly. This prediction was tested in flocks of nutmeg mannikins (Giraldeau and Livoreil, 1998). As predicted, scrounger use decreased as the proportional share of a patch monopolized by food finders increased (Fig. 2.9). In addition to documenting tactic payoffs in groups with different producer-to-scrounger ratios, Mottley and Giraldeau (2000) also allowed birds to move freely between the two sides of their apparatus. As predicted, fewer individuals occurred on the scrounger side when the producer of a patch monopolized a larger share of each patch.

A producer's advantage has also been documented in a field study of capuchin monkeys in Argentina. Fruits were distributed on platforms in the forest, and after each discovery the amount of food obtained by the finder was noted before the scroungers arrived (di Bitetti and Janson, 2001). As would be expected, the producer's advantage increased when foragers were more spread out and when the number of fruits was smaller. The producer's advantage was not related to sex or dominance status although the overall amount of food consumed at the patch increased with dominance status. In another field study involving primates, scrounging was more prevalent when chacma baboons exploited larger food patches (King et al., 2009). However, these large food patches were rare and tended to occur in a risky habitat, which may predispose individuals to share. In ring-tailed coatis foraging on fruits

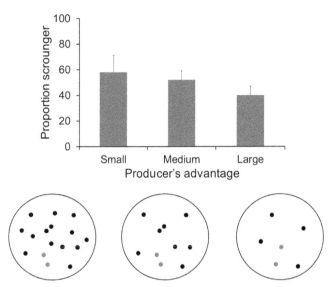

FIGURE 2.9   Producer's advantage in the producer-scrounger game: scrounger use decreases when the producer's advantage varies from low to high. Changes in the producer's advantage were induced by changing the number of food items per patch. Food items monopolized by a producer are shown in red. Error bars show one standard deviation. *Adapted from Giraldeau and Livoreil (1998).* (For interpretation of the references to colour in this figure legend, the reader is referred to the online version of this book.)

from naturally occurring trees, individuals that arrived first enjoyed greater foraging success (Hirsch, 2011), which is compatible with the occurrence of a producer's advantage.

Increasing the producer's advantage should help food finders maximize their food intake rate. However, in a group, scroungers are difficult to avoid and may even evict subordinates from a patch (Held et al., 2000; Liker and Barta, 2002) or obtain a disproportionate share of the resources, as was the case in capuchin monkeys. To increase the producer's advantage, individuals may try to maintain large inter-individual distances and avoid giving signals that a patch has been discovered (di Bitetti and Janson, 2001). Similar avoidance strategies have been documented in pigs where the subordinate member of a pair tended to avoid searching for food when a scrounger was nearby or foraged faster to increase its share of each patch (Held et al., 2002; Held et al., 2010). Nutmeg mannikins also abandon food patches they produce more quickly when scroungers arrive, allowing them to locate new patches sooner and exploit resources alone longer (Beauchamp and Giraldeau, 1997). It should be clear from this discussion that the producer's advantage should not be viewed as a fixed attribute, but rather as the outcome of food exploitation strategies, which are, to some extent, under the control of food finders. PS models assume that the producer's advantage explains variation in scrounger use. However, one can make the case that food finders may manipulate scrounging to some extent by adjusting the size of the producer's advantage.

2.5.2.3. Foraging Requirements

Risk-sensitive PS models predict that an increase in patch encounter rate should increase scrounger use so as to reduce the probability of an energy shortfall. A different version of the risk-sensitive PS model makes the opposite prediction when patches are small relative to group size. A laboratory experiment with European starlings aimed to examine the effect of patch encounter rate on scrounger use in the context of minimizing the chances of an energy shortfall (Koops and Giraldeau, 1996). Unfortunately, in this experiment, changing the encounter rate with food patches also changed forager density and the size of the producer's advantage, making it difficult to isolate the effect of encounter rate. This is important because forager density, on its own, can influence scrounger use (Barta et al., 2004; Monus and Barta, 2008). Therefore, it is essential to manipulate the rate of encounter with food patches while controlling for forager density and other confounding variables.

Manipulation of food requirements offers a unique opportunity to examine predictions from risk-sensitive PS models. In a further experiment with their starlings, Koops and Giraldeau also manipulated foraging requirements by altering food deprivation levels. However, the results were quite inconclusive as scrounger use was relatively unaffected by foraging requirements. A more recent experiment with nutmeg mannikins investigated risk-sensitivity in a more direct fashion (Wu and Giraldeau, 2005). The experimenters determined that hopping with the head down, which was associated earlier with producing in this species, generated a more variable food intake rate over a fixed period of time than hopping with the head up, the scrounger-associated posture. In addition, calculations revealed that the producer tactic would reduce the probability of energy shortfall when foraging requirements are higher. Unfortunately, differences in the risk of energy shortfall proved quite small between the two tactics in the actual experiment, and the relative use of the two tactics was not strongly influenced by changes in foraging requirements.

Risk-sensitive PS model predictions have also been tested in other species. In captive house sparrows, scrounging provided more stable foraging gains than producing, echoing the results with nutmeg mannikins (Lendvai et al., 2004). With low energy levels, individuals increased their use of scrounging early in the day, suggesting the adoption of a risk-averse strategy to reduce the risk of energy shortfall. In a further experiment, dominant house sparrows in a flock also increased their use of scrounger when on a negative energy budget (Lendvai et al., 2006). Similar experiments with zebra finches have provided inconsistent results. In one study, birds in poorer condition scrounged more, as did the sparrows in the above study (Mathot and Giraldeau, 2010a). However, in a more recent study, producing generated a more, rather than less, stable food intake rate over a fixed time period, and birds with higher foraging requirements actually produced more (David and Giraldeau, 2012). Nevertheless, taken together, the above results suggest that scrounging has attributes of a risk-averse strategy and may respond to some extent to manipulation in foraging requirements.

2.5.2.4. Spatial Position

The prediction that producer use should be more frequent at the edges of a group was tested in flocks of nutmeg mannikins by assigning some birds to play producer and determining their position relative to the centre of the group (Flynn and Giraldeau, 2001). In these flocks, producers tended to occur at the edges of the groups and flocks with proportionately more scroungers were also more compact, as predicted by the Barta et al. (1997) spatial model of scrounging. The model assumes that each individual only uses one tactic. However, some of the scroungers in the experiment could not be prevented from producing, which may have affected their relative position in the group. The inevitable increase in local forager density as scroungers aggregate temporarily in the same area constitutes a more problematic issue when testing the model. While such an increase is predicted by the model, it may also, to some extent, represent a residual after-effect of patch exploitation. Perhaps the solution is to test these predictions in the absence of food.

Similar findings regarding the spatial distribution of producers and scroungers have been documented in other species, including free-living European tree sparrows (Monus and Barta, 2008), chacma baboons (King et al., 2009), and capuchin monkeys (di Bitetti and Janson, 2001). Although these results provide support for the above spatial model of scrounging, it is important to keep in mind that many factors other than producing and scrounging can influence spatial position in a group, including level of hunger and dominance status (Krause, 1994a). In capuchin monkeys, subordinate individuals may be forced to the edges of the group to provide protection against predators to those more inside the group (di Bitetti and Janson, 2001). In addition, producing may be more difficult in the centre of a group due to increased competition, forcing central individuals to scrounge more. Testing the spatial location prediction can be fraught with ambiguity.

2.5.2.5. Learning

Whether individuals involved in a PS game use a specific learning rule has not yet been documented. However, using a learning rule to achieve the stable equilibrium frequency of scrounger should produce tell-tale patterns that can be tracked by researchers. For instance, individuals should gradually adjust scrounger use to current environmental conditions and, perhaps more tellingly, past patterns of tactic use should continue to exert an influence on current tactic use because of memory lag. To address these issues, some nutmeg mannikins were trained in a context that favoured high or low scrounging behaviour. All these individuals were then tested under the same condition. The use of a

learning rule should lead to gradual changes in scrounging reflecting past preferences. Indeed, the efficiency of scrounging improved when individuals encountered new foraging contingencies. Crucially, individuals with prior experience
with the current scrounging conditions adapted more rapidly (Morand-Ferron
and Giraldeau, 2010). A similar finding was documented in house sparrows.
Young birds were trained under conditions that differentially reinforced the use
of scrounger. When tested subsequently in a shared aviary, young birds with
positive reinforcement for scrounging showed higher levels of scrounging, suggesting that past experiences with tactic payoffs can determine current tactic use
(Katsnelson et al., 2008). In a further experiment with the same species, adult
birds that were rewarded for using companions as cues to locate food patches
on a given day used scrounger to a greater extent the following day in a novel
environment, suggesting again that differential success associated with the use
of a foraging tactic can influence the choice of tactics subsequently (Belmaker
et al., 2012).

I note that different learning rules predict different trajectories for scrounger
use through time and also different stable equilibrium frequencies. Testing these
quantitative predictions represents an important next step.

2.5.3. Phenotypic Limitations

Asymmetric players are expected to use different tactics at equilibrium and
to experience different payoffs. Stable equilibrium frequencies have not been
tested quantitatively in an asymmetric PS game, but asymmetry among players
appears a common feature in many species. This is an important consideration
because symmetric and asymmetric PS games predict different outcomes, as
described earlier.

Several factors have been related to specialized scrounger use in animals, including dominance status, personality traits, and competitive ability
(Table 2.1). In this compilation, I excluded cases of kleptoparasitism involving
single prey items, and concentrated instead on cases where food patches contain
multiple items. I also excluded studies that focused exclusively on outcomes at
food patches. Indeed, the observation that dominant foragers exclude subordinates from a patch does not indicate that dominance status is related to tactic
use overall since both low- and high-ranking individuals may find food patches
equally. In the remaining cases, I looked for evidence that individuals with different phenotypic attributes allocate more time to producing or scrounging.

The most obvious finding from this compilation is that subordinates act as
food finders for the benefits of more dominant companions in many species.
Why this is not always the case may be partly related to differences in food
patch characteristics from one study to another. In the first empirical study of
producing and scrounging, dominance was in fact a poor predictor of tactic
use in house sparrows (Barnard and Sibly, 1981). Other studies with this and
other species, however, uncovered a strong association between scrounger use

TABLE 2.1 Phenotypic Attributes Correlated with Scrounger Use in Animal Species

Species	Trait	Correlation with Scrounger Use*	Reference
Birds			
House sparrow	Dominance	0	Belmaker et al. (2012)
House sparrow	Dominance	+	Liker and Barta (2002)
House sparrow	Dominance	+	Lendvai et al. (2006)
House sparrow	Dominance	0	Barnard and Sibly (1981)
Pigeon	Dominance	0	Giraldeau and Lefebvre (1986)
Dark-eyed junco	Dominance	+	Baker et al. (1981)
Dark-eyed junco	Dominance	+	Caraco et al. (1989)
Nutmeg mannikin	Dominance	0	Giraldeau et al. (1990)
Zebra finch	Dominance	0	Giraldeau et al. (1990)
Zebra finch	Dominance	0	Beauchamp (2006)
Zebra finch	Dominance	0	Biondolillo et al. (1997)
White-throated sparrow	Dominance	0	Wiley (1991)
Harris's sparrow	Dominance	+	Rohwer and Ewald (1981)
Barnacle goose	Dominance	+	Stahl et al. (2001)
Bald eagle	Dominance	+	Hansen (1986)
Mexican jay	Dominance	0	McCormack et al. (2007)
Zebra finch	Food-finding ability	–	Beauchamp (2006)
Barnacle goose	Shyness	+	Kurvers et al. (2010)
House sparrow	Ability to learn	–	Katsnelson et al. (2011)
Mammals			
Tufted capuchin	Dominance	0	Di Bitetti and Janson (2001)
Ring-tailed coati	Dominance	+	Hirsch (2011)
Fish			
Striped parrotfish	Dominance	+	Clifton (1991)

*The following symbols indicate the sign of the correlation: positive (+), negative (–), or absent (0).

and dominance status. Barnard and Sibly (1981) used patches with few food items and scrounging rarely involved interactions between patch finders and scroungers. In the other studies, scroungers could displace food finders, leading to an association between scrounger use and dominance status.

As far as I know, the all-or-nothing pattern of scrounger use in relation to dominance status has not been empirically documented. Instead, scrounger use typically increases with dominance status. Dominance may not be the only trait associated with scrounger use, as indicated by the fact that food finding ability and personality traits also correlate with scrounger use. It appears that the ability to explore, which may allow individuals to find food faster or to learn individual tasks more easily, can facilitate the use of producer.

Phenotypic limitations have many consequences for group members in the PS game. In particular, some individuals may be more successful than others in the same group even at the stable equilibrium. Furthermore, some individuals may be less able to adjust scrounger use than others to changing environmental conditions. For instance, shy foragers in a group may perform rather poorly in an environment that requires a greater use of producer.

Phenotypic differences associated with personality traits or physical attributes may persist over time and create consistent patterns of tactic use. Empirical studies have indeed documented consistency in tactic use in different contexts in bird species (Beauchamp, 2001a; Morand-Ferron et al., 2011). However, the fact that success with one tactic in one context predisposes an individual to use the same tactic in another context, as we have seen earlier, complicates the interpretation of the results. Such carry-over effects can potentially explain consistency in tactic use. Disentangling the relative contribution of learning carry-over effects and phenotypic limitations represents a challenge for future studies.

2.5.4. Targets of Scrounging

PS models assume that scroungers should join all food patches produced by others regardless of the identity of the food finders. Preference for patches discovered by a particular type of individuals should be avoided because it would lead to deviation from the stable equilibrium frequency of tactic use. However, interactions at a food patch between scroungers and food finders are not necessarily neutral. A scrounger, for instance, may evict the food finder to obtain a larger share of resources and may, therefore, target lower-ranking companions when scrounging. Therefore, individuals may increase their share of resources at a patch by selectively avoiding or selectively choosing to scrounge from patches exploited by specific individuals. Evidence for targeted scrounging has been documented in many species and involves factors such as pair bond (Beauchamp, 2000a), dominance status (McCormack et al., 2007), and sex (King et al., 2009). In general, the type of individuals exploiting a patch together can determine whether the patch is shared amicably or not (Bugnyar and Kotrschal, 2002; Ha and Ha, 2003).

The effect of kinship, in particular, has attracted much attention with respect to targeted scrounging. Showing a preference for patches discovered by kin can make sense because food finders may show a greater tolerance for scrounging kin (Mathot and Giraldeau, 2010b). Nevertheless, the evidence has been mixed on whether kinship influences scrounging choices. Some studies have reported no effect of kinship on scrounging (Ha et al., 2003; King et al., 2009; McCormack et al., 2007). House sparrows, on the other hand, are less likely to use aggressive scrounging when joining patches produced by relatives or tended to avoid joining patches discovered by kin, suggesting strategies to reduce the cost of scrounging for closely related companions (Toth et al., 2009). A similar finding was documented in zebra finches where tolerance for scrounging was higher in groups composed of more closely related companions (Mathot and Giraldeau, 2010b).

Evidence for targeted scrounging certainly poses a problem for PS models which all assume random joining. Including selective joining and also selective use of aggression will add a new layer of complexity in future PS models.

2.6. CONCLUDING REMARKS

The literature on producing and scrounging has provided a wealth of theoretical and empirical studies. This burgeoning literature reflects the importance of exploitative strategies in foraging behaviour research. In Chapter 1, I showed how foraging in groups can increase foraging efficiency and how competition among foragers can reduce expected benefits. Scrounging can also be viewed as a cost of foraging in groups. Perniciously, scrounging is expected to be more prevalent in large groups, those very groups that should provide more, rather than fewer, benefits to individual foragers. Determining the factors that influence scrounging in groups should thus remain high on the agenda of behavioural ecologists. Scrounging studies, once the confine of laboratory research with birds, have expanded in scope in recent years. Scrounging has now been investigated in the field and involves a broader range of species. Restrictive assumptions in the basic PS model have also been relaxed subsequently, which should make it easier to examine predictions in a broader range of conditions and species.

The greatest weakness in the producing and scrounging literature appears to me to be the relative lack of evidence for the incompatibility assumption. This is also a weakness for information-sharing models as well, which make the opposite assumption. If some compatibility between search modes does exist, a new approach will be needed allowing for simultaneous use of the two search modes with some form of cost compensation. This possibility was addressed early on in the PS literature by considering an opportunist tactic that allows simultaneous use of the two search modes. However, subsequent PS models have neglected this tactic. I also note that, empirically, the lack of a behavioural marker for tactic use will make it difficult to determine what proportion of time is allocated to the opportunist tactic since the end result of joining a patch can be a manifestation of different tactics.

Beyond the issue of incompatibility, the two research frameworks may not be as diametrically opposed as first thought. In fact, complicating matters, many predictions from PS models can be accommodated by the information-sharing framework by incorporating constraints on joining. For instance, in a spatially explicit setting, foragers may not be able to perceive scrounging opportunities beyond a given radius of detection (Ruxton et al., 1995). This simple adjustment would bring the stable equilibrium frequency of scrounging much closer in the two frameworks. In addition, foragers will be less likely to reach small patches in time (those with a large producer's advantage), mimicking the prediction that scrounging should decrease as the producer's advantage increases. Joining in this type of environment will increase with group size and when patch encounter rate is low, just as predicted by PS models. The actual radius of detection may also be considered a target of selection. For instance, it may turn out that subordinate group members should evolve a lower detection radius than more dominant companions, explaining why dominance status correlates with tactic use. At the qualitative level, predictions from the two research frameworks may thus be hard to distinguish. A challenge for future theoretical work must be to provide unique predictions for each research framework.

At the mechanistic level, learning based on the costs and benefits experienced by individuals when using the different tactics is thought to underlie variation in scrounger use. An alternative view has been proposed by Sumpter (2010) who suggested that joining increases nonlinearly with the number of individuals already exploiting a patch. For example, a large patch will take longer to deplete, and the number of foragers present should increase over time (Sumpter, 2010). As the number of individuals in the patch increases and passes a response threshold, yet more individuals should join the patch. This leads to the prediction that scrounging should increase with patch size, just as predicted in traditional PS models. In this scenario, natural selection works on the functional form of the relationship between joining and the number of foragers present at a patch rather than tweaking how individuals respond to past failures and successes when joining. This modelling leaves aside details of patch detection and potential incompatibility between searching for food and joining. However, the model proposes that what matters is how individuals respond to the number of companions foraging in a patch. This is certainly an avenue worth investigating in future work.

Part B

Prey

Antipredator Ploys

3.1. INTRODUCTION

In the upper Bay of Fundy, on Canada's eastern coast, semipalmated sandpipers, a small shorebird, aggregate in large numbers during fall migration to feed on abundant mudflat amphipods. Individuals spend two to three weeks in the Bay area to accumulate the fat needed to fuel their nonstop flight to coastal areas of South America (Hicklin, 1987). Roosting flocks on the shore during high tide can exceed 100,000, and feeding flocks in the thousands are common during low tide. Such aggregations attract the attention of many birds of prey, including peregrine falcons, which nest on nearby cliffs. Roosting flocks are harassed frequently by falcons, perhaps explaining why sandpipers often stay on the wing over the ocean during high tide, investing considerable energy (Dekker et al., 2011).

Sandpipers deploy several lines of defences against predation attempts by falcons. The sheer number of anxious eyes scanning the horizon certainly increases the chances of locating falcons before they come fatally close. Indeed, the ability of sandpipers to detect an approaching falcon is uncanny. In addition to falcons, sandpiper flocks also attract many bird watchers who, despite all their fancy equipment, always spot a fast approaching falcon long after the sandpipers raise the alarm. Even though this interrupts feeding at a crucial time of the year, sandpipers allocate a sizable proportion of time to vigilance against predators, highlighting the importance of early detection.

Not all birds in the flock detect the approaching predator, and yet the whole flock becomes alerted very rapidly. The benefits of early detection through vigilance would not translate into an advantage for the flock if information about

Social Predation. http://dx.doi.org/10.1016/B978-0-12-407228-2.00003-2

67

a detected threat failed to pass quickly through the flock. Rapid information transfer ensures that individuals that fail to detect the predator directly can flee sooner than if they waited to detect the predator on their own. In fact, birds are so attuned to fright reactions by companions in the flock that false alarms are quite common (Beauchamp, 2010b). Detectors sometimes make mistakes in identifying a threat like, say, an inoffensive crow that from afar looks like a falcon, and their quick departure triggers a wave of escape across the whole group, suggesting that individuals react to the alarm cues of their companions rather than to the alleged threat itself.

The sheer number of targets for the predator to choose from represents another line of defence available to sandpipers. Falcons can only capture one prey at a time during an attack. Therefore, the risk of predation for any individual can be effectively diluted by the mere presence of many alternative targets. Obviously, if larger groups attracted proportionately more falcons, the dilution advantage would be somewhat reduced. If predators attack the first individual in the group that they encounter during an attack, then the presence of companions can also reduce risk by adding layers of protection between an individual and the predator. In this case, risk can be manipulated by pursuing more advantageous positions in the group. For example, positions that are not in the direct line of attack of a falcon approaching from the outside of the group offer more protection.

Once an alarm has been raised in the group, individuals in some species simply flee as fast as possible to safety. In other species, individuals adopt defensive formations aimed at dissuading predators. Sandpipers cannot hide anywhere on the open mudflats. In addition, such small birds cannot put up much of a fight against a much larger predator like the falcon. However, sandpipers cluster in tight airborne formations that move in erratic fashion, alternatively flashing their upper darker parts and their white belly. From a distance, waves of flashing pass through the group very rapidly, and such displays may make it more difficult for the predator to target any single prey in the group. That such formations serve as a line of defence is readily evident since the chances of capture increase dramatically for any individual that becomes separated from the swirling flock.

This example with sandpipers illustrates the various lines of defences available to a prey species that lives in groups. Some defences are deployed early in the predation sequence, like detecting the predator at the onset of attack. Other defences come into play later in the sequence when the predator attempts to capture fleeing prey. Caro (2005) describes the many ways by which birds and mammals may decrease their risk of predation at different stages of the predation sequence. Several of these mechanisms apply only to prey in groups, such as the dilution and confusion effects. I will focus here on the antipredator ploys available to animals that live in groups. Several of the ploys described above for sandpipers are also used frequently in other species, and I will describe these, including collective detection, dilution, and confusion. I will leave two aspects to separate chapters in view of the large literature devoted to them: antipredator vigilance (Chapters 4 and 5) and the selfish-herd hypothesis (Chapter 6). Most of the themes developed here apply to both single- and mixed-species groups. However, group formation may provide unique benefits to participants in mixed-species groups, and these will be covered in Chapter 8.

3.2. ANTIPREDATOR PLOYS

3.2.1. Collective Detection

The concept of collective detection is an old one: more than 100 years ago, Galton noted that in a herd of cattle many eyes and ears are available to detect predators, and crucially that an alarmed individual soon warns all the others (Galton, 1883). Detecting a predation threat can be divided into two aspects: vigilance and then the response in the group once a threat is detected. Here, I am concerned with the latter. The role of vigilance in early detection of predation threats will be discussed in the following chapter.

In influential models of antipredator vigilance, detection by one individual in the group is expected to alert all remaining group members instantaneously, which means that early detection allows everyone in the group to escape before the predator closes in (Lima, 1990). This information transfer about predation threats from detectors to non-detectors in the same group is called collective detection. In terms of survival, the benefit to collective detection is the difference between the probability of surviving an attack for a non-detector, when at least one other companion detects the predator, and the probability of surviving when no group member detects the predator soon enough. Without collective detection, early detection only benefits early detectors, and the remaining group members must rely on other means to evade predation.

3.2.1.1. Signals Used for Collective Detection

What cues can be used to transfer information about predation threats from detectors to non-detectors within the same group? Chemical alarm substances are known to be produced in many taxa, including fish (Krause, 1993) and invertebrates (Machado et al., 2002), in which they are known to provide information about predation risk. However, the release of these compounds requires mechanical damage of the skin by a predator, and it is therefore unlikely that chemical communication can work before the predator captures an individual, as is required by collective detection. Nonetheless, such cues might alert prey to recent local successful predation, which might usefully alert them to high-risk areas. This is somewhat similar to collective detection, except here prey that have recently been in the same area can warn others about heterogeneous predation risk in space rather than about specific impending attacks.

Acoustic cues have many desirable properties for communicating alarm within a group. Such cues are typically loud and rapidly transmit over large distances. In many species of birds and mammals, detectors of predation threats emit alarm calls that very quickly alert non-detectors in the same group (Hoogland, 1979; van der Veen, 2002). Another example of an auditory cue that communicates alarm is the plop produced by Iberian green frogs (*Rana perezi*) jumping into the water in response to a predation threat (Martín et al., 2006).

Noises produced by an escaping bird, including whistles or whirring sounds produced by a rapid take-off, are also used to transfer information about threats (Coleman, 2008; Hingee and Magrath, 2009). The intensity of the sounds produced during an alarm probably acts as a cue. But how do we know that it is not simply a visual cue associated with the rapid departure, rather than an auditory cue, that is triggering the subsequent alarm? In these two studies, the authors played back sounds emitted by alarmed birds and observed escape responses by foraging individuals in the absence of visual signals. The relative contribution of visual and auditory cues of alarm has also been examined in common voles (Gerkema and Verhulst, 1990). Voles that were prevented from hearing escape noises from alarmed companions often failed to react to an alarm raised. Furthermore, at large distances between detectors and non-detectors, only auditory cues were effective in transmitting alarm.

Visual cues can also be used to communicate alarm. In contrast to sound production, however, visual cues have a shorter range, especially in cluttered habitats, and they can only be perceived when the receiver of information is facing in the general direction of the sender. Changes in locomotor behaviour in response to predation threats are thought to visually communicate alarm in many species. For instance, in mammals, the rapid movements associated with flight contrast with slow foraging movements and provide information about danger (Berger, 1978; Caro, 1986). In some bird species, rapid alarm flights not associated with intention movements, which are commonly displayed in many species prior to casual departures, communicate alarm to others (Davis, 1975). In a marine insect, the arrival of a fish predator leads to changes in velocity and frequency of turning for peripheral foragers, which act as an alarm signal for the remaining group members (Treherne and Foster, 1981). A similar effect was found for prey fish foraging in a habitat that precludes visual detection of the predator for most individuals in the group (Handegard et al., 2012).

Animals can respond to both the quality and number of visual alarm cues. Although changes in locomotor activity in only one group member may be ambiguous with respect to causation, several flight reactions in rapid succession can strongly signal a threat (Cresswell et al., 2000; Lima, 1995b; Marras et al., 2012).

Mechanical perception plays an important role in some species. Alarm signals can propagate through the substrate and influence companions independently of correlated visual or auditory cues. Agitated movements in the water in response to threat detection can transmit as waves, which can be detected rapidly by all group members (Vulinec and Miller, 1989). Foot drumming in alarmed fossorial mammals transmit alarm through vibration in the ground, which can be detected by individuals out of visual or auditory range (Randall, 2001). Spiders can perceive vibrations on the silk of their colony webbing produced by the evasive movements of companions, allowing individuals to react more quickly to an approaching predator (Uetz et al., 2002). Mechanical communication probably varies in effectiveness depending on the nature of the substrate that carries vibration, and transmission was probably restricted to immediate group members in the above examples.

Threat detection may also be associated with alert postures or behaviour patterns that communicate alarm to the rest of the group. In a fish species, alerted individuals flick their fins to transmit information about predation threats to nearby companions (Brown et al., 1999). In another fish species, the startle responses of individuals exposed to a predator transmit alarm to companions that cannot see the predator (Magurran and Higham, 1988). Rats detect alarm when their neighbours become silent, a cue that a threat is being investigated (Pereira et al., 2012). In yet another example, common voles exposed to a bird of prey model often freeze, and this cue alerts companions that cannot see the predator (Gerkema and Verhulst, 1990).

Alarm calls have probably evolved to play a role in transmitting information about predation threats, but many visual and mechanical cues are probably inadvertent, the simple by-product of individual flight responses. As such, such cues cannot be considered intraspecific signals of danger (Maynard Smith and Harper, 2003). Even if a case can be made that such cues act as danger signals, the cue may have evolved to provide information in other contexts. For instance, fin flicking in fish may be aimed at predators rather than companions (Brown et al., 1999), and may have been subsequently co-opted for intraspecific signalling. A similar argument can be made for ink production in squids, which can act as both a predator deterrent and a cue to conspecifics about a threat (Wood et al., 2008).

When the behaviour of an individual has been shaped because of the alerting effect it has on others (the cue has become a signal), it is natural to ask what the benefit to the signaller might be from warning others. This benefit may be kin selected if it increases the survival of related individuals (Sherman, 1977). Alternatively, there may be selfish benefits. For example, a ground-feeding bird that spots an approaching falcon may benefit from encouraging flockmates to take flight at the same time, so as to benefit from the confusion effect (see "How the Confusion Effect Works" later in the chapter). However, it is important to bear in mind that there can be costs to such signalling. One natural cost may be energetic expense and another may be an increase in personal predation risk. This might happen if issuing the call attracts the attention of the predator, or requires the signaller to delay its own fleeing until the signal is sent. Such considerations are important to understand why some cues evolve into shaped signals.

3.2.1.2. Effectiveness of Collective Detection

To be effective, collective detection must enable non-detectors to escape more rapidly than would be the case without the signal from detectors. The effectiveness of collective detection signals has been evaluated using the speed of transmission of information in the group following detection, or the length of time needed by a non-detector to react to the information provided by a detector.

In some species, the speed of transmission of information within the group is much higher than the speed at which a predator approaches the group, suggesting that individuals that are alerted last can still enjoy an early warning effect. In their study of a marine insect, for instance, Treherne and Foster (1981) noted that agitation by peripheral foragers upon detection of a predator was transmitted very quickly through the group so that individuals could initiate avoidance behaviour before they became directly aware of the predator. This phenomenon was dubbed the 'Trafalgar effect', referring to the transfer of information using signals sent from ship to ship, which provided an early warning to Admiral Nelson's fleet of an imminent attack long before the French and Spanish armada was in sight. This effect has also been documented in a fish species (Godin and Morgan, 1985).

Ultimately, the early warning provided by collective detection should lead to enhanced survival. However, few studies have documented the fitness consequences of collective detection. In one example, predator wasps experienced lower success during their attacks on a social spider species because the evasive responses of spiders provided an early warning to non-detectors located farther away on the colony webbing (Uetz et al., 2002). Similar studies are much needed to evaluate the fitness benefits of collective detection.

An advantage of collective detection can also be evident if the delay in reaction of a non-detector after initial detection is shorter than the delay that would be expected if the non-detector was alone or did not receive the warning signal. Such an effect was documented in Iberian green frogs (Martín et al., 2006) (Fig. 3.1), in one species of bird (Fernández-Juricic et al., 2009), and in one species of fish (Webb, 1982).

Collective detection may not always be effective. In one avian species, for example, delays to responding by non-detectors were never shorter than what would be expected for solitary individuals, perhaps because this territorial species does not typically live in groups (Fernández-Juricic et al., 2009). Reaction to the sight of a simulated predator did not influence nearby unaware companions in groups of house finches (Roth et al., 2008), in sharp contrast to previous findings in fish (Magurran and Higham, 1988; Mathis et al., 1996), mammals (Gerkema and Verhulst, 1990), and one amphibian species (Crane et al., 2012). It is perhaps the case that fright reactions in birds provide fewer cues of threat

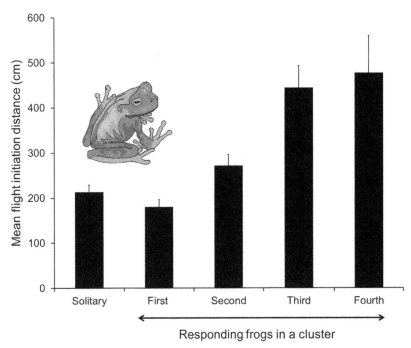

FIGURE 3.1   Evidence of collective detection: Iberian green frogs respond earlier to a threat when in a cluster by initiating flight at a greater distance. Mean flight initiation distance of a solitary frog is contrasted with that of the first, second, third, and fourth individuals responding in a cluster. Error bars show one standard deviation. *Adapted from Martín et al. (2006).* (For interpretation of the references to colour in this figure legend, the reader is referred to the online version of this book.)

detection than in other species, but it is difficult to be conclusive given the paucity of research on this topic in birds.

Perhaps the most spectacular failure to document collective detection comes from a study of dark-eyed juncos fleeing from a simulated attack (Lima, 1995b). Lima targeted individual birds in foraging flocks with a small ball launched from a chute. The ball moved noiselessly down the chute and could only be seen by the targeted birds. Despite their quick flushing to safety, targeted birds failed to entice others to flee as well. Indeed, when only one bird was targeted, only about 10% of non-detectors flushed to safety. However, when two birds were targeted, close to 70% of non-detectors flushed to safety. The obvious interpretation of these results is that ball-induced departures failed to provide the sort of cues that would indicate danger to others. However, in a further experiment, a stuffed bird of prey induced just as many flushes in non-detectors as did the ball, suggesting that the ball was just as effective as the model bird of prey. Nevertheless, only 10% of the non-detectors flushed in response to either type of stimuli, suggesting that these stimuli were not very effective in triggering alarm in the group. It would thus seem that flushes to safety are ambiguous cues of alert for these birds, and can be mistaken for casual, non-threat-related departures.

3.2.1.3. Collective Detection When Cues Are Ambiguous or Detectors Are Wrong

Collective detection implies that non-detectors will react as quickly as possible to alarm cues provided by detectors. However, is this always the best strategy? The junco study presented earlier raised the possibility that alarm cues may be ambiguous. In this case, non-detectors are left wondering whether or not to respond to alarm cues. In some situations, instead of blindly initiating a flight response, the best response may be to scan the surroundings to verify the reliability of the signal prior to investing in energetically costly fleeing. Which is best will depend on the relative costs of responding to false alarms versus the predation risks of delaying flight until scanning reveals that an attack really is in progress (Lima, 1994a; Proctor et al., 2001).

Another possibility is that detectors provide unambiguous cues of alarm to the rest of the group, but that the information that the detectors have acquired from the environment is not completely accurate. For instance, detectors may mistake harmless environmental signals for real threats and raise the alarm needlessly, causing what is known as a false alarm. False alarms are quite common in many species of birds (Beauchamp, 2010b; Crook, 1960; Kahlert et al., 1996) and mammals (Blumstein et al., 2004b; Hoogland, 1981). In one study, for example, alarms attributed to harmless species or to unknown causes represented up to 75% of all alarms given (Cresswell et al., 2000). Similarly, in semi-palmated sandpipers roosting on shore during high tide, there was, on average, about one false alarm every minute (Beauchamp, 2010b).

False alarms reduce foraging time and are energetically costly. When classification errors by detectors are common, non-detectors may be selected to respond only when more than one detector raises the alarm (Beauchamp and Ruxton, 2007b). Indeed, in redshanks, a shorebird species, non-detectors reacted more quickly when more detectors of a predation threat were initially involved (Cresswell et al., 2000) (Fig. 3.2). Such a delay in responding to an alarm has also been documented in large groups of weaverbirds, suggesting that individuals are attempting to assess the cause of alarm before responding (Lazarus, 1979). A delay makes more sense in a larger group since other lines of defences, such as risk dilution and confusion (explored later), are available to protect individuals should the threat be real (Ydenberg and Dill, 1986).

The above discussion suggests that the reaction to alarms raised by detectors may actually be modulated depending on whether detectors are reliable and/or transmitting unambiguous information about a potential threat. Such adaptive adjustments in reaction speed need to be taken into account when assessing the effectiveness of collective detection. In some cases, slow reaction may actually make more sense than the almost instantaneous reaction postulated in collective detection models.

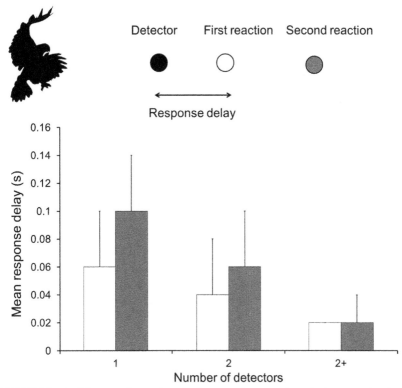

Detector   First reaction   Second reaction

Response delay

FIGURE 3.2   Avoiding superfluous alarm flights: more detectors of attacks by sparrowhawks in groups of redshanks lead to faster response time by non-detectors. Median response delays are illustrated for the first bird to depart following detection (white bars) and for the second bird (grey bars). Error bars show 95% confidence intervals. *Adapted from Cresswell et al. (2000).*

3.2.1.4. Factors That Influence Collective Detection

As collective detection relies on the transmission of visual, auditory, or mechanical cues of alarm to other group members, any environmental variable that interferes with information transfer will influence collective detection. Obstacles to collective detection are mostly known for visual cues. However, it would be interesting to see whether increases in, for instance, environmental noise levels can interfere with collective detection based on auditory cues like alarm calls or escape noises. Obviously, a greater distance between neighbours will decrease the effectiveness of cues transmitted as waves or vibrations, because the detection range for these cues must be quite limited.

For ball-induced departures in juncos, Lima and Zollner (1996) showed that visual barriers between foragers decreased the effectiveness of collective detection: far fewer flushes to safety occurred after initial detection when a small wall prevented birds from seeing the early stages of flushes (Lima and Zollner, 1996). One difficulty with the use of visual barriers between foragers is that the perception of group size may be altered at the same time. Foragers may not consider birds that are not visible as part of their group, and therefore they focus their attention on visible companions, which may detract from their ability to monitor what is happening elsewhere.

Increasing the distance between companions can also be used to examine the limits of collective detection. In the same study, Lima and Zollner (1996) found that increasing inter-individual distances also reduced the effectiveness of collective detection. Monitoring neighbours probably becomes more difficult when individuals are farther apart because fewer companions fall within the field of view at any one time. In addition to monitoring difficulties, obstacles like rocks or patches of vegetation may block the view more frequently when foragers are farther apart, reducing the effectiveness of collective detection (Metcalfe, 1984). However, this was not the case in the Lima and Zollner study as individuals foraged in an open habitat. A similar decrease in the effectiveness of collective detection with distance has been documented in common voles (Gerkema and Verhulst, 1990) (Fig. 3.3). Obviously, at some point, distant companions will not be treated as part of the group. This may be the point when auditory cues provided by distant companions become imperceptible or when neighbours are too difficult to monitor visually. At that point, collective detection ceases to provide benefits to group foragers. Indeed, common voles failed to react to alarms raised by neighbours 0.1 m away when only visual cues could be used (Gerkema and Verhulst, 1990). In European starlings, neighbours 5 m away exert little influence on vigilance, and orientation between birds becomes more random (Fernández-Juricic et al., 2004c). In general, in birds at least, the probability of detecting a flushing response decreases very rapidly with distance, especially for smaller species with lower visual acuity (Fernández-Juricic and Kowalski, 2011). Consistent with the point of view that distance between neighbours affects information transfer, waves of flushing in groups tend to be slower when inter-individual distances increase (Beauchamp, 2012a; Hilton et al., 1999; Quinn and Cresswell, 2005).

3.2.1.5. Ontogenetic Changes in Collective Detection and Learning

Do animals have to learn how to react to alarm cues provided by companions? This is an important issue as it suggests that the effectiveness of collective detection may vary according to age. Few studies have examined ontogenetic changes in collective detection abilities. In one fish species, the social transmission of fright responses starts about 30 days post-hatching, suggesting that the recognition of alarm cues is not effective in young individuals (Nakayama et al., 2007). Incomplete maturation of the senses would delay responses to alarm signals, yet schooling, which involves use of much of the same senses, begins a full two weeks prior to the onset of fright recognition. It thus appears unlikely that development of the sensory apparatus is limiting the response to alarm cues. It would be interesting to extend such findings to other taxa such as birds and

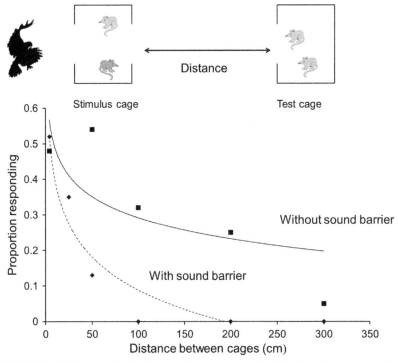

FIGURE 3.3   Limits to collective detection: fewer common voles flee when alarmed companions are farther away especially when a sound barrier prevents the transfer of auditory information. The experimental set-up comprised a stimulus cage where voles were exposed to a model predator and a test cage where voles could detect alarm signals from stimulus companions but not the predator. *Adapted from Gerkema and Verhulst (1990).*

mammals, where young and inexperienced individuals join adult groups at the end of the breeding season. A failure to react to the alarm cues provided by other group members would certainly reduce chances of survival during an attack.

In addition to developmental changes in collective detection abilities, the quality of the information provided by frightened individuals may also change with age. Young and inexperienced individuals may respond with alarm to a much broader range of stimuli than adults, many of which are harmless to adults (Cheney and Seyfarth, 1990). It would, therefore, pay individuals to evaluate the reliability of the signaller when responding to alarm signals (Blumstein et al., 2004b).

3.2.2. Encounter-Dilution

While collective detection relates to the ability to detect predators and escape early, encounter and dilution effects act to reduce the individual risk of attack in the group. Although both encounter and dilution effects can act alone, the two mechanisms may work together and form what is known as attack abatement (Turner and Pitcher, 1986).

3.2.2.1. Encounter Effect

The encounter effect is a form of predator avoidance; it works by decreasing the rate of encounter between predators and prey. To benefit from the encounter effect, individuals of a social species must encounter predators less often when in groups than when alone. If a predator can capture all individuals in a group during an attack, the encounter effect will work only if groups are less likely to be detected than solitary foragers. If a predator can only capture one prey in the group, the encounter effect will work if the rate of encounter with a group of size N is less than N times the rate of encounter with a solitary forager (Inman and Krebs, 1987). Notice that in this case, antipredator benefits arise from a combination of encounter and dilution effects.

3.2.2.1.1. How the Encounter Effect Works

The encounter effect hinges on changes in the likelihood of detecting prey as a function of group size. To illustrate how this works, imagine prey distributed along the periphery of a circle and a predator located in the centre scanning the periphery to locate individual prey. In the simplest case, the sensory apparatus of the predator has enough resolution to detect each prey separately. The area scanned by a predator at any one time is called a location. With the above spatial resolution, each location includes at most one individual prey. If the predator randomly selects one location to look for prey, the probability of being targeted for an individual prey should be the same whether N prey are distributed randomly or aggregated in one group of N. Now, if the resolution of a predator is coarser, a location may contain more than one individual. It is when a location contains more than one prey that detection issues come into play. For instance, it may be easier to detect a location with several prey, each increasing the conspicuousness of the location to the predator (Cullen, 1960; Vine, 1973). Spatially aggregated individuals are thus likely to encounter a predator more frequently (Hassell, 1978). In this simple scenario, the distance between predator and any prey, whether alone or in groups, is always the same. As discussed below, this assumption is probably unrealistic and larger groups may actually be harder to locate.

Aggregation may still offer protection to prey even if predators can detect larger groups more easily, for at least three reasons. First, while a group of N individuals may be more conspicuous than any of the N solitary individuals scattered in the habitat, it is unlikely that the group is N times more conspicuous (Inman and Krebs, 1987). For instance, a predator approaching from the outside may not be able to detect all individuals in the group as some may be hidden from view. Therefore, we may expect encounter rate to increase with group size but less than proportionately, which should thus reduce the per capita capture rate if the predator cannot capture all individuals in the group upon attack.

Second, the rate of encounter with aggregated prey may be more variable over a fixed time horizon because prey are distributed over fewer locations. A run of bad luck in locating prey in groups may force the predator to look elsewhere, thus reducing encounter rate for prey in groups (Taylor, 1984). In fact, it has been suggested that this potential disruption in the search process of predators caused by spatial clumping has been a major factor driving aggregation behaviour in dense locust populations (Reynolds et al., 2009).

Finally, while groups may be more conspicuous than solitary individuals, when seen from the same distance, the number of groups in a finite population must always be smaller than the corresponding number of solitary individuals. Therefore, the average distance between a searching predator and prey in groups will tend to increase with group size. For a simulated visual predator, combination of these two effects (conspicuousness and sparseness) did indeed result in an overall decrease in the rate of encounter with prey as group size increases (Ioannou et al., 2011).

Is it possible to conceive of a situation where detection risk increases faster than linearly with group size? Consider, for example, fish shoals targeted by human fishermen whose sonar can only detect large shoals against background scattering. Thus, human fishermen may select against fish forming large groups. It is not known to what extent natural predators also suffer from an inability to detect small groups of prey, but by analogy with fishermen, marine mammals or other species using sonar-like detection may be an appropriate group to explore biased encounters.

3.2.2.1.2. Empirical Evidence for the Encounter Effect

What is the evidence for the encounter effect? Many studies have reported the rate of attack on groups of different sizes, but this is not equivalent to encounter rate between predators and prey. This is because predators may not attack all groups that they encounter. For instance, profitability may vary as a function of group size because, say, large groups are more effective in evading attacks, and predators may thus skew their attacks to groups of specific sizes (Cresswell, 1994; Krause and Godin, 1995). Returning to the human fishermen example, economic arguments may explain why groups of a certain size are preferred. Even if small shoals could be detected as easily as large ones, it is likely that fishermen would ignore these since the (time and/or fuel) costs of deploying the fishing nets may be approximately the same for shoals of all sizes, but the rewards from a catch increase with shoal size. Thus, there will be a critical shoal size below which, even if detected, the shoal would not worthwhile catching. In short, while the encounter effect only involves detection ability, realized attack rate combines detection and the choice to attack.

Data on the encounter effect are sparser than those on attack rates. In a well-known study of the encounter effect, larger pupal groups of a stream-dwelling insect attracted more predators (Wrona and Dixon, 1991). However, this may arise because predators prefer to attack prey in larger groups, or remain in the vicinity of larger groups longer, and not because they encounter such groups more frequently. The converse can also be true. Larger groups of an aposematic butterfly species were attacked less frequently than solitary individuals (Finkbeiner et al., 2012), again suggesting differences in encounter rates with groups of different sizes. However, butterfly predators have probably learned to avoid brightly coloured, non-palatable butterflies, and a large aggregation of these butterflies makes the unpalatability signal even stronger, thus discouraging predation. Again, in this study, it is not clear whether groups of different sizes were encountered at different rates.

Many studies have encountered this difficulty in disentangling encounter rate from attack choices. For example, tracks in the snow from larger groups of elk mingle more often with wolf tracks, which the authors interpreted as an increased risk of encounter with predators (Hebblewhite and Pletscher, 2002). However, it may be the case that wolves can detect elk in small and large groups just as easily, but avoid chasing elk in smaller groups because of lower profitability. In a laboratory experiment with one fish predator, the rate of attack on prey fish was independent of school size (Morgan and Godin, 1985), suggesting, at first glance, that groups of different sizes were equally conspicuous to the predator. The authors noted changes in the positioning of the predator after visual detection of the prey, but did not report the time needed to detect prey in groups of different sizes, which would really indicate whether small and large groups were equally detectable.

In yet another study of encounters between predators and prey, the rate of attack by wasps on spider aggregations increased with group size in a non-linear fashion, which the authors interpreted as a perceptual constraint whereby larger colonies may not appear necessarily much larger than smaller ones (Uetz and Hieber, 1994). Although preferences by wasps to attack detected groups of different sizes were not examined, the non-linear change in attack rate with group size may indicate real differences in the detection rate of groups of different sizes. In a totally different system, parasitoids seeking insect larvae hosts to deposit their eggs landed more often on leaves with more larvae (Low, 2008), indicating either that larger aggregations are more detectable or that parasitoids avoid landing on leaves with fewer larvae.

In all of the above studies, it was not possible to isolate the encounter effect unambiguously. Detailed studies from the perspective of the predator are needed to assess the encounter effect. To date, few studies have addressed the encounter issue directly. We need more detailed studies like those detailed below on sticklebacks and their swarming prey. A first study found that the time needed by the sticklebacks to approach swarms of *Daphnia* decreased with prey group size (Ioannou et al., 2009), indicating that larger groups are more conspicuous to the sticklebacks. This is true in the particular case where the distance between the sticklebacks and swarming *Daphnia* is constant regardless of prey group size. In a second study, the sticklebacks took longer to visually detect larger groups of *Daphnia* (Ioannou et al., 2011) when the same total number of prey were distributed in groups of different sizes. In this case, the average distance between predator and prey increased with the size of the aggregation, making rarer, larger aggregations more difficult to find despite their greater conspicuousness when viewed at the same distance (Fig. 3.4).

Recent work in the laboratory involves subjects looking for arbitrary signals occurring alone or in clusters of different sizes. The main advantage of this approach lies in teasing apart detection of these clusters from the choice to attack, since subjects are rewarded when detecting any signals. In separate studies focusing on a bird species and human subjects, the rate of detection of arbitrary signals that were difficult to locate against the background increased but eventually levelled off with group size (Jackson et al., 2005; Riipi et al., 2001). An increase in cluster size for prey that are difficult to locate increases conspicuousness for visually searching predators. Difficulty in locating single items against a similar background may explain why parasitoids, in the above example, landed more often on leaves with more larvae.

In view of the scarcity of reliable data, we cannot yet draw many conclusions on the existence and nature of the encounter effect. It is essential that researchers control for the number of groups as well as the size of these groups

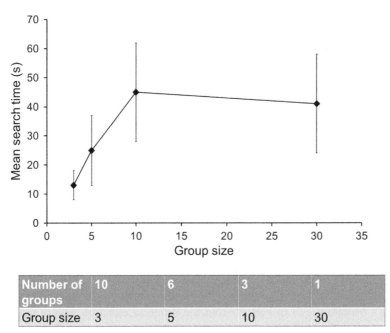

Number of groups	10	6	3	1
Group size	3	5	10	30

FIGURE 3.4   Group size and conspicuousness to predators: three-spined sticklebacks take longer to detect a group of prey as group size increases. Total number of prey was maintained constant at 30: an increase in group size means that fewer groups were present in the same volume of water. *Adapted from Ioannou et al. (2011).* (For colour version of this figure, the reader is referred to the online version of this book.)

because conspicuousness and sparseness can operate independently to determine the rate of encounter between predators and prey. All work so far has been
conducted with visually searching predators. It also remains to be seen whether
predator species using different sensory modalities also encounter groups of
different sizes at different rates.

3.2.2.2. Dilution Effect

Once an attack has been launched, individuals in a group can benefit from a
dilution of predation risk due to the presence of many alternative targets in the
group (Bertram, 1978). Obviously, dilution can only work if the predator cannot capture all prey in the group upon attack. To properly document a dilution
effect, it is necessary to measure number of attacks per individual per unit time
as a function of group size. A finding that attack rate is independent of group
size is not sufficient to infer that attack rate per individual will be equal to $1/N$,
where N represents group size. This is because some individuals may be more
likely to be attacked than others, depending, for instance, on their phenotypic
attributes or their position in the group. In addition, documenting a dilution
effect alone is not sufficient to determine that grouping provides antipredator
benefits because large groups may attract proportionately more predators, thus
reducing or even negating the dilution effect.

   In many invertebrate and vertebrate prey species, the proportion of individuals in a group that are captured per unit time during an attack has been found
to decrease as a function of group size (Foster and Treherne, 1981; Jensen and
Larsson, 2002; Lucas and Brodeur, 2001; Morgan and Godin, 1985; Sorato
et al., 2012; Turchin and Kareiva, 1989; Uetz and Hieber, 1994; Wrona and
Dixon, 1991). Returning to the stream-dwelling insect example, the proportion
of dead pupae in an aggregation showed a pronounced decrease with pupae
group size (Wrona and Dixon, 1991) (Fig. 3.5), suggesting a strong dilution
effect.

3.2.2.2.1. How the Dilution Effect Works

Empirical studies suggest that the dilution effect can arise through several
mechanisms, including predator satiation, increased prey handling time and
interference between predators. Predator satiation imposes an upper limit
on the number of prey that a predator can capture over time, which means
that a larger proportion of individuals can avoid predation in larger groups if
satiation takes place before all prey are consumed (Rasmussen and Downing,
1988). Mobile prey can benefit from the satiation effect by simply moving
away before the predator is hungry again. Satiation can even work for sessile
prey if sated predators leave the prey patch before depletion (Low et al.,
2012), perhaps because being in prey patches exposes the predator to its own
enemies. Long prey handling time also allows members of a group to escape
to safety before the predator is ready to attack anew. This mechanism can only

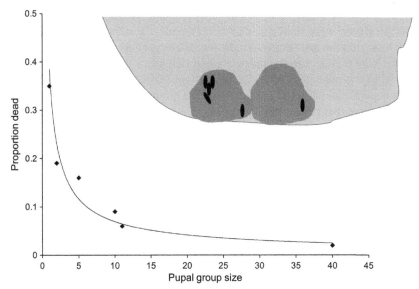

FIGURE 3.5 Evidence for risk dilution: attacked caddisfly pupae in larger groups suffer a lower risk of mortality. Inset illustrates a cross-section of a stream. Pupae (dark spots) aggregate on stones at the bottom of streams and the co-occurrence of their triclad predators was noted for each aggregation on a stone. A power function was fitted to the data. *Adapted from Wrona and Dixon (1991).* (For colour version of this figure, the reader is referred to the online version of this book.)

work for mobile prey. For example, it takes a peregrine falcon several minutes to consume one semipalmated sandpiper, allowing other birds in the attacked flock plenty of time to relocate elsewhere. Finally, interference between several predators attracted to the same prey aggregation can also reduce the probability of attack for prey (Hassell, 1978). Several predators in the same patch may jostle for position or attack one another, distracting them from prey exploitation.

The challenge is to rule out alternative explanations for the observed decrease in per capita attack rate in larger groups. Dilution need not be invoked, or may not work alone, if larger groups are more efficient in detecting and avoiding attacks (see the preceding section) or confuse the predator (see "Confusion" later in the chapter). In some cases, alternative mechanisms can easily be ruled out. Collective detection is not an issue when prey are immobile or cannot detect predators. For instance, flotillas of a marine insect at the surface of water cannot detect attacks by fish predators from below (Foster and Treherne, 1981). Similarly, encased pupae of stream-dwelling insects are immobile and cannot detect or flee from predators (Wrona and Dixon, 1991). Another possibility is that large groups of prey disorient or confuse a predator in pursuit, leading to a reduction in capture rate (Allen, 1920b). Confusion can effectively be ruled out for immobile prey. For mobile prey, it is necessary to determine whether predator efficiency varies according to prey density.

Dilution is typically viewed as dividing risk equally among all group members. However, some individuals in the group may be more at risk than others and, here, dilution does not work equally for everyone. For example, predation risk is often higher at the edges of a group (Krause, 1994a), and so peripheral individuals do not benefit from the full dilution effect. This will also be the case for any particular subset of individuals in the group that are inherently more at risk, like juveniles or mothers with young. It is, therefore, important to incorporate such particular phenotypic attributes when modelling the benefits of dilution (Dehn, 1990; Rieucau et al., 2012).

3.2.2.2.2. Collective Detection and the Dilution Effect

The concept of collective detection and the dilution effect are intimately related. The relationship is complex, as is shown by recent modelling studies described below. In this section, I explore how collective detection ability can influence the range of the dilution effect in both space and time.

With respect to space, we have seen that an increase in inter-individual distances decreases the efficiency of collective detection. Now, how does group density influence the dilution effect? Intuitively, dilution must be less effective if individuals in a group are farther apart. However, in contrast to collective detection, it is not the ability to monitor neighbours that plays an important role in the dilution effect. Consider the following example with two groups visually separated from one another. Collective detection will not work as individuals from the two groups cannot detect one another. However, dilution can still be effective if a predator can detect both groups when deciding to attack a particular individual (Jackson et al., 2006). A predator that fails to capture prey upon attack of one group will relocate elsewhere, say at distance x. Therefore, any prey within a circle of radius x from the last unsuccessful encounter will effectively be spared. All individuals within that circle are effectively yoked in terms of predation risk even if individuals are not necessarily in direct contact.

Dilution range in space is a function of the ability of prey to transfer information about threats in the group. Once a predator is detected, information spreads rapidly to all within the range of communication (visual, auditory, or mechanical). The detected predator must move beyond the alarm communication range of the group to encounter an unalarmed prey again, which effectively sets a space limit for the dilution effect. Alarm communication range can be rather short: centimetres on a social spiderweb versus metres for vocal alarm calls. While the range of visual alarm signals is probably limited in a closed habitat, it can be surprisingly large in open habitats. For instance, European starlings can detect a falcon attack on a neighbouring flock located kilometres away (Carere et al., 2009). Obviously, falcons would have to move more than a few kilometres to encounter unwary starlings.

The previous argument is that dilution can be defined in space in relation to the range of alarm communication. Bednekoff and Lima (1998) illustrate the case where the range of dilution varies in the time dimension in relation to the effectiveness of collective detection (Bednekoff and Lima, 1998b). Their models explore predation risk for pairs of prey upon attack by a single predator, the simplest situation that incorporates both risk dilution and collective detection. An attack is successful if the predator can get within striking distance unde-tected by both prey. Beyond a fatal point, the predator will catch one of the two individuals before they can reach cover. If collective detection works perfectly, only one individual needs to detect the predator before it gets fatally close to allow each one to escape. Imperfect collective detection imposes a slight delay between initial detection by one group member and reaction by the other. There-fore, the predator must be detected slightly sooner when collective detection is imperfect to ensure that both prey escape safely. If collective detection involves a sizable time delay, late detection may mean that only the detector can escape safely.

If the predator chooses a target near the beginning of its attack, predation risk is diluted among the two prey whether or not collective detection is perfect, because the ability to detect the predator is the same for each individual. Now, consider what happens if the predator chooses a target after the last point for individual detection. With late targeting, the predator will pursue at random any individual that remains behind. If only one prey has detected the predator early, the companion that has failed to detect the predator before the last point for detection will shoulder the full risk of predation. Here, the risk of being targeted cannot be assumed to be divided equally. The risk will in fact increase the sooner the companion detects the predator and the longer it takes for col-lective detection to work. Therefore, an increase in the delay to react to alarm signals from a companion reduces the efficiency of collective detection and also the benefits from dilution since the burden of predation will be shouldered for a longer time period alone by the non-detector.

Another consequence of late targeting is that, counterintuitively, foraging with a vigilant companion can actually be riskier. At first sight, a more vigi-lant companion should increase safety by allowing earlier detection. However, early departure of the more vigilant individual when an attack occurs effectively cancels any dilution benefits. It might even be that foraging with vigilant indi-viduals is more risky than foraging alone, if being in a larger group increases the rate of predation on the group, and the variation in vigilance means that the low-vigilance individual does not benefit from dilution of risk.

In the previous section on collective detection, I noted that the speed of transmission of information about predation threats was often much higher than that of the approaching predator. In such cases, the delay in reaction within the group will be small and dilution will work for detectors and non-detectors alike. When this is not the case, dilution will be less effective for non-detectors, especially when collective detection is slow.

Testing the targeting options considered by Bednekoff and Lima remains challenging. Evidence for early targeting comes from the reaction of individual prey birds that are not in the direct line of attack by a predator: instead of trying to escape to safety, these birds often freeze or fly in a perpendicular direction (Lima and Bednekoff, 2011). Preference for less vigilant foragers in a group may also be an indication of early targeting (FitzGibbon, 1989; Krause and Godin, 1996a). However, in this case, predators may target the last individuals to leave the group, which will be on average those that tend to be less vigilant. More information about predator targeting behaviour for predators is needed to assess the interaction between dilution and collective detection. At the very least, the above discussion indicates that the two mechanisms are intimately related.

3.2.2.2.3. Disentangling Dilution and Collective Detection

Although a case for dilution acting alone can easily be made for immobile prey or prey that cannot detect predators, it remains a challenge to rule out alternative hypotheses for mobile prey that can detect predators. In particular, much attention has been devoted to disentangling the effects of dilution versus collective detection. These two mechanisms often make the same predictions regarding antipredator behaviour. Both mechanisms imply that larger groups are more protected: through the presence of alternative targets for the predator, in the case of risk dilution, and through enhanced predator detection, in the case of collective detection. In response to a decrease in perceived predation risk, individuals in large groups are thus expected to invest relatively less in antipredator defences. The finding that antipredator vigilance decreases with group size, for instance, can be explained by risk dilution acting alone, by collective detection acting alone, or by both acting together (Roberts, 1996).

Roberts (1996) suggested that one way to disentangle the effects of both mechanisms is to study antipredator defences in settings involving just one mechanism. For instance, risk dilution may be kept constant by comparing antipredator defences in groups of the same size. Variation in collective detection can then be induced by altering the ability to detect predation threats, say, by blocking the field of view for some members of the group. Another possibility is to keep collective detection constant and vary risk dilution among groups. To this end, Ruxton and I investigated antipredator vigilance in groups of semipalmated sandpipers foraging on a mudflat during migration (Beauchamp and Ruxton, 2008). Peregrine falcons launch their attacks on sandpiper groups from the wooded area bordering the mudflat. Since attacks all originate from one general direction, flocks that spread along the receding tideline have a riskier side: the one facing the wooded cover. Collective detection in all flocks of sandpipers can be assumed constant. This is because attacking falcons can be detected from anywhere in the flock, and no visual obstacles prevent information transfer about threats in the open mudflat habitat of sandpiper flocks. Risk dilution, however, is expected to vary depending on the spatial location of individuals within the flock. Birds on the side facing the woods are in the direct line of attack and can only dilute predation risk with close neighbours. Birds on the opposite side of the flock benefit from the presence of several layers of protection. We expected, therefore, that birds on the riskier side of the flocks would invest more in antipredator defences, such as vigilance, at the expense of foraging. The results showed that foraging intensity was indeed lower on the riskier side of the flocks (Fig. 3.6). The possibility that birds on different sides of the flocks have different phenotypic attributes, which could explain any differences in foraging behaviour, was dismissed because birds readily move from one part of the flock to another as the flock tracks the receding tide.

Another way to disentangle the effects of risk dilution and collective detection is to statistically evaluate the relative contribution of each factor (Dehn, 1990). The assumption is that the relationship between antipredator defences and group size will have a different shape depending on whether collective detection acts alone, risk dilution acts alone, or collective detection and dilution act together. Fitting these different predicted relationships to the data allows one to determine which mechanism predominates. So far, this statistical approach has been used in three species of mammals. In two species, the data suggest that both dilution and detection probably act together (Dehn, 1990; Rieucau et al., 2012). In the other, detection acting alone provided a better

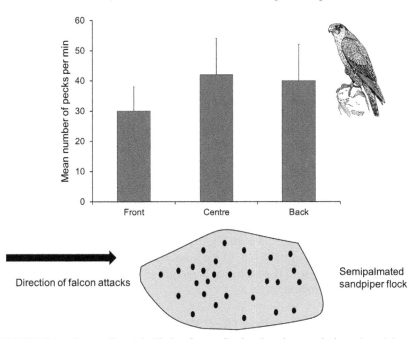

FIGURE 3.6 Disentangling risk dilution from collective detection: semipalmated sandpipers spend more time foraging, at the expense of antipredator vigilance, in the more protected areas of a flock. In flocks of sandpipers, collective detection is constant and risk dilution varies depending on the location of the individuals in the flock. Attacks from peregrine falcons (inset) originate from the wooded cover. The riskier side of a flock is thus at the front. Error bars show one standard deviation. *Adapted from Beauchamp and Ruxton (2008).*

fit to the data (Fairbanks and Dobson, 2007). The statistical procedure makes assumptions about how collective detection and dilution work in groups. For instance, predators are assumed to attack groups of different sizes with the same probability, and all individuals in the group are equally likely to be targeted by the predator. It is important to assess such assumptions before proceeding with the statistical approach.

3.2.2.3. Attack Abatement

Attack abatement represents the combination of encounter and dilution effects. The main point here is that it is difficult to demonstrate antipredator benefits of grouping without a detailed knowledge of how each of these two effects influences predation risk. Neither mechanism acting alone can guarantee a decrease in per capita attack rate for prey in groups. Dilution, for example, will only provide benefits if the encounter rate with predators does not increase proportionately with group size. This is the case in pupae of a stream-dwelling insect where dilution of risk in larger groups compensated for the higher probability of occurrence of predators in larger groups (Wrona and Dixon, 1991).

3.2.3. Confusion

By contrast to collective detection and encounter-dilution effects, confusion acts later in the predation sequence, when the predator closes in on a group of prey. Confusion involves a reduction in attack success when attempting to capture an individual from a fleeing group. A role for confusion as an antipredator ploy has long been suspected. Early naturalists noted that the scattering of group members upon attack may reduce capture rate by dividing the attention of the predator (Allen, 1920b; Grinnell, 1903; Miller, 1922). For instance, repeated failures of a loon to capture sardines in a large school lead Allen (1920) to suggest that the bird was confused by the availability of so many potential targets. Similarly, reduced capture rate by aquatic predators when chasing prey in larger groups is thought to reflect distraction of attention from the targeted prey by the presence of several non-target companions (Landeau and Terborgh, 1986; Lavalli and Spanier, 2001; Neill and Cullen, 1974). In a particularly telling example, predatory bass captured prey only 11% of the times when their silvery minnow prey occurred in groups of 15 (Landeau and Terborgh, 1986), while they never failed when minnows occurred singly.

3.2.3.1. How the Confusion Effect Works

Confusion is a line of defence that works by exploiting the cognitive limitations of the predator (Krakauer, 1995). Confusion reduces the ability of the predator to select and/or pursue a target in the fleeing group. As a consequence of confusion, the number of captures per contact with prey decreases, resulting in increased protection for aggregated prey. Pinpointing exactly which aspects of prey behaviour confuse the predator requires detailed experimental work. Starting with the observation that three-spined sticklebacks prefer to attack stragglers in a swarm of *Daphnia* (Milinski, 1977a), which suggests that capturing prey is more challenging in a larger group, Ohguchi (1981) designed a series of elaborate experiments to determine exactly how confusion arises in the stickleback/*Daphnia* system. The *Daphnia* were held in test tubes to prevent attacks, allowing the experimenter to focus on the factors determining how the sticklebacks select a target and maintain contact with the target (Ohguchi, 1981). The experimenter controlled the number and density of test tubes and could move these tubes at different speeds and in different directions to simulate the movement of a swarm. Many factors reduced the number of predation attempts on individual *Daphnia*, including higher movement speed, movements in the same direction, increased swarm size and density, and increased uniformity in the appearance of swarm members. Cross-movements by the prey also constituted a strong predictor of predation attempts. I shall return to the appearance of group members in Section 3.2.3.2. The other factors indicate that flight trajectories and speed are crucial in determining predation attempts in this species. Adjustment in flight speed can be an asset for solitary as well as aggregated prey. However, movements in the same direction or criss-crossing paths can only provide benefits to aggregated prey, and may therefore be a crucial component of the confusion effect.

Another approach to understanding the confusion effect is to model predator cognitive abilities (Krakauer, 1995; Tosh et al., 2006). By mimicking properties of the predator's nervous system, from perception of environmental stimuli to the corresponding mapping of information in the brain, artificial neural networks can be used to determine which factors constrain predation performance when hunting aggregated prey (Box 3.1). Such models are particularly suited to the investigation of the confusion effect, as this effect depends on the predator's ability to maintain contact with a prey target surrounded by many non-target individuals.

In the artificial neural network model implemented by Tosh et al. (2006), which follows the procedure detailed in Box 3.1, predator accuracy, measured in terms of errors in mapping the precise location of the target prey, first decreases and then increases with prey group size (Fig. 3.8). Alleviation of the confusion effect takes place when the target prey are more conspicuous than nontargets. Prey are more difficult to target in homogenous-looking groups than heterogeneous groups. Predators are also more accurate when targeting isolated prey. Finally, in contrast to the results reported earlier with swarms of *Daphnia*, group density does not influence predator accuracy. All these effects have been tested with human subjects using a mouse cursor to capture one object on a computer screen containing a large number of fast moving, identical-looking objects. Human experimental data corroborated all the above theoretical predictions (Tosh et al., 2006).

BOX 3.1 Artificial Neural Networks

Artificial neural networks provide a framework for investigating and modelling mechanisms underlying behavioural patterns (Haykin, 1994). By specifying the actual structure and functioning of neural networks, such models can fill the proverbial black box linking behavioural responses to environmental stimuli. Artificial neural networks have been used in many fields, from cognitive psychology to engineering. Animal behaviour students have been rather late in adopting artificial neural networks in their toolkit (Tosh and Ruxton, 2010). Recent examples include the use of artificial neural networks to investigate vocal learning in primates (Pozzi et al., 2010) and division of labour in social insects (Duarte et al., 2011). Artificial neural networks typically consist of three layers: an input layer, an output layer, and a hidden layer that connects these two layers (Fig. 3.7).

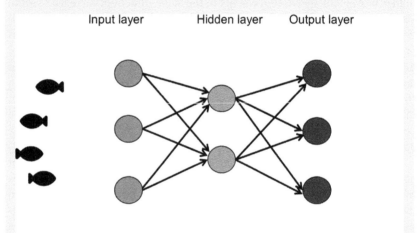

FIGURE 3.7   Structure of a typical neural network model. Stimuli in the environment, here the location of prey fish, are perceived by a layer of input receptors and passed on as signals to a hidden layer of interneurons. These interneurons pass the information to output neurons, which can then provide the necessary feedback to perform behaviour, such as to attack a particular fish in the school. Such a network has many parameters controlling the flow of information along the pathways. Arrows indicate connection weights that can be positive or negative, thus altering the strength of the signals along different paths. Values for these parameters are obtained after a series of training sessions. The trained network can then be used to determine what behaviour will be used in response to new environmental stimuli. (For colour version of this figure, the reader is referred to the online version of this book.)

Here, I will illustrate the basic components of artificial neural networks for the specific case of the confusion effect in predator–prey interactions. An artificial neural network for the confusion effect asks if and how a specific neural network configuration can recognize a particular prey in a group of prey (Krakauer, 1995; Tosh et al., 2006). The individual target in a group activates input units (say, retinal cells in the eye), which in turn activate interneurons linking input units to an internal

BOX 3.1 Artificial Neural Networks—cont'd

representation (the map) that preserves the spatial organization of the group. Good mapping of the target location would eventually lead to more coordinated predation behaviour, although this motor aspect is not usually examined in artificial neural networks.

The devil is in the wiring scheme between the input units, interneurons, and output units, and in the strength ascribed to the signals connecting various layers. Different wiring schemes can be investigated, including schemes where each input unit links directly to a unique location on the map or to several adjacent regions. Information flows from input units to the map and information along the way is pooled at the level of each interneuron and each location on the map.

Units in the input layer are activated by the perception of environmental stimuli—in this case, the position of the target prey. The level of activation is typically bounded between 0 and 1. Interneurons integrate information from each input layer unit to which they are connected. Integration is a weighted sum of all such activations. The integration value is then transformed using a transfer function so as to produce an output to the next layer. Similarly, units in the output layer integrate information from each interneuron to which they are connected, again using weights and a transfer function to provide the final signal processed by the network. In the case of prey detection, this is the mapping of the precise location of the target.

Artificial neural networks are initially trained to recognize specific environmental stimuli (typically called training stimuli). This learning phase determines the specific weight values to be used to modulate signal strength along the pathways of the network. The learning phase proceeds by comparing the output from a particular configuration of the network to a target output, and then by tweaking parameter values of the artificial neural network to achieve the desired output. Different algorithms are available to tweak the network to achieve the desired output, the most common being back propagation. Since different algorithms may produce different results, a sensitivity analysis should be done at this stage.

The testing phase uses the trained artificial network. The network is exposed to various novel environmental stimuli that differ from the training stimuli, the aim being to examine how the network can generalize from slightly different inputs. In the particular case of the confusion effect, researchers can modify, for example, both the number and density of individuals in prey groups.

Artificial neural networks have been criticized for being misleadingly simplistic (Dawkins and Guilford, 1995). Simple models, it has been argued, may not capture the real properties of sensory systems, and their results may be an artefact of the way the networks are build. This type of criticism is not unique to artificial neural networks, and time and again models have been criticized for making simplifying assumptions. Nevertheless, one thing is clear: we should be confident that the predictions that arise from artificial neural networks follow from the underlying biology subsumed in the model, rather than from the arbitrary choices made during model implementation. Recent work emphasizes that network performance can be influenced by many factors, including network topology and details of network training (Franks and Ruxton, 2008; Tosh and Ruxton, 2007). It is, therefore, important to understand what part of the output genuinely reflects biology rather than simple implementation choices made by modellers.

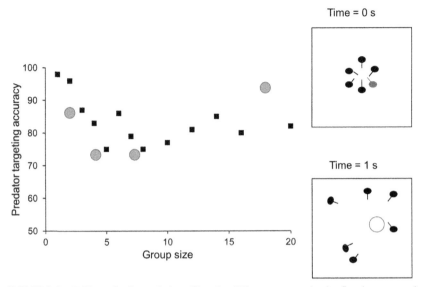

FIGURE 3.8   Evidence for the confusion effect: the ability to target a stimulus first decreases and then increases with the number of targets provided on a computer screen to human subjects. The insets on the right-hand side illustrate a computer screen with objects that could be tracked using a mouse pointer. At the start, the objects are all positioned near the centre and one is identified as the target. The objects then disperse and the viewer has to capture the target with the mouse-operated cursor. Different sizes of object clusters were tested. Observed targeting accuracy over many trials with many subjects is shown in the graph on the left-hand side. Green dots show values that fell outside the 95% confidence intervals based on predicted values obtained from an artificial neural network. *Adapted from Tosh et al. (2006).* (For interpretation of the references to colour in this figure legend, the reader is referred to the online version of this book.)

Interestingly, as predicted by the model, increasing the density of a group did not make it more difficult for human subjects to capture a target at any of the group sizes tested in the experiment. This result implies that the compaction of groups upon predatory attacks (see Chapter 6) may be beneficial for reasons that are not related to the confusion effect. The U-shaped pattern between confusion and group size is also unexpected in the sense that, at first sight, confusion should become more problematic the larger the group. One difficulty in relating the empirical results with human subjects to theoretical predictions is that the human subjects faced a moving array of objects, possibly criss-crossing each other as they moved away from the centre of the screen, while the artificial neural network mapped unmoving objects. Human subjects may concentrate harder as the task becomes more difficult, as was probably the case when isolating a target from a larger group, explaining, perhaps, the increase in accuracy when group size increases. Because the network maps objects from groups of different sizes with the same effort, it is possible that the effect of group size on the confusion effect reflects different mechanisms in human subjects and artificial neural networks.

In a similar study using human subjects, but this time facing an array of objects moving from one side of the screen to the other, confusion increased with group size and showed no signs of decreasing in larger groups (Ruxton et al., 2007). This result raises the possibility that the degree of confusion may be related not only to group size but also to the general configuration of prey movements.

3.2.3.2. Oddity Effect

Studies using artificial neural networks have identified contrast between the target and other non-target individuals in the group as a main factor influencing predator accuracy. Oddity alleviates the confusion effect by making the target more conspicuous with respect to the rest of the group. Returning to the bass attacking minnows, the inclusion of one or two odd individuals in the group, following a dyeing treatment, greatly increased attack success (Landeau and Terborgh, 1986). In another experiment using groups of minnows with different proportions of small and large individuals that all looked the same except for their size, the bass attacked individual minnows of the minority size proportionately more often (Theodorakis, 1989), suggesting that variation in size as well as colouration within the group can induce oddity. The oddity effect was also noted in the above experimental study of confusion using human subjects (Ruxton et al., 2007).

If oddity does indeed increase conspicuousness, one would predict that animals would attempt to reduce oddity when under attack. The tendency for frightened individuals to join individuals of the same size (Krause, 1994b; Theodorakis, 1989), or to abandon groups when looking differently for the rest of the group (Wolf, 1985), can all be interpreted in the light of reducing the oddity effect.

Oddity may pertain not only to the appearance of individuals but also to their behaviour. For instance, individuals in swarms of *Daphnia* adopted a more homogenous speed when frightened, which may be adaptive because predators prefer to target individuals swimming at a different speed than the rest of the group (Szulkin et al., 2006). The key point, therefore, is that predator confusion can be more easily overcome when prey appearance or behaviour varies within the group.

Confusion and oddity should act together to promote homogeneity of behaviour and appearance within a group under threat of predation. Nevertheless, many groups are composed of heterogeneous individuals or even visually distinct species (see Chapter 8). Such within-group heterogeneity may be maintained for several reasons. It may, for example, be preferable to be an odd individual in a group that provides some form of protection than a solitary individual exposed to a high risk of predation (Landeau and Terborgh, 1986).

A further possibility has been identified recently using an artificial neural network model (Tosh et al., 2007). This new approach has shown the importance of conspicuousness with respect to the background, not just the other individuals in the group. The model considers two species that differ in conspicuousness against the background. Even when rare, the less conspicuous species in a group may evade predation because the combination of the two types of species confuses the predator and reduces attack rate on the more cryptic species. Confusion and oddity would predict higher predation success on odd individuals regardless of their conspicuousness against the background. By contrast, the artificial neural network model predicts that odd individuals that are more cryptic than the rest of the group would evade predation. A laboratory experiment with three-spined sticklebacks and *Daphnia* prey examined these predictions. Results indicate that more cryptic *Daphnia* prey were never attacked more often than their more conspicuous companions, even when rare in the group (Rodgers et al., 2013) (Fig. 3.9). These results suggest that the oddity effect, which is a form of conspicuousness in relation to the rest of the group, can be overcome in heterogeneous groups when it interacts with the degree of conspicuousness against the background. It remains to be seen whether this evolutionary force has acted to shape the composition of naturally occurring groups.

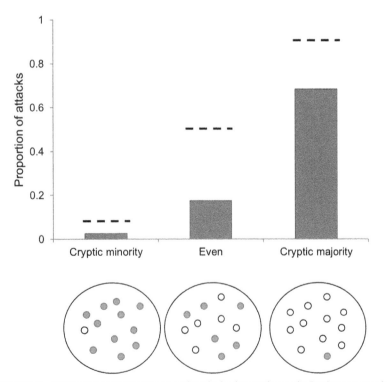

FIGURE 3.9 Oddity and conspicuousness against the background: cryptic *Daphnia* are attacked less often than predicted when rare or common in the population of prey, suggesting a deflection of attacks toward more conspicuous group members. The random expectation is shown as a broken line. Cryptic *Daphnia* were dyed to match the colour of the water. *Adapted from Rogers et al. (2013).* (For colour version of this figure, the reader is referred to the online version of this book.)

3.2.3.3. Protean Displays

Work with *Daphnia* swarms has indicated that the criss-crossing of paths when fleeing creates more confusion for the predator. In this case, predators may have difficulty maintaining their original target as individuals flee. Seemingly erratic fleeing paths adopted by animals upon attack have been referred to as protean displays, and it has been postulated that such displays confuse the predator (Chance and Russell, 1959). Although erratic displays can be performed by solitary individuals, the effect is thought to be greatly enhanced when fleeing involves many individuals simultaneously (Humphries and Driver, 1970). The erratic nature of the fleeing path is critical, because a predator may learn to overcome the confusion effect if prey animals always flee in the same pattern.

To show that protean displays can serve as an antipredation device requires experimental work in which the fleeing paths of individuals are controlled. Ohguchi (1980) enclosed individual *Daphnia* in test tubes and manipulated the speed and direction of fleeing movements. The criss-crossing of paths reduced the ability of a fish predator to attack successfully, thus suggesting a deterrent effect of protean displays. The mouse-cursor paradigm, described above, has been extended to explore whether human subjects take longer to capture a moving object if the fleeing path of the target is less predictable (Jones et al., 2011). As predicted, more unpredictable fleeing paths did indeed reduce capture efficiency. However, the protean effect acted independently from the confusion effect, which is related to group size, since unpredictability reduced capture efficiency to the same extent in small and large groups. Therefore, protean displays, as least with human subjects, would be equally effective in deterring predation for solitary and aggregated prey. To be considered an antipredator ploy for prey species living in groups, protean displays should be more effective in larger groups.

In contrast to solitary prey, however, prey in groups can coordinate their effort to produce group-level displays whether protean in nature or not that may be efficient in deterring predation. Collective displays by groups under predation threats have been noted in many species. Familiar examples include wheeling and flashing in avian flocks (Carere et al., 2009), turning schools of fishes (Gerlotto et al., 2006), wing vibrating in bees (Kastberger et al., 2008), and simultaneous head shaking in caterpillars (Cornell et al., 1987). That such displays can be effective in reducing attack success is suggested by recent work with European starlings fleeing from peregrine falcons (Procaccini et al., 2011) (Fig. 3.10). Rapid turning in flocks upon attack propagated through the flock as a wave. Such waves tended to originate from the position of the attacking bird of prey and radiate through the flock. The formation of waves in fleeing flocks reduced predator success, as would be expected if such displays confused the predator. It is also believed that waves of shimmering in colonies of giant honeybees upon attacks by hornets reduce predation success (Kastberger et al., 2008), although a reduced rate of attack may simply mean that a predator that has been detected abandons attacks before confusion can even play a role (Tan et al., 2012).

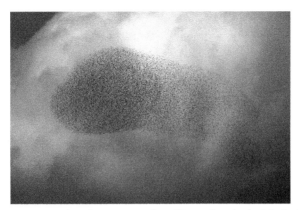

FIGURE 3.10   A murmuration of European starlings. Waves of turning in tight flocks of starlings propagate as waves, which are thought to deter predation from birds of prey. *Photo credit: Muffin. This figure is reproduced in colour in the colour section.*

3.2.3.4. Confounding Factors

In their review of the literature, Jeschke and Tollian (2007) documented confusion in 16 out of 25 predator–prey systems involving both invertebrate and vertebrate species. Although confusion plays a role for both tactile and visual predators, it is more common with visual predators when their prey are agile (Jeschke and Tollrian, 2007). Confusion effects thus appear to be quite general in scope. However, the authors pointed out that exactly how confusion effects arise is poorly understood and that alternative factors must be ruled out in order to conclude that confusion is responsible.

Morgan and Godin (1985), in their laboratory study of predation on fish schools of different sizes, ruled out confusion by noting that all transitions in predatory behaviour, from the number of attacks per encounter to the number of captures per attack, failed to vary with group size, indicating that the predator was equally efficient in handling schools of different sizes. Not all measures of predator success, however, can be used to document confusion. For instance, the number of captures per contact with prey may decrease when group size increases simply because of a limit on the number of prey that a predator can handle per unit time (i.e. functional response) (Jensen and Larsson, 2002).

Computer experiments with human subjects offer a way of teasing apart the various factors that can contribute to the confusion effect by effectively controlling the behaviour of all individuals. Using human subjects aiming a mouse-cursor at moving objects on a computer screen, Ruxton et al. (2007) determined that the time needed by subjects to capture a moving object increased with the number of moving objects on the screen, thus demonstrating the confusion effect. Objects on the screen showed no coordination, in contrast to the aforementioned collective displays. Therefore, the confusion effect can arise, at least in human subjects, from the sheer number of environmental stimuli, and does not require systematic fleeing patterns, although coordination may still act to strengthen this effect.

A recent experiment with computer-generated prey targeted by a live predator fish addressed similar issues (Ioannou et al., 2012). The behaviour of computer-generated prey was manipulated by changing their attraction to one another and the straightness of their paths on a computer screen, producing a wide range of group size and path tortuosity within groups. In contrast to a random targeting strategy, real fish preferentially attacked prey in small groups and those that moved in less coordinated fashion (Fig. 3.11). Because the computer-generated prey are unaware of the presence of the fish, collective detection could not explain why larger groups were attacked less often. The selfish-herd effect can also be ruled out because prey could not hide behind one another on the flat computer screen. The authors argued that the results are best explained by the confusion effect. This type of experiment with real predators and manipulated prey, while restricted to predator–prey systems amenable to laboratory studies, offers real opportunities to examine in great details prey selection processes in predators.

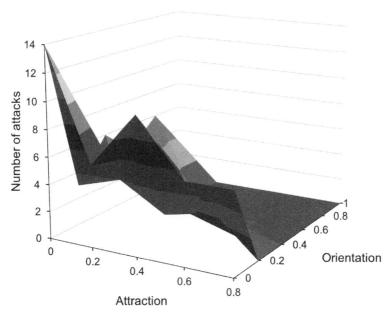

FIGURE 3.11 Confusion effect in a simulated habitat: predatory fish targeted disproportionately computer-generated prey that show less tendency to aggregate (low attraction) and less tendency to move straight (low orientation), as shown by the total number of attacks from fish over several trials with different combinations of orientation and attraction among prey. Prey with large values of attraction and large values of orientation tended to move in coordinated fashion. *Adapted from Ioannou et al. (2012).*

3.2.4. Swamping

Predator swamping arises when a predator faces constraints in the number of prey it can consume per unit time (Darling, 1938). Such constraints may be related to the time needed to handle or digest prey. For example, if handling prey takes considerable time, other individuals in the aggregation can escape predation. Mobile prey can simply move away from the predator before it is ready to attack again. If the rate of prey consumption is low, sessile prey can evade predation by passing through to a less vulnerable life stage or grow to a size that precludes predation.

Swamping has been invoked as an antipredator mechanism to explain mast fruiting in plants and reproductive synchrony in animals (Ims, 1990), but the concept can also be applied to the formation of animal groups. Predator swamping, viewed this way, is similar to predation risk dilution, which, as we saw above, can only work if the predator cannot capture all individuals in the group. Predator swamping can work, in theory, for prey that are spatially dispersed but available at the same time, such as the seeds of trees in a large forest that are produced simultaneously. Risk dilution in animal behaviour research focuses on aggregations of prey in both time and space, and typically the time scale of interactions between predators and prey is short enough that prey cannot evade predation by, say, growing or passing through a different life stage. Because the term 'swamping' has a broader meaning and has been used for prey that are essentially inanimate (e.g. seeds or eggs), I would suggest avoiding this term to describe typical predator–prey systems in animal behaviour research and refer to 'risk dilution' instead.

3.2.5. Group Defence

The various antipredator ploys described in the preceding sections paint a picture of passive prey that offer little resistance to predators. I have considered cases where prey animals try to reduce encounter rate with predators, share predation risk with each other, or flee upon detection of a threat. However, prey animals in groups can also rely on more active means to deter predation. I will now explore three such ways in which prey strike back: mobbing, group defence, and predator inspection.

3.2.5.1. Mobbing

Mobbing has been defined as the use of visual and/or acoustic displays typically aimed at a predator that poses little risk. Counterattacks and rescue attempts are not generally viewed as mobbing (Crofoot, 2012). Mobbing may include direct contact with predators, but it usually involves harassment from a distance (Curio, 1978). Mobbing, thus defined, implies a mobile prey species that can detect predators from a distance. While mobbing typically involves diurnal species, which can detect a predator more easily at a distance, it can also work at night for species that are able to maintain group cohesion (Schuelke, 2001). The very act of mobbing entails time and energy costs, and sometimes increases immediate predation risk by bringing prey and predator closer (Tórrez et al., 2012). However, mobbing offers several potential benefits, including moving the predator away, at least temporarily (Ishihara, 1987), and in some cases, even killing it (Cowlishaw, 1994).

Mobbing is usually considered a stressful experience for a predator (Consla and Mumme, 2012; Flasskamp, 1994), and it typically triggers an evasive response to reduce harassment (Cowlishaw, 1994; Pettifor, 1990). Therefore, a mobbing group, as opposed to a single mobbing individual, should be more effective in deterring predation. Evidence for a deterrent effect of mobbing in groups comes from the observation that an increase in the number of mobbing individuals reduces attack success rate of predators in two colonial nesting species of birds (Andersson, 1976; Robinson, 1985). In Formosan squirrels (Tamura, 1989) and spectral tarsiers (Gursky, 2006), the duration of mobbing increases with the number of mobbing individuals, presumably increasing its effectiveness. However, it may also be the case that an increase in the duration of mobbing indicates that the predator is not easily intimidated, and mobbing is actually less effective when it lasts longer. Risk dilution represents a further advantage of mobbing in groups since each individual in the group is less likely to be targeted by the retaliating predator (Brown and Hoogland, 1988; Poiani, 1991). As long as there is some benefit to mobbing by one individual, it is quite clear that joining a mobbing group will be attractive to nearby companions. This is perhaps why mobbing activities, including calls, entice companions and even other species to join (Forsman and Mönkkönen, 2001; Hurd, 1996; Iglesias et al., 2012).

Mobbing groups often include family members (Griesser, 2009; Gursky, 2006; Poiani, 1991), which raises the possibility that kin selection may play a role in the formation of these groups. It is also possible that adults might encourage their offspring to take part in mobbing as a low-risk way of giving them experience of predators (Cornell et al., 2012). Certainly, the presence of close kin members provides an extra incentive to joining a mobbing group. Generally speaking, a benefit to group living may be the increased number of individuals available to mob potential predators.

3.2.5.2. Defensive Formations

Rather than mobbing potential predators, several species adopt defensive formations when faced with a predation threat. A few examples include outward-facing radial formations in groups of wildebeest (Creel and Creel, 2002) or spiny lobsters (Herrnkind et al., 2001), or inward-facing radial formations in groups of sperm whales (Pitman et al., 2001) or cowtail stingrays (Semeniuk and Dill, 2005). In such formations, weapons, such as horns or tails, project outward and more vulnerable parts, or even more vulnerable group members such as offspring, stay inside the group. It is difficult to prove the adaptive value of defensive formations because such formations can rarely be induced under controlled conditions. Nevertheless, observational studies have examined attack success rates for such formations and shown some benefits to the behaviour. For instance, wild boars suffered fewer attacks from packs of wolves when adopting an outward-facing radial formation (Jędrzejewski et al., 1992). Similarly, coyotes were less likely to attack mule deer that bunched together rather than spread out after an encounter (Lingle, 2001). Interestingly, white-tailed deer, a closely related species, failed to bunch upon attacks by coyotes and preferred flight to group defence to avoid predation (Lingle, 2001). Such interspecies differences in the tendency to use defence formations indicate that economic considerations probably play a role in the choice of strategies to elude predators. Essentially, a defensive formation should only be attractive if prey are willing to maintain it longer than the predator is prepared to wait for it to break up. In addition, the ability to outrun a predator is probably important in determining whether prey animals take a stand or flee upon attack. Indeed, a comparative analysis in artiodactyls (even-toed ungulates) revealed that bunching was more likely in species that live in larger groups and that tend to be larger (Caro et al., 2004). However, the relative size and number of their predators was not included in the analysis, making it difficult to determine whether relative speed of escape played a role in the evolution of bunching responses.

One study manipulated the occurrence of defensive formations by randomly allocating individually tethered spiny lobsters to dens with and without companions (Butler et al., 1999). Solitary individuals suffered a higher rate of mortality than those in groups. While this finding may simply indicate a risk dilution effect, the authors note that no survival advantages accrued to smaller individuals in groups, which are probably unable to defend themselves against predators. Group defence was thus proposed as a benefit of aggregation in larger juvenile spiny lobsters. Similarly, individually tethered Caribbean spiny lobsters suffered fewer bites from fish predators when their rosette formation contained more individuals (Lavalli and Herrnkind, 2009).

Prey in groups can also build or use structures that are efficient in deterring predators. These structures play a similar role as defensive formations, but use natural substrate rather than bodies to reduce capture rate. For instance, spiders in larger groups can build larger protective structures made of leaves, which reduce the ability of their predators to reach inside the colony (Unglaub et al., 2013). Protection from external structures implies that animals in groups must spend a significant amount of time in the same location. This is the case in many species of burrow-digging rodents. Burrow digging in one group of rodents has been linked to social group size (Ebensperger and Blumstein, 2006), suggesting the intriguing possibility that the presence of several group members may allow rodents to dig tunnels that provide more protection while also reducing the individual burden of digging.

3.2.5.3. Predator Inspection

Mobbing may ensue after members of a group encounter a predator. However, in some circumstances, a subset of individuals leave the group, approach the predator, stop, and then return to the group. In contrast to mobbing, such behaviour rarely involves aggressive behaviour toward the predator. This behaviour has been labelled predator inspection, and is known in many schooling species of fish and in other animals as well (Caro et al., 2004; FitzGibbon, 1994; Magurran, 1986).

The function of predator inspection appears to be gaining information about the location and motivation of the predator, which is then passed along to companions in the group. It may be that most predators that prey encounter are currently uninterested in attacking, and inspection may be a way of identifying such situations and saving costs of increased vigilance or fleeing in response to a predator that does not actually pose a current threat. Like mobbing, predator inspection appears, at first sight, ill-advised. Rather than risk death by approaching a potential predator, a group member should rely, instead, on the various means of protection afforded by the group, such as risk dilution and confusion. Early research with fish indicated that predator inspection was more likely when the predator was not moving, and thus less of a threat, and when many individuals joined the inspection party, suggesting that inspectors assess very closely the risk associated with approaching a predator (Pitcher et al., 1986). Nevertheless, predator inspection closes the distance between predator and prey, and has been shown to be riskier than remaining behind in the group (FitzGibbon, 1994; Milinski et al., 1997). Non-inspectors in the group obtain all the benefits of inspection without paying the costs, suggesting that inspection must provide unique benefits.

In a particularly revealing study about the value of predator inspection in guppies, fish predators were less likely to initiate an attack and kill fish that inspected them than those that remained behind (Godin and Davis, 1995), suggesting that predator inspection deters predation (Fig. 3.12). A similar function of predator inspection was suggested for European minnow (Magurran, 1990) and Thomson's gazelle (FitzGibbon, 1994). However, interpretation of such results is complicated by the possibly that inspectors may often be intrinsically higher-quality individuals than non-inspectors (discussed later in this section). In addition, there are few studies on the potential benefits of predator inspection, and some studies have failed to uncover any benefits (Milinski et al., 1997).

Predator inspection often involves more than one individual (FitzGibbon, 1994; Pitcher et al., 1986), and so each individual can dilute the risk of attack by the predator (Milinski et al., 1997). However, dilution cannot explain why individuals inspect predators in the first place; dilution can only reduce the costs associated with inspection, and benefits must be sought to explain why such risks are taken.

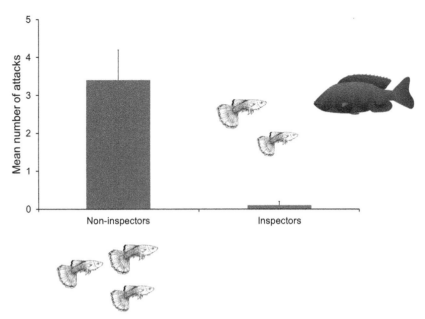

FIGURE 3.12   Risk of predator inspection: inspecting guppies are actually safer from attacks despite being closer to the predator. Error bars show one standard deviation. *Adapted from Godin and Davis (1995).* (For colour version of this figure, the reader is referred to the online version of this book.)

So how exactly does predator inspection deter predation? There are several possible mechanisms. First, predator inspection can escalate to mobbing, and thus move the predator away, as noted earlier. Second, approaching inspectors may confuse the predator, which may then prefer to attack non-inspectors (Curio, 1978). This hypothesis only works for species that approach predators as a group. In the guppy study, the predator approached inspectors whether they occurred in small or larger groups, contradicting the confusion hypothesis. Third, inspection may be a signal aimed at the predator to indicate that it has been detected and that an attack is likely to be futile (perception advertisement). The predator would then prefer to attack non-inspectors that are less wary. This hypothesis can only work before the inspectors have returned to the group and if non-inspectors are sufficiently far away not to have noticed the predator. It would also seem that perception advertisement could be done from the safety of the group by, say, making loud noises or using visually conspicuous displays. Finally, in the same vein, how close an individual approaches a predator can be a signal of quality that effectively discourages the predator from attacking those that come closest (quality advertisement). Support for this mechanism comes from the observation that predator inspection is condition-dependent in some species. For instance, larger three-spined sticklebacks and those in better condition are more likely to approach predators (Külling and Milinski, 1992). Further study is needed to assess the benefits of predator inspection.

Predator inspection in groups, regardless of the benefits that ensue, offers an opportunity to study how individuals coordinate their efforts to approach the predator. The leader in inspection groups faces a greater risk than the others behind (Milinski et al., 1997), and laggards can be considered defectors in a cooperative venture. A large body of literature has focused on predator inspection as a means to investigate the evolution of cooperation in animals (Dugatkin and Wilson, 2000; Milinski, 1987).

3.2.6. Aposematic Displays

Unpalatability is often associated with conspicuous colouration that warns a predator of the unpleasant experience that may follow after an attack (Ruxton et al., 2004). Warning signals may thus reduce predation rate by decreasing contacts with non-naïve predators. This is a strategy used, for example, by brightly coloured poisonous frogs and butterfly larvae. Aposematic species are often gregarious (Cott, 1940), raising the possibility that aggregation in such species may provide additional antipredator benefits.

How warning signals are able to evolve has been the subject of much speculation. A rare, conspicuous form of a cryptic prey species that has just evolved would face the double whammy of being rare and conspicuous, thus attracting the attention of predators unaware of their defences (Beatty et al., 2005). Aggregation has been thought to facilitate the evolution of aposematism. The argument is that an attack on a prey with warning signals living in a group would teach the predator to learn to avoid other similarly coloured individuals, thus sparing the rest of the group, which may consist of relatives sharing the same trait (Fisher, 1930). Aggregation would thus provide antipredator benefits by enhancing signals of unpalatability to predators.

Research on the association between aposematism and aggregation has proceeded along two distinct lines. First, comparative analyses have been used to determine whether the evolution of aposematism is indeed associated with gregariousness. In lepidopteran larvae, warning signals evolved prior to the apparition of gregariousness, contradicting the hypothesis that gregariousness would appear first to facilitate the subsequent evolution of aposematism (Sillen-Tullberg, 1988). As a cautionary note, in these studies, human perception served as the basis to classify species as cryptic or aposematic, but whether the real predators of such prey have a similar perception remains an open question. Similar issues have been raised in studies investigating mimicry in animals (Cuthill and Bennett, 1993). It is also interesting to note that there were few transitions to a gregarious lifestyle in these species. Given that aposematism has evolved many times, this also suggests that gregariousness is not necessarily a prerequisite for the evolution of aposematism. To resolve the issue of the ordering of events in the evolution of aposematism, a dataset with more transitions to a gregarious lifestyle is needed.

The second approach uses experiments to determine the vulnerability to predation of aposematic prey species living alone versus in groups. An experimental

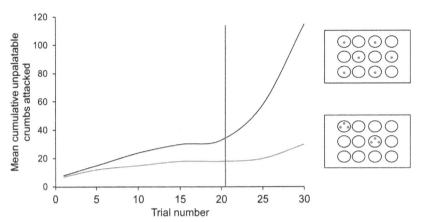

FIGURE 3.13 Enhancement of the value of aposematic signals in groups: chicks learn to discriminate unpalatable food crumbs more rapidly when crumbs are clumped. A stronger aversion is indicated by a lower cumulative number of unpalatable food crumbs taken over several trials. Chicks were presented with food crumbs presented singly (black line) or in clumps (red line) as shown in the inset. After the 20th trial, all crumbs were palatable. *Adapted from Gagliardo and Guilford (1993)*. (For interpretation of the references to colour in this figure legend, the reader is referred to the online version of this book.)

approach can tackle mechanistic questions regarding how warning signals interact with gregariousness to influence predator deterrence. In a particularly clever experiment to address these issues, chicks were allowed to forage on a tray with food crumbs, some of which were brightly coloured and unpalatable (Gagliardo and Guilford, 1993). Chicks learned more quickly to avoid unpalatable crumbs when the same number of crumbs were presented in clumps rather than singly, and persisted longer in avoiding these items in trials where the crumbs were now all palatable (Fig. 3.13). The authors suggest that ingestion of unpalatable food represents a more memorable experience when viewed simultaneously with other similar stimuli. Enhanced learning of warning signals associated with such prey aggregation has been documented in subsequent studies as well (Mappes and Alatalo, 1997; Reader and Hochuli, 2003; Skelhorn and Ruxton, 2006). I would like to note that predators may also learn to associate unpalatability with various traits of the prey species, including non-aposematic features such as body shape or size. It would be interesting to determine whether aversion is stronger when associated with warning signals rather than with other traits, as would be expected if gregariousness and aposematism are indeed associated.

What happens to aposematic individuals when rare in the population, as would presumably be the case in the early stages of the evolution of aposematism? In the previous study, the chicks received an equal mixture of palatable and unpalatable food items. The oddity effect, reviewed earlier in this chapter, involves a greater risk of attack for rare individuals that are distinct from the rest of the group. More generally, rare morphs of a prey may be attacked proportionately more

(Allen and Greenwood, 1988). Therefore, such selection against novelty may provide a further burden for the evolution of warning signals. In an experiment using great tits as predators of artificial prey, with which they had no previous experience, attack rate was indeed higher on aposematic prey when they were rarer in the population (Lindström et al., 2001). The encounter-dilution effect may also have played a role in this experiment. The rate of encounter with brightly coloured prey may increase due to their greater conspicuousness, but since not all prey are taken from a group, per capita predation rate becomes lower for prey in groups. The dilution effect may be especially strong for unpalatable prey because a bad experience with one individual prey probably reduces the chances that the predator remains in the same patch, thus protecting the rest of the group (Riipi et al., 2001). This mechanism would also work for species without warning signals but faced with predators that cannot capture all individuals in the group.

The comparative analysis approach in lepidopteran larvae suggests that aposematism can precede the evolution of gregariousness, but it remains to be shown whether this conclusion also pertains to a range of taxa. The experimental approach indicates that there are benefits of aggregation for aposematic prey. Most experimental studies have been carried out using a few select avian species as predators. Including other species may help broaden the conclusions to species that differ in sensory ecology. In the end, it may not be surprising that in different taxa, with different types of predators, the causal relationship between gregariousness and aposematism may run in different directions.

3.3. ARE ANTIPREDATOR PLOYS EFFECTIVE?

The list of antipredator ploys available for prey in groups is long. Some ploys act early in the predation sequence, such as collective detection, while others act later, like confusion and group defences. In view of the arsenal available for grouped prey, what is the evidence that aggregation increases safety? In the previous section on encounter-dilution effects and group defences, several studies found that per capita predation risk decreases with group size. Caro (2005) reviewed the literature in birds and mammals and found only one study among fifteen that did not report a decrease in individual predation risk as group size increases. A formal review of the literature is beyond the scope of this book. Instead, I prefer to focus on cases not reviewed by Caro, where individual risk failed to decrease with group size, to illustrate how antipredator ploys may break down.

Surprisingly, several studies have reported an increase in per capita mortality risk as group size increases, especially in larvae of several species of insects and in juveniles of coral reef fish species (Table 3.1). It is important to remember that a greater prevalence in such taxa may simply reflect the fact that predation risk in relation to group size has been investigated more often in these species. Insect larvae and juvenile coral reef fishes tend to be relatively sedentary during growth, and mortality can be measured over weeks, making such species

TABLE 3.1 Studies That Fail to Provide Evidence for a Decrease in Per Capita Predation Risk as Group Size Increases

Species	Range of Group Sizes	Findings	References
Fish			
Yellow damselfish	2–5+	Larger groups may encounter more predators; higher competition for enemy-free space	Brunton and Booth (2003)
Yellow-tailed damselfish	3–6	Competition for enemy-free space (shelter space is limited)	Holbrook and Schmitt (2002)
Six bar wrasse	1–2	Higher number of predators in larger groups	Shima (2001)
Spiny chromis damselfish	13–170	No dilution effect upon attack	Connell (2000)
Insect Larvae			
Baltimore checkerspot	250–2500	Large groups attract more predators	Stamp (1981)
Pipevine swallowtail	4–12	Large groups may be encountered more frequently; dilution may be less effective in larger, denser groups	Fordyce and Agrawal (2001)
Cabbage butterfly	5–100	Increased rate of parasitism in larger groups	Le Masurier (1994)
Black slug moth	45–541	Larger groups may be attacked more frequently	Reader and Hochuli (2003)
Pine sawfly	5–10, 10–40	Larger groups attract more ant predators	Lindstedt et al. (2006)
Pine sawfly	10–50	Larger groups attract more avian predators	Lindstedt et al. (2010)
Nymphalid butterfly	1–30	–	Denno and Benrey (1997)
Processionary caterpillar	46–233	More predation on larger groups	Pescador-Rubio (2009)

TABLE 3.1 Studies That Fail to Provide Evidence for a Decrease in Per Capita Predation Risk as Group Size Increases—cont'd

Species	Range of Group Sizes	Findings	References
Japanese giant silk moth	3–18	No dilution effect upon attack	Iwakuma and Morimoto (1984)
Other Species			
Caribbean spiny lobster	1–3	Larger groups may encounter predators more frequently	Childress and Herrnkind (2001)
Elk	1–30+	Small groups encounter predators less often	Hebblewhite and Pletscher (2002)
Brambling and chaffinch	10–10,000	Large groups are attacked more frequently and more successfully	Lindstrom (1989)

quite amenable to study. Birds and mammals, in contrast, range widely, forage in groups of varying sizes, and live much longer, making it more challenging to document changes in predation risk with group size. Nevertheless, it may be the case that a breakdown of antipredator ploys happens more frequently in species with restricted mobility.

From the table, one of the most common reasons why per capita mortality risk fails to decrease with group size is that the increase in the rate of encounter with predators in larger groups is not compensated by risk dilution. In fact, in some species studied, risk dilution did not operate at all, because predators could consume all prey in the group. Competition for enemy-free space, areas where individuals are protected from predators, also increases with group size in some fish species, increasing the proportion of individuals exposed to predation in larger groups as these safe areas are often limited in size. Overcoming confusion, a factor that is only relevant to mobile prey, has also been proposed as the reason why attack success rate did not decrease with group size in one avian species.

Why would species persist in joining groups if individual risk does not decrease in larger groups? Obviously, living in groups may provide other types of benefits, such as increased foraging efficiency (Chapter 1), which can compensate for compromised personal safety. Nevertheless, from the point of view of antipredator ploys, social prey species may be selected to invest relatively less in costly antipredator defences (Daly et al., 2012). Reduced investment in defences makes it possible to allocate resources to other fitness-enhancing traits, such as foraging, and recoup losses in personal safety due to reduced investment in defences.

I present two examples to illustrate the trade-off between personal safety and other fitness-enhancing traits. First, individuals in many species of birds and mammals reduce their investment in antipredator vigilance in larger groups, making groups more vulnerable to predation than would be expected from the pooling of so many eyes and ears attuned to predator detection (see Chapter 4). However, the time saved from a reduction in individual vigilance can be reallocated to foraging or resting, which provides many benefits. The second example is the finding that insect larvae of one species defend themselves less readily when in groups than when alone (Daly et al., 2012). Gains from a reduction in defence costs can be reallocated to individual growth, for instance, which is a crucial trait when competing for limited resources in a group. As these two examples show, animals may tolerate a certain level of predation risk, which allows individuals in groups to reallocate resources to other fitness-enhancing traits, thus yielding an overall advantage of living in groups despite a weak relationship between safety from predators and group size.

Comparative analyses of species that live in groups of different sizes can also shed some light on the relationship between survival and group size in animals. In insects, larvae of gregarious species are less likely to die in early development than those of solitary species (Hunter, 2000). Although a formal analysis using phylogenetic information is currently lacking, these results suggest that personal safety increases with group size in this set of species, although the pattern is not universal, as noted in Table 3.1. Intriguing results have emerged recently from comparative analyses in birds and mammals relating longevity to group size. Life-history theory predicts that increased protection in large groups should select for an increase in longevity by reducing extrinsic sources of mortality. However, controlling for known correlates of longevity, such as body size, three different studies failed to document an increase in longevity in species living in larger groups (Beauchamp, 2010c; Blumstein and Møller, 2008; Kamilar et al., 2010). Various factors may explain this unexpected result. As noted above, the association between living in groups and personal safety may not be as strong as predicted. Second, the fluid nature of group sizes in many species of birds and mammals makes it difficult to quantify sociality. Finally, individuals in solitary species may compensate for the lack of protection afforded by companions by adopting niches that reduce contact with predators (Lind and Cresswell, 2005).

3.4. CONCLUDING REMARKS

The greatest challenge in the study of antipredator ploys remains the disentangling of the effects of the various mechanisms available to increase individual safety in groups. In many species, more than one mechanism plays an important role, and several mechanisms often make the same predictions.

Although a large body of evidence suggests that living in groups reduces predation risk, this relationship is far from universal, and in some cases, a weaker association may occur because limited resources are allocated to other fitness-enhancing traits like foraging. We have some evidence for fitness benefits associated with the encounter-dilution effect in largely immobile species, but survival consequences in other species and in relation to other mechanisms, such as collective detection and confusion, are currently lacking.

A further difficulty is that variation in predation risk may reflect differences in group size as well as differences in phenotypic attributes among individuals in groups of different sizes. For instance, individuals in small groups may be in poorer condition, exaggerating the real effect of group size on predation risk. This represents a potential problem for observational studies such as those conducted on birds and mammals. Experimental studies are particularly valuable in this regard and have been conducted in fish and other invertebrate species, as reviewed earlier, and would be welcome in a broader range of species.

Antipredator studies typically concentrate on one specific ploy, but it is probably the case that more than one ploy plays a role in deterring predation. It would be interesting to determine for a given species the best combination of ploys for deterring predation, and under which environmental conditions certain combinations work best. Obviously, some species are restricted in the number of ploys available to them. Sessile species, for instance, can only deter predation using encounter-dilution effects and group defences. But even in such cases, is it better to invest more in group defences or in factors that reduce encounter rate with predators? The behavioural ecology approach, which focuses on the fitness consequences of different strategies in particular environments, may be particularly well suited to investigating the adaptive value of different combinations of antipredator ploys.

As a final consideration, which will be explored in greater depth in Chapter 9, it is important to remember that the best course of action for deterring predation is also dependent on what the predators can do. Predators may be able to learn or evolve over several generations, thereby altering the effectiveness of antipredator ploys. Predicting the best antipredator ploy needs to take into account the fact that predator behaviour can change in response to the deployment of antipredator defences by prey.

Chapter 4

Antipredator Vigilance: Theory and Testing the Assumptions

4.1. INTRODUCTION

Among primates, humans are exceptional in the degree to which activities are carried out on two feet rather than the usual four. The evolutionary origin of hominin bipedalism remains murky, but several selection pressures are thought to have played a role in fostering an upright posture. In particular, the vigilance hypothesis argues that standing on two feet broadens the visual horizon in cluttered habitats, facilitating the detection of predators and competitors (Dart, 1959). Many other four-legged species, including non-human primates (Videan and McGrew, 2002) and rodents (Bednekoff and Blumstein, 2009), carry out bipedal vigilance, suggesting that the ability to scan visually from a vantage point may be beneficial. Consistent with the view that better vigilance matters, sentinels in many avian and mammalian species also seek vantage points (Clutton-Brock et al., 1999b). Although recent archaeological

Social Predation. http://dx.doi.org/10.1016/B978-0-12-407228-2.00004-4

evidence suggests that factors other than vigilance played a more prominent role in ushering in hominin bipedalism (Niemitz, 2010), advantages related to enhanced vigilance may have helped maintain bipedalism once evolved.

Vigilance represents an allocation of time by an animal aimed at detecting significant environmental stimuli. An animal may be vigilant for various reasons, one of which was already explored in Chapter 1, namely, the detection of food sources discovered by conspecifics. Here, I shall focus on the detection of potential threats related to the presence of predators and competitors—two types of threats suggested earlier by the vigilance hypothesis for hominin bipedalism. The detection of predation threats, any signal that indicates the presence of predators nearby, has been the main focus of research on vigilance. Vigilance, in this context, represents an antipredator ploy like those investigated in Chapter 3. I devote two full chapters to vigilance considering the extensive literature that focuses on this theme. In this chapter, I examine the most influential models of vigilance, and then review the empirical evidence supporting their assumptions. The following chapter covers the empirical evidence for two key predictions.

The level of vigilance maintained by an individual should be under strong selection pressure. Not detecting a food item may not make a large dent in the survival of an individual, but failing to detect a nearby predator will greatly increase the risk of death. Therein lies the interest of studying vigilance. The obvious solution of allocating a large amount of time to vigilance is not feasible for most animals. As the amount of time available to perform activities is necessarily finite, an increase in time spent vigilant must detract from time spent on other fitness-enhancing activities, such as foraging and resting. In addition, maintaining vigilance for a long period of time may be associated, paradoxically, with a decrease in the probability of detecting important environmental signals due to habituation or fatigue (Dimond and Lazarus, 1974). For these reasons, a major task for animals consists in allocating the right proportion of time to vigilance to increase survival.

A solitary forager can only reduce the burden of vigilance by foraging in places or at times with reduced predation risk. For social animals, joining a group provides a very effective solution to the problem of individual vigilance. The previous chapter described the various ways living in groups can reduce individual predation risk. In particular, animals in groups can benefit from an increase in predator detection ability and from a dilution of predation risk. Here, I show how living in groups can allow individuals to spend less time vigilant without incurring a higher risk.

4.2. WHAT VIGILANCE IS AND HOW IT IS MEASURED

From a mechanistic point of view, vigilance represents the ability to detect a particular stimulus at a given moment in time. Vigilance should be viewed as a mental state of alertness aimed at detecting these stimuli. Resources are thus allocated in the brain to direct the senses, including eyes and ears, and to interpret the information gathered over a given period of time. Consequently, vigilance should be measured from brain activity measurements. Obviously, this is not practically feasible when investigating vigilance in the field. Vigilance, therefore, is typically measured using behavioural postures, which are visible expressions of alertness. In most studies, vigilance is thought to occur when animals raise their head to scan their surroundings. A meerkat on the lookout for predators on a high perch provides a classic example of antipredator vigilance (Fig. 4.1).

Head position was indeed the defining feature of vigilance in the very first model of vigilance in animals (Pulliam, 1973). Although using head position solves the issue of actually measuring an observable state, such postures may be poorly related to actual vigilance as might be revealed by brain activity measurements. For instance, an animal seemingly scanning its surroundings may be tuned out, experiencing only a marginal probability of detecting important stimuli. Assumptions about what the eyes can see are implicit in any measurement of vigilance related to behavioural postures, but little is known about the visual detection ability of many animals (Fernández-Juricic et al., 2004a). For instance, species with a large field of view may still be able to detect predation threats while foraging head down (Guillemain et al., 2002; Tisdale and Fernández-Juricic, 2009; Wallace et al., 2013). Reliance on postures also assumes that vigilance is mostly performed visually, but acoustic vigilance may be a common feature in animals (Ridgway et al., 2006).

Unfortunately, linking vigilance and postures mostly restricts studies of vigilance to birds and mammals, the two taxonomic groups where measurable changes in postures appear related to visual scanning. It is thus necessary to rely on indirect indices to infer vigilance in other species. In fish, for instance, vigilance has been related to indirect behavioural indices like skittering (Magurran et al., 1985). In lizards, non-vigilance was inferred when the animals closed their eyes, but this is certainly a definition of vigilance based on the absence of evidence (Lanham and Bull, 2004). I am not aware of vigilance studies based

FIGURE 4.1   A meerkat on the lookout. *Photo credit: Kevin Ryder. This figure is reproduced in colour in the colour section.*

on postural measurements in invertebrates even though a state of alertness must surely exist. For example, vigilance in squids has been inferred indirectly from changes in skin colour indicative of threat detection (Mather, 2010). Other indirect measurements of vigilance include the amount of food left behind in a patch (Kotler et al., 2010), but such measurements assume that there is an inverse correlation between the time spent vigilant and foraging, which is probably true although the relationship may not be exactly one to one.

The posture approach to defining vigilance implies that there is a certain level of incompatibility between vigilance and other activities. Therefore, prey animals must trade-off antipredator vigilance and the time, energy, and attention devoted to other activities, such as foraging and sleep. Individual allocation of time and effort to antipredator vigilance has thus generally been viewed as a balancing act between competing activities (Caraco, 1979b; Lima, 1987b). In this context, two types of vigilance have been proposed (Blanchard and Fritz, 2007): routine or induced vigilance. Routine vigilance occurs when an individual monitors its surrounding during spare time, while induced vigilance can be characterized by a strong reaction to a predation threat. Routine vigilance represents the main focus of antipredator vigilance. Unlike routine vigilance, induced vigilance may disrupt the foraging process, as animals are expected to dedicate their full attention to monitoring a threat. Relatively cost-free, routine vigilance may thus be considered low-level alertness, but induced vigilance typically involves a large reduction in time spent foraging. In view of the varying costs associated with each type of vigilance, it would not be surprising if factors like group size influence the two types of vigilance differently. This discussion highlights the importance of paying attention to what the animals are actually looking at during vigilance. Some researchers associate vigilance with active monitoring of a threat (Yaber and Herrera, 1994) or distinguish between low vigilance and high vigilance, the latter presumably aimed at imminent threats (Koboroff et al., 2008; Lazarus, 1979). Practically, it should be a priority to identify the type of vigilance recorded in a study.

Estimates of vigilance are typically gathered from time budgets. In a time-budget study, the percentage of time allocated to vigilance during a predetermined time period is determined for a number of identified individuals. Vigilance has also occasionally been estimated using the percentage of individuals in the group that are vigilant at predetermined time fixes. Three measures of vigilance are routinely reported in a time-budget study after pooling results from all focal individuals: time spent vigilant, which is the average percentage of time allocated to vigilance during a bout of observation; vigilance frequency, which is the average number of vigilance bouts initiated per unit time; and vigilance duration, which is the average duration of all vigilance bouts. Although these different measures tend to be correlated with one another, they are not interchangeable. For instance, the duration of vigilance bouts appears less sensitive to variation in group size in birds than their frequency (Beauchamp, 2008c).

Solely investigating time spent vigilant may hide crucial information about vigilance. Indeed, to achieve a given percentage of time spent vigilant, individuals can use a few long scans or many short ones (McVean and Haddlesey, 1980). The question of why some combinations of vigilance duration and frequency may be preferred has received little attention in an evolutionary context (Sirot and Pays, 2011). Vigilance is probably more than a simple pattern of raising and lowering the head. Recent studies indicate that different head movements, while actually scanning the surroundings, can be used to optimize detection or monitor different targets, such as predators or companions (Fernández-Juricic et al., 2011; Wallace et al., 2013). These results highlight the contribution of head movements in predator detection and recognition. To conclude, simplifying assumptions have been made to measure vigilance, and more work is definitely needed to assess their relevance in any study system.

4.3. THEORETICAL BACKGROUND

In a letter to Mr Barrington in 1772, Gilbert White stated that "birds may crowd together to dispel some degrees of cold and a crowd may make each individual safer from the ravages of birds of prey and other dangers" (Hindwood, 1937). Galton in 1871 noted that "a lone beast is not only defenceless but easily surprised. A herd of cattle is always on the alert and at any moment some eyes, ears or noses will command all approaches. The start or a cry of alarm by one beast is a signal to all other companions. The protective senses of one individual are multiplied by a large factor and animals in groups thereby receive maximum security as a minimum cost of restlessness" (Galton, 1871). These early naturalists articulated the notion that being in a group would provide more protection to any individual than being alone. This enhanced safety arises from the pooling of information acquired from the senses of all companions in the group, a principle referred to nowadays as the many-eyes effect. The many-eyes effect was the basis for the first formal model of vigilance proposed by Pulliam 40 years ago (Pulliam, 1973). By making explicit assumptions about vigilance in animals, this first model was therefore crucial in moving the field of research on vigilance from purely descriptive to hypothetico-deductive. Many of the assumptions made by Pulliam have been re-examined over the years, and these will be discussed in detail subsequently.

4.3.1. Pulliam's 1973 Model

Pulliam considered a predator that attacks a group of prey animals. The predator breaks cover at random times irrespective of whether individuals in the group are vigilant or not. From the time where the attack is launched, the predator needs time T to get close enough to the prey animals to make a capture. Prey animals obtain resources while head down, at which time the predator cannot be detected. Animals interrupt their foraging and raise their heads to scan their surroundings, which allows them to detect the predator before it gets fatally close. Scans are assumed to be short and constant in duration. Prey animals interrupt their feeding at random times following a Poisson process. Therefore, a feeding bout with the head down will be interrupted at a fixed rate λ.

Intervals between successive scans are referred to as interscan intervals. Culled over many bouts of vigilance, the distribution of interscan intervals is expected to follow a negative exponential distribution as long as the rate of interruption is constant. In such a distribution, most interscan intervals are short with a few long intervals (Fig. 4.2).

This is similar to the distribution of lightbulb life expectancy or the time to the next telephone call for not too busy a scientist. How long a lightbulb will last before breaking down, which is the same idea as the interruption of feeding to scan, follows a negative exponential distribution with parameter λ, which provides the rate at which lifetime decays.

If the feeding bout were considered a series of discrete choices between remaining head down and raising the head, the fixed rate parameter could be thought of as the probability of raising the head at any moment a choice is made. Consider a series of coin flips for a well-known analogy. In a series of coin flips, the fixed rate parameter may be likened to the probability of getting a head

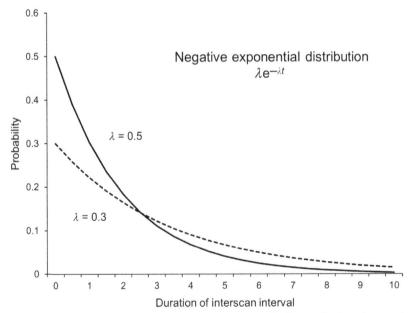

FIGURE 4.2 A simple model of vigilance: a forager interrupts feeding at a fixed rate λ to scan the surroundings for possible threats. The interval between two scans is called an interscan interval. In this simple model, the duration of an interscan interval is independent of the duration of the previous intervals. The distribution of interscan intervals follows a negative exponential distribution ($\lambda e^{-\lambda t}$). A higher value of λ induces a faster decay in the duration of interscan intervals.

after one flip, which is 0.5 in this case. This probability is independent of how many heads or tails were obtained previously. If more than one coin is flipped at the same time, getting a head with one coin has no bearing on the results for any other coins. Returning now to the scanning dynamics, the random process implies that the initiation of a scan is independent of the length of time spent head down. In addition, since choices to interrupt feeding by each individual are dictated by separate but identical random processes, interruptions in foraging occur independently among individuals in the same group.

After launching its attack, the predator will capture any solitary forager that fails to scan before time T, an event which occurs with probability $e^{-\lambda T}$. Now, consider the case of N individuals foraging together in a group. The dangerous situation for an individual happens when no one in the group raises its head during the approach by the predator. Should one individual spot the predator before the critical time, all individuals can escape to safety. Since each individual scans independently, the situation where no one scans before the critical time arises with probability $e^{-N\lambda T}$. Therefore, the probability of detection by at least one individual in the group is the complement, namely, $1 - e^{N\lambda T}$ (Fig. 4.3).

For a solitary forager, the probability of detection intuitively increases when the time needed by the predator to get fatally close increases. This increase is progressively smaller as attack time gets longer, simply because the chances of not raising the head become vanishingly small the longer the predator takes to get fatally close. For a group member, the probability of detection, for any attack time, increases with group size. When group size becomes large enough, the probability of detection gets very close to 1. In such a case, it is not even remotely possible for the predator to remain undetected during the attack, and all foragers can escape unharmed every time.

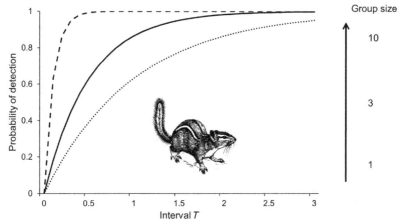

FIGURE 4.3   Predator detection and group size: the probability of detecting a predator by prey animals increases when the time (T) needed by the predator to get fatally close increases and when group size increases.

Detection is almost inevitable in large enough a group. Group members could thus reduce the rate at which they interrupt feeding and still enjoy an advantage over remaining alone. Therefore, the model makes two testable predictions: (1) the probability of detection should increase with group size and (2) individual investment in vigilance should decrease with group size. Pulliam's model makes several assumptions regarding the vigilance process, including independent scanning among individuals and instantaneous randomness in the initiation of scans. Empirical evidence regarding these assumptions will be examined in the next section.

Weaknesses became readily apparent in Pulliam's model in the years following publication. Survival in Pulliam's model only varies as a function of the probability of detection. However, several other factors can reduce predation risk in a group. As mentioned in the last chapter, encounter-dilution effects as well as confusion can greatly influence the probability of attack and capture of an individual and should be taken into account to get a broader perspective on the management of risk in a group. The model also fails to specify why prey animals should initiate scans randomly. One could easily imagine other scenarios for the temporal organization of scans, including interruptions at regular intervals (Bednekoff and Lima, 2002; Scannell et al., 2001). The model does not take into account the benefits that arise from foraging. As such, the model fails to specify why individuals should not spend all their time vigilant. The trade-off between vigilance and foraging now forms the basis of all models of vigilance. Attack rate by predators was assumed to be independent of group size, which is not consistent with many empirical findings (Beauchamp, 2008a; Cresswell, 1994; Hebblewhite and Pletscher, 2002).

Perhaps the greatest weakness in the model concerns the question of cheating. In a Pulliam group, an individual would greatly benefit by reducing its own investment in vigilance at the expense of others. Cheaters would get the benefit of increased detection, provided by the remaining group members, without having to pay the full cost of investing time in vigilance. The time thus saved could be used to obtain more resources and thus increase survival. Modelling vigilance requires an approach that takes into account the possibility of cheating. Some of the above issues were addressed in a subsequent model by Pulliam and collaborators.

4.3.2. Pulliam et al.'s 1982 Model

A 1982 model by Pulliam et al. represents quite an improvement over the previous attempt by Pulliam in 1973 (Pulliam et al., 1982). Here, foragers can still benefit from joining a group by improving their chances of detecting a predator before it gets fatally close. However, increased vigilance now comes at a cost as it reduces the rate at which food is gathered. Individuals with a low food intake rate are forced to spend more time in the open, where they are exposed to more attacks by predators. Therefore, the core of this model includes the fundamental trade-off between vigilance and foraging. One would expect individuals to adopt an intermediate level of vigilance, which provides safety without sacrificing food intake rate. Survival varies as a function of the expected number of attacks from predators. However, as individuals must remain in the foraging patch until a fixed food requirement is met, starvation is not an issue for survival.

This new model incorporates cheating, because the optimal investment in vigilance for any individual takes into account the scanning behaviour of other group members. To illustrate the concept of cheating, consider the following two extreme patterns of vigilance. The strategy of investing heavily in vigilance, when adopted by all group members, is not stable because a less vigilant cheater would get more food and enjoy the same safety as the others. Similarly, the strategy of investing lightly in vigilance is also unstable, as a more vigilant cheater may lose only a small amount of food but experience a large increase in safety.

Game theory is perfectly suited to examining phenomena where payoffs are frequency dependent. I already discussed the use of such models in the context of producing and scrounging (Chapter 2). Pulliam et al. used a game-theoretic approach to address the issue of cheating. The authors uncovered two possible solutions in the vigilance game. The cooperative solution emerges when the vigilance level that maximizes fitness is adopted by all individuals. However, this solution can easily be invaded by cheaters. When all individuals adopt the selfish solution, choosing an alternative level of vigilance can never provide higher fitness.

Using field estimates of attack rate and duration, the authors compared observed scanning rates in groups of different sizes to those predicted under the two types of solutions in yellow-eyed juncos, a small granivorous bird. It turns out that the cooperative solution provided a better fit to the observed scanning rates than the selfish solution (Fig. 4.4). I note that the selfish solution entails that all juncos should forego vigilance when in groups. Therefore, the only means of protection for juncos in groups would be risk dilution, downplaying any benefits related to collective detection. I would argue that support for the cooperative solution in this species is not really indicative of a preference to cooperate. Rather, I think that the selfish solution identified by the model is too extreme. As I will discuss later, vigilance may provide direct benefits to an individual during an attack. If the selfish solution involved any non-negligible allocation of time to vigilance, it may turn out to be closer to the observed values than the cooperative solution.

What happens to animals when they become satiated before the end of the day is problematic in this model. Satiated animals are expected to leave their groups to hide in safe places away from predators, thus reducing the number of attacks to which they are exposed. However, the model assumes that group size remains constant, which means that individuals leaving the group must be replaced immediately by companions so as to maintain the same group size. This assumption may be questionable in animal groups with stable or slowlychanging membership (Beauchamp and Ruxton, 2007a).

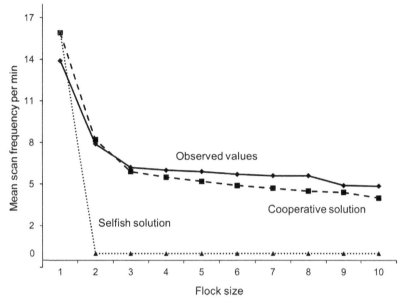

FIGURE 4.4   Vigilance and group size: the number of scans per minute decreases as a function of flock size in yellow-eyed juncos. Observed values are contrasted with the cooperative and selfish solutions to the vigilance game derived from the 1982 vigilance model of Pulliam et al.

4.3.3. McNamara and Houston's Model

A model by McNamara and Houston (1992) is by far the most sophisticated attempt to capture the essence of vigilance. This model formalizes the notion that direct detection of a threat, rather than indirect detection through the reaction of companions, can increase the chances of survival of a forager. This potential advantage was first considered in an earlier phenomenological model of vigilance (Packer and Abrams, 1990). As I explained in the previous section, the model by Pulliam et al. (1982) assumed no advantage to direct detection. Nevertheless, direct detection can give a head start when escaping and increase the chances of escaping safely.

The McNamara and Houston model predicts that the magnitude of the group-size effect on vigilance should depend on the relative advantage of direct versus indirect detection (Fig. 4.5). In particular, the stable level of vigilance should increase when those that detect an attack directly enjoy a considerable reduction in the probability of death upon attack over those that detect the threat indirectly. Increased reliance on personal vigilance to detect attacks also means that the decrease in vigilance with group size is shallower. This result certainly illustrates the importance of considering the advantage of personal vigilance when modelling vigilance.

McNamara and Houston also considered the role of confusion on vigilance. The hurried departure of several individuals in a group may confuse

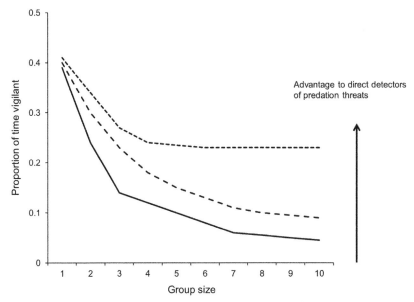

FIGURE 4.5   Vigilance and detection: the evolutionary stable level of vigilance decreases more slowly with group size when the probability of death upon attack by a predator is larger for non-detectors than detectors. *Adapted from McNamara and Houston (1992).*

the predator and thus reduce the probability of death upon attack (see Chapter 3). Crucially, this probability should be a decreasing function of group size. When incorporating the confusion effect, the stable level of vigilance decreases more slowly with group size, which is an unexpected finding not discussed by the authors. At first sight, confusion should provide extra safety and allow individuals to decrease their vigilance even more in larger groups. I note that the authors only modelled confusion in the situation where the probability of death upon attack was similar for detectors and non-detectors alike. I wonder whether this counterintuitive prediction would also prevail when personal vigilance is beneficial. Interestingly, the fit to Pulliam's own data on vigilance in yellow-eyed juncos was better when considering the confusion effect.

Scan duration has always been assumed to be short and, more importantly, constant in duration. However, scan duration can vary quite extensively in groups of different sizes (McVean and Haddlesey, 1980). McNamara and Houston suggested that the probability of failing to detect a predator may be a decreasing function of scan duration. Indeed, a very short scan may be quite useless in detecting a predator. However, very long scans will provide little extra safety and delay the resumption of foraging. The optimal scan duration will balance these conflicting demands. It turns out that optimal scan duration should decrease when the proportion of time spent vigilant decreases, a trend that has been documented in many avian studies (Beauchamp, 2008c). Clearly, the assumption that scans are constant in duration needs to be revisited.

Oddly enough, previous models of antipredator vigilance predicted that vigilance should be independent of predation risk, the very factor driving vigilance in the first place. McNamara and Houston's approach predicts that vigilance should increase with predation risk when there is a risk that feeding will be interrupted. Pulliam et al. (1982) assumed no interruption in feeding so that the whole day was available to accumulate resources. Therefore, the risk of starvation was non-existent. However, many factors can alter the duration of the foraging time horizon, including changes in the weather or external disturbances. When foraging can end prematurely, starvation becomes more likely if feeding rate is too low, and the stable level of vigilance ought to increase with predation risk.

In a further development, the authors included in the model the level of internal energy reserves available to a forager. Intuitively, a forager with more reserves should be less likely to sacrifice safety to increase foraging. The model thus predicts how vigilance levels should vary as a function of state. The results show that vigilance should generally increase with the level of reserves, but the effect interacts with group size (Fig. 4.6). In small groups, vigilance is predicted to increase rapidly with the level of reserves before reaching a plateau. By contrast, in large groups, vigilance is state-independent until reserves reach a high level. In a large group, individuals with low reserves are protected by

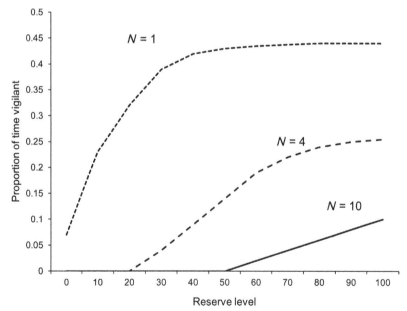

FIGURE 4.6 Vigilance and energy reserve levels: the evolutionary stable level of vigilance increases when individuals have more internal energy reserves but the pattern interacts with group size. *Adapted from McNamara and Houston (1992).*

risk dilution and collective detection and can afford to invest more heavily in foraging to build up reserves, which will increase their chances of avoiding starvation. This is not the case for individuals in small groups where dilution and collective detection effects are less effective. As the level of reserves increases, individuals become less likely to sacrifice safety to reduce the likelihood of starvation, and thus invest more in vigilance.

As animals build up reserves through foraging, the stable level of vigilance should also vary as a function of food intake rate. In particular, the model predicts that vigilance should decrease when food intake rate increases, controlling for group size and level of reserves. Because food intake rate is usually a function of food density in the habitat, vigilance levels should generally decrease with food density. In view of these novel predictions, incorporating state dependence has certainly been a remarkable improvement over previous models. State-dependent solutions for the vigilance game may explain why vigilance levels can vary from one individual to another within the same group and in response to variation in the level of reserves and food intake rate. It is unfortunate that the mathematical complexity of McNamara and Houston's model has limited its appeal to empirical researchers. The richness of their predictions certainly deserves more empirical scrutiny.

4.4. VALIDITY OF THE ASSUMPTIONS

The models explored thus far make several assumptions about the vigilance process. It is important to assess the relevance of these assumptions before evaluating the predictions. In the next two sections, I explore the assumption that group size drives changes in vigilance and that there is a trade-off between vigilance and foraging. If time spent vigilant fails to compete with the time allocated to other activities, such as foraging, vigilance is not expected to vary at all with group size. I also explore the assumption that direct detection of predation threats is advantageous.

Other assumptions are related to the nature of the scanning process. All models assume that individuals interrupt foraging to scan at random times dictated by a Poisson process. The Poisson process entails that the initiation of a scan will occur at the same rate regardless of how long the animal has spent head down. This feature has been labelled instantaneous randomness (Bednekoff and Lima, 1998a). With a Poisson process, the distribution of interscan intervals should follow the negative exponential distribution (Fig. 4.2). In a negative exponential distribution, longer intervals are progressively less likely and the distribution has a mode at the smallest possible value and no hump (as is the case in a normal distribution). Sequential randomness, a corollary of instantaneous randomness, implies that the duration of an interscan interval should be independent of the duration of previous intervals. Finally, scanning is assumed to be independent among foragers in the group because otherwise the detection ability of each forager cannot really add up.

4.4.1. Vigilance Varies as a Function of Group Size

This fundamental assumption often fails to be stated explicitly. All models of vigilance imply that variation in group size alone drives changes in vigilance. In an observational study, however, the possibility arises that unmeasured variables correlated with group size may actually explain much of the variation in vigilance. Two approaches have been used to determine whether group size drives vigilance. The first approach consists in manipulating group size while maintaining other variables constant, which ensures that any variation in vigilance can only be attributed to group size. This experimental approach is typically restricted to the laboratory since it is not feasible to randomly allocate individuals to groups of different sizes in the field unless individuals show a limited range of movements.

Doubts are often expressed as to the validity of studying vigilance in the laboratory where predation risk is absent. Nevertheless, I point out that animals in the laboratory may not know that predation risk is non-existent. In fact, the group-size effect on vigilance is no less apparent in the laboratory than it is in the field (Beauchamp and Livoreil, 1997; Blumstein et al., 1999; Gosselin-Ildari and Koenig, 2012; Rieucau et al., 2010; Vahl et al., 2005). Therefore, group size can exert a strong influence on vigilance levels even in the laboratory, which lends support to the assumption that group size drives changes in vigilance.

The second approach can be used in the field and consists in documenting variation in vigilance as a function of rapid changes in group size. Given that vigilance is documented before and immediately after a real-time change in group size, it can be safely assumed that all remaining variables have not changed save for group size. As an example, isolated birds decrease their vigilance immediately after hearing broadcasted calls simulating the presence of companions (Radford and Ridley, 2007). Similarly, preening birds alter their vigilance straightaway after the departure or the arrival of one group member (Roberts, 1995). While all the above results do not dispel the notion that group size drives changes in vigilance, any discussion of causality must entertain the possibility that variables other than group size may be involved.

4.4.2. Trade-Off between Vigilance and Foraging

At first, vigilance was thought to be entirely incompatible with the acquisition of resources. This is because vigilance and other behavioural patterns are carried out with postures deemed suitable to only one activity at a time. This assumption probably reflects the fact that early models of vigilance focused on birds that feed head down, a position that presumably occludes their field of vision to some extent. However, recent research with birds, as well as other species, indicates that vigilance and alternative activities are not necessarily mutually exclusive. For example, animals may be vigilant to some extent even when asleep (Rattenborg et al., 1999; Ridgway et al., 2006). With their laterally placed eyes with large fields of view, many bird species can maintain vigilance when feeding head down (Bednekoff and Lima, 2005; Fernández-Juricic et al., 2008; Lima and Bednekoff, 1999). Birds can also monitor their surroundings while searching or handling food in a head-up posture (Barbosa, 2002; Kaby and Lind, 2003). Similarly, rats can monitor overhead stimuli when looking down (Wallace et al., 2013) and mammalian herbivores can be vigilant while processing their food head up (Cowlishaw et al., 2004; Fortin et al., 2004b). Primates often feed in an upright position, allowing an overlap between vigilance and feeding (Treves, 2000).

Overall, the trade-off between foraging and vigilance may not be very costly when animals can overlap different processes while foraging (Baker et al., 2011; Fortin et al., 2004a) or monitor their surroundings even when head down. Similarly, faced with easy foraging tasks, animals may be able to perform vigilance at little or no cost (Lawrence, 1985; Teichroeb and Sicotte, 2012). These results suggest that in many animal species, vigilance and foraging may not be as incompatible as once thought, although few would argue that vigilance is typically cost-free.

4.4.3. Randomness in Scanning

4.4.3.1. Instantaneous Randomness

Several studies have examined the distribution of interscan intervals to examine the assumption of instantaneous randomness in the initiation of scans. Initial attempts found a good fit between the empirical distribution of interscan intervals and the negative exponential distribution, or its discrete equivalent, the geometric distribution (Bertram, 1980; Caraco, 1982; Studd et al., 1983). Nevertheless, the fit proved quite poor in other studies (Elcavage and Caraco, 1983; Pöysä, 1987; Sullivan, 1985). Generally, very short or very long intervals were less frequent than expected (Hart and Lendrem, 1984; Lendrem et al., 1986). Obviously, there must be a minimum interval dictated by the speed with which an animal can raise its head to scan from a head-down position. However, this minimum value has been shown to vary with group size, suggesting that an adaptive process rather than a biological constraint is at play (Elcavage and Caraco, 1983; Lendrem, 1983; Sullivan, 1985). The relative rarity of long scans suggests an increase in the probability of initiating a scan after a long interscan interval (Lendrem et al., 1986).

Recent studies have expanded the range of species investigated, which was typically limited to small avian species, and found rather drastic departures from instantaneous randomness. In birds, humped distributions of interscan intervals have been documented in downy woodpeckers (Sullivan, 1985), farmed ostriches (Ross and Deeming, 1998), Caribbean flamingos (Beauchamp, 2006a) and greater rheas (Carro and Fernandez, 2009). In flamingos, the distribution of interscan intervals definitely shows a humped pattern for both small and large groups (Fig. 4.7). Flamingos have few natural enemies and disturbances are typically human-related. A random scanning pattern, which is expected to detect fast-approaching predators, may not be relevant in this species. A more regular scanning pattern, which would tend to produce a humped distribution, may be all that is needed to detect such disturbances.

The negative exponential distribution also proved a poor fit to the distribution of interscan intervals in three mammalian species (Pays et al., 2010). A humped distribution was particularly apparent for the eastern grey kangaroo and European roe deer, the two species in the study experiencing the lowest predation risk. In the waterbuck, which faces more substantial predation threats, the distribution of interscan intervals showed no hump but still did not fit the negative exponential distribution well. Together with the flamingo results, it thus appears that low predation risk is associated with more regular scanning patterns. Ideally, this idea should be tested using different populations of the same species exposed to different levels of predation risk to rule out species-specific factors that may promote non-random scanning patterns.

The above empirical studies rejected the assumption of instantaneous randomness but did not address the processes that may actually produce a humped distribution of interscan intervals. Lendrem et al. (1986) proposed a time-dependent process of scan initiation. In a time-dependent process, the initiation of a scan becomes more likely when interscan intervals are longer. This time-dependent process reduces the likelihood of very short as well as very long scans, and can produce a humped distribution of interscan intervals when the

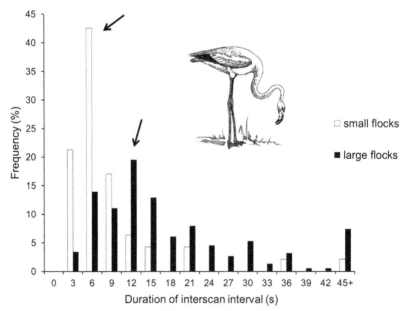

FIGURE 4.7 Distribution of interscan intervals: the frequency distribution of interscan intervals shows a hump (identified with an arrow) in small and large flocks of Caribbean flamingos. *Adapted from Beauchamp (2006).*

rate of feeding interruption is low. In flamingos and the other species mentioned earlier, scan initiation may thus be time dependent. The problem here is that this model fails to explain why time dependence should arise in the first place. As such, the model provides a phenomenological explanation for the results rather than a solution based on first principles.

Basic knowledge about the factors playing a role in the duration of interscan intervals is needed to build a truly predictive theory of their distribution. The duration of interscan intervals may be influenced by factors acting during prey and predator detection. Here, I just provide some pointers for future work without getting into too much detail. In terms of prey detection, the ability of a forager to locate items may decline if the foraging bout is interrupted too early or too frequently, favouring longer interscan intervals. This may be the case for prey items that are camouflaged or difficult to detect (Fritz et al., 2002). For other types of prey, it may be preferable to accumulate a large mouthful with one handling period rather than many smaller mouthfuls each interrupted by a scan and a handling period, which would again select for longer intervals. In terms of predator detection, the benefit of early interruption may be negligible when predation risk changes slowly. However, it may be costly to sustain long feeding periods without scanning, as predation risk may have changed since the last scan (Sirot and Pays, 2011). For instance, a predator may have moved nearby undetected during the feeding bout. The duration of interscan intervals may also reflect the outcome of a game between predators and prey in which prey animals favour widely variable interscan intervals to plant uncertainty in the mind of their observant predators. If prey scanned at regular intervals, an observant predator may launch attacks when prey are predictably vulnerable (Scannell et al., 2001). Combining all these various costs and benefits will help us better understand why the distribution of interscan intervals shows a particular shape.

4.4.3.2. Sequential Randomness

Sequential randomness refers to the lack of predictability in the duration of successive interscan intervals. Two methods have been used to examine this assumption. The first method uses spectral analysis to uncover periodicity in the rises and falls of interscan intervals collected over long sequences of feeding and vigilance. Claims that vigilance sequences in various animal species show statistically significant periodicity, or, in other words, that the duration of an interscan interval can be partially explained by the duration of previous intervals (Ferriere et al., 1996), have been controversial. Reanalysis of the original data, for instance, revealed much fewer significant results than previously reported (Suter and Forrest, 1994). In addition, short-term correlation between interscan intervals may also be driven by responses to external events, such as the arrival of companions or the detection of a threat (Ruxton and Roberts, 1999), which would tend to produce temporal clusters of short interscan intervals. Overall, the evidence from spectral analysis regarding the sequential randomness assumption is rather weak (Bednekoff and Lima, 1999).

A simpler method classifies interscan intervals as short or long, and seeks runs of such intervals in vigilance sequences. Longer runs than expected by chance would indicate that several consecutive intervals are similar in duration, thus inducing sequential predictability. Statistically significant runs have been documented in a number of vigilance sequences in flamingos (Beauchamp, 2006a) (Fig. 4.8), greater rheas (Carro and Fernandez, 2009), and captive black tufted-ears marmosets (Barros et al., 2008), suggesting a certain level of predictability in successive interscan intervals.

While avoiding methodological hurdles associated with spectral analysis, the simpler method loses valuable information by characterizing each interscan interval in a sequence as short or long. In addition, such tests cannot take into account other factors that may influence the level of predictability over time, such as group size. In their analysis of vigilance randomness in three mammalian species, Pays et al. (2010) also determined the correlation between successive interscan intervals, controlling for group size and other potentially confounding factors. In all three species investigated, the duration of an interscan interval was positively correlated with the duration of the previous interval, suggesting again predictability in the organization of vigilance sequences.

What are the consequences of failing to meet the assumption of random scanning? As suggested earlier, any regularity in scanning behaviour may be exploited

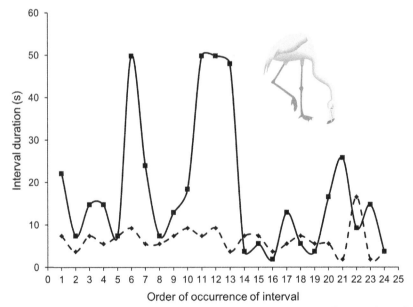

FIGURE 4.8   Temporal variability in interscan intervals: temporal sequences of interscan intervals in two Caribbean flamingos illustrating high (solid line) and low (dashed line) temporal variability. *Adapted from Beauchamp (2006).* (For colour version of this figure, the reader is referred to the online version of this book.)

by stalking predators who could time their attacks based on the behaviour of their prey (Bertram, 1980). It is not clear to me whether predators can actually perceive deviation from randomness to time their attacks. A stalking predator with only a limited amount of time to launch its attack may be hard pressed to detect slight deviation from randomness. Many predators rely on surprise attacks rather than stalking (Cresswell, 1996). In this case, a regular pattern of scanning by the prey may be more appropriate since the goal is to detect rather than confuse the predator by unpredictable scanning behaviour (Bednekoff and Lima, 1998a). The consequences of more regular scanning for predator detection ability and the magnitude of the group-size effect on vigilance have not been investigated in a group setting. I surmise that regular scanning may synchronize vigilance among group members and reduce the effectiveness of collective detection.

4.4.4. Advantages to Detectors

Whether the prey animal is vigilant or not at the time of the attack was not considered a factor in the probability of escaping safely in the early models of vigilance. Results from McNamara and Houston's model presented earlier certainly indicate that an advantage to direct detection influences vigilance patterns.

Two mechanisms may explain why vigilant individuals at the time of an attack may survive better. First, it has been suggested that predators may initially target less vigilant individuals in the group (FitzGibbon, 1989). Second, less vigilant foragers may escape more slowly, increasing their chances of being targeted. Slower escape for individuals foraging head down has been documented in fish (Krause and Godin, 1996a) and in birds (Elgar, 1986; Hilton et al., 1999).

Determining which animal in a group is the initial target of an attack is challenging, especially when prey individuals are close together and when the predator attacks from far away. The escape advantage of vigilant animals is easier to address empirically. In a particularly convincing study, Lima used small balls rolled down a chute as a proxy for predation attempts. Each ball could be aimed at a specific bird in the group (Lima, 1994b). Targeted birds flushed to safety as if they were attacked by a real predator, and no nearby foragers appeared to detect the rolling ball before the initial flush. After a targeted bird flushed to safety, Lima noted how long nearby vigilant and non-vigilant birds took to flush. In two species of birds, vigilant birds escaped sooner than birds feeding head down (Fig. 4.9). While the difference in escape time between the two classes of birds was small, about 0.1 s, this difference may allow a swooping bird of prey to approach much closer to a bird before flushing. The results gathered so far suggest that direct detection is indeed advantageous.

4.4.5. Synchronization and Coordination of the Vigilance

The assumption that the initiation of scans is independent among group members is fundamental. Without full independence, it would not be possible to

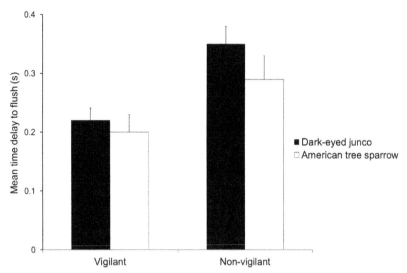

FIGURE 4.9 Response delay and vigilance: the delay to flush after the sudden departure of a companion is shorter for vigilant than non-vigilant neighbours in flocks of two emberezid sparrows. Error bars show one standard error. *Adapted from Lima (1995).*

consider that all individuals contribute equally to predator detection. In mathematical terms, this assumption forms the basis for including group size in the calculation of the probability that at least one individual in the group will detect the predator in a given time interval. For collective detection to work, individuals cannot be oblivious to their companions. A certain awareness of others is critical to determine group size, so as to adjust vigilance accordingly, and to detect threat-related information from the behaviour of companions. However, paying attention to the actual vigilance of others should not be a consideration according to game-theoretic models of vigilance. Any unilateral deviation in vigilance will lower fitness when vigilance in the group is at equilibrium. Therefore, there is no reason why an individual may wish to monitor the vigilance of companions: the cost of deviating from the stable level of vigilance means that all individuals should adopt the same level of vigilance. The existence of a stable level of vigilance implies that vigilance should be independent among individuals in the group.

Nevertheless, independent scanning means that by chance many foragers, or few, or none at all may be vigilant at the same time. Independence would thus seem to lead to temporary episodes where too much or too little safety prevails. Ideally, collective detection should be maintained at the same level by coordinating vigilance efforts, effectively ensuring that at least one individual is vigilant at any one time. Coordination of vigilance implies monitoring of vigilance to ensure that a minimum vigilance is maintained at all times. The use of sentinels carrying out vigilance duties for the whole group provides the simplest solution to achieve coordination (Radford et al., 2009). Sentinel behaviour

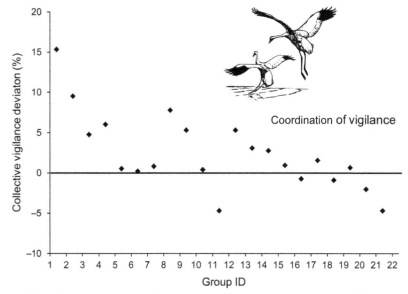

FIGURE 4.10   Coordination of vigilance: pairs of adult common cranes coordinate their vigilance. When coordination occurs, observed collective vigilance is higher than predicted under independent scanning. The solid line indicates independent scanning and the area above the line represents instances of coordination. *Adapted from Ge et al. (2011).*

appears most common in kin-related groups and is rarely observed in casual groups lacking kin ties.

The prohibitive cost of monitoring the vigilance of neighbours was thought to explain why coordination of vigilance was rare in groups without sentinels (Ward, 1985). Recent models have examined this issue more thoroughly. Their results indicate that coordination may be adaptive when in small groups, when direct detection of threats is unreliable, and when collective detection is efficient (Fernández-Juricic et al., 2004b; Rodriguez-Girones and Vasquez, 2002). One recent study found evidence for coordination of vigilance in a species without sentinels (Ge et al., 2011). In common cranes, individuals forage in small family groups typically consisting of an adult pair with one or two offspring. Observations were carried out in the buffer zone of the Yancheng National Natural Reserve in China, where disturbance by people was quite high. Collective vigilance proved indeed quite high in the buffer zone: the probability that at any one time at least one adult scanned was about 0.6. The predicted level of collective vigilance under the assumption of independent scanning was calculated using the percentage of time spent scanning by each adult. The observed level of collective vigilance was actually greater than predicted, indicating that common cranes tended to avoid periods where no adult was scanning (Fig. 4.10).

Coordination of vigilance makes sense for common cranes in the Reserve. This species forages in small kin groups exposed to high level of human intrusion. A relatively open habitat and the close distance between foragers facilitate communication about threats within the group. Coordination of vigilance reduces the number of times the two adults in a family are head down at the same time, which probably helps detecting disturbances more quickly. It is not known whether coordination of vigilance also extends to larger groups, and when visual barriers, such as high vegetation, reduce the effectiveness of collective detection.

Monitoring vigilance may prove beneficial in other contexts as well. Vigilance models identify the stable level of vigilance, but not how this level is reached. As a group starts to forage, vigilance may be high initially before gradually decreasing over time (Beauchamp and Ruxton, 2012a; Sirot and Pays, 2011). Monitoring vigilance may be important in these initial steps to determine how quickly the group is converging on the stable level of vigilance.

Recent models of vigilance have considered the possibility that vigilance may not be independent among foragers in a group (Bahr and Bekoff, 1999; Beauchamp et al., 2012; Jackson and Ruxton, 2006; Rands, 2010; Sirot, 2012; Sirot and Touzalin, 2009). In particular, Sirot and Touzalin (2009) raised the possibility that individuals in a group may copy the vigilance of their neighbours. Copying vigilance will lead to more bouts where all or no individuals are vigilant at the same time. Collective vigilance, when copying occurs, should be less than predicted under independent scanning. Any reduction in collective vigilance beyond the level predicted by independent scanning appears at first sight counterproductive since one evolutionary purpose of joining a group is to provide extra safety. However, as I indicated earlier, direct detection of threats ensures a more rapid escape and reduces the odds of capture. As the group escapes, laggards are thus quite likely to be drawn from the pool of non-vigilant foragers. If predators target laggards preferentially, an increase in the number of vigilant neighbours should increase the probability that an individual remains or becomes vigilant. Extra vigilance when more neighbours are vigilant acts to decrease the relative risk of attack when predators select laggards preferentially. Similarly, when many neighbours are not vigilant, a decrease in vigilance may pay off since the risk of being targeted will now be diluted equally among many foragers. Copying may tend to produce alternating bouts through time where many or few foragers are vigilant at the same time. Such temporal waves of vigilance have been found to be a robust phenomenon in a further model of vigilance copying, but as would be expected less likely when fewer neighbours are available to share information (Beauchamp et al., 2012).

Copying vigilance may also arise if individuals use one another as a real-time source of information about predation risk. An individual may thus become more vigilant when more neighbours are vigilant because their high level of alertness signals impending danger (Eilam et al., 2011; Sirot, 2006). Here, the benefit of copying is not related to predator targeting behaviour but rather to the need to obtain a better estimate of predation risk. Copying in this context may lead to the propagation of false information, which may explain why false alarms are common in animal groups (Giraldeau, 2002; Sirot, 2006). The evolutionary stability of rules to pool information from companions was not investigated by Sirot (2006), and as such it is not clear if copying to acquire information about predation risk represents a stable strategy.

Empirical evidence for vigilance copying has increased dramatically over the last few years in birds and in mammals (Beauchamp, 2009b; Ge et al., 2011; Michelena and Deneubourg, 2011; Öst and Tierala, 2011; Pays et al., 2012). In a telling example, individual eastern grey kangaroos were more likely to be vigilant when more of their neighbours were vigilant in the previous time period (Pays et al., 2009). Evidence for temporal waves of vigilance has also been documented in sleeping gulls (Beauchamp, 2011a).

Two studies, however, failed to find evidence for vigilance copying. In one study, the addition of hungry birds in a flock of dark-eyed juncos, birds which were visibly less vigilant, did not lead to a reduction in vigilance in their companions as would be expected if birds monitored the vigilance of flock members (Lima, 1995a). In the other study, pairs of zebra finches maintained the same level of vigilance whether or not they could monitor the vigilance of their partner (Beauchamp, 2002b). Here, birds were expected to alter their vigilance level to compensate for the possibility that their companion was less vigilant. Overall, the evidence certainly suggests that vigilance may not be independent among group members.

Obviously, other reasons may explain why vigilance tends to be synchronized in animal groups. Short-term correlation in vigilance among foragers may be generated by individual and independent reaction to external sources of disturbances (Ruxton and Roberts, 1999). It is therefore imperative to document putative cases of copying when external disturbances are absent or controlled. Synchronization of vigilance may also arise as a side effect of synchronization of foraging activities. Recent work suggests that both vigilance and foraging may be copied (Fernández-Juricic and Kacelnik, 2004; Michelena and Deneubourg, 2011). In some cases, synchronization of foraging activities can effectively be ruled out, as was the case in sleeping gulls (Beauchamp, 2011a).

4.5. CONCLUDING REMARKS

Models of antipredator vigilance have made remarkable strides since the early work of Pulliam. It is now possible to examine changes in vigilance in response to variation in a host of ecological factors, such as patch richness and energy levels. Many predictions from more recent models await empirical scrutiny.

The occurrence of copying, which appears to be a common feature in many animal groups, certainly complicates matters for vigilance models. Vigilance for any individual now becomes a function of the behaviour of many neighbours. The simple pooling of individual detection ability, used in earlier models of vigilance, must be adapted accordingly. The elegant analytical solutions that characterized earlier models are likely to be replaced by more cumbersome numerical solutions of complex equations. Although the foundation for vigilance models including copying has been laid, much theoretical work remains to be done. It is not clear, in particular, how copying should vary with factors such as group size and predation risk.

Although many empirical studies have questioned the validity of several assumptions of vigilance models, including instantaneous and sequential randomness and independence of vigilance among group members, this should not be viewed as a failure of the theory to investigate vigilance. Early models in any field of endeavour tend to make simplifying assumptions, which are often challenged by subsequent empirical studies. The key in the field of animal vigilance, however, has been the ability to generate ever more complex models more in line with the empirical findings. In particular, recent models of vigilance include the possibility of cheating, and the notion that detectors of threats can have an advantage over indirect detectors. Fortunately, some of the key predictions of vigilance models, such as early detection and the group-size effect on vigilance, have been found to be robust. I examine the evidence for these predictions in the following chapter.

Chapter 5

Antipredator Vigilance: Detection and the Group-Size Effect

5.1. INTRODUCTION

In the previous chapter, I laid the theoretical foundation for vigilance research in animal behaviour and reviewed key assumptions. In this chapter, I examine the empirical evidence for two influential predictions, namely, that living in groups allows individuals to detect predators more quickly and reduce their investment in antipredator vigilance. The second prediction has fostered one of the most active research areas in animal behaviour over the last four decades. In particular, I will explore the magnitude of the group-size effect on vigilance, present alternative explanations for this effect, and discuss how ecological factors shape and potentially confound the relationship.

5.2. INCREASED DETECTION IN GROUPS

The many-eyes effect or, more generally, the pooling of information about threats gathered from the many senses available in a group, predicts that the

Social Predation. http://dx.doi.org/10.1016/B978-0-12-407228-2.00005-6

detection of predation threats should occur more quickly as group size increases. If this were not the case, living in groups for a prey species would provide no advantages in detecting threats, which means that group members could only rely on dilution and the selfish-herd effects to reduce their risk of attack.

To actually show that detection ability varies in groups of different sizes is challenging because predation attempts are rarely witnessed and control over the distance at which predators launch their attacks on a group is usually lacking (but see FitzGibbon, 1990). Most studies thus rely on trained predators (Kenward, 1978), or on simulated attacks using model predators (Powell, 1974) or humans (Altmann, 1958). Determining when detection actually takes place represents another difficulty, as overt reaction to a threat may only become apparent after detection.

The earliest research on the relationship between detection ability and group size was performed with birds. In a classic study, a trained goshawk was used to attack wood pigeons in groups of varying sizes from a standard distance (Kenward, 1978). The results showed that as the number of pigeons increased in the group, the median detection distance also increased, as would be expected from the many-eyes effect (Fig. 5.1).

Similar results have been obtained in many other species of birds (Boland, 2003; Cresswell, 1994) and mammals (FitzGibbon, 1990; Hoogland, 1981). In many of these empirical studies, the detection ability at the group level is probably underestimated in large groups. This is because each individual in a large group typically expends less time on vigilance than when alone or in small

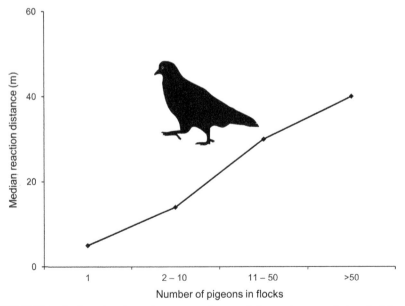

FIGURE 5.1   Predator detection and prey group size: wood pigeons detect attacks launched from a standard distance by a trained goshawk sooner in larger groups, as revealed by the median (m) detection distance. *Adapted from Kenward (1978).*

groups (see below). Consequently, collective vigilance increases more slowly with group size than expected on the basis of individual vigilance, implying that the corporate ability to detect threats will probably fail to reach its full potential in large groups. In order to properly assess the magnitude of the detection effect, it is thus important to maintain collective vigilance at the same level in groups of different sizes or provide a statistical control.

An additional problem is that overt responses to the presence of a predation threat may also depend on group size (Ydenberg and Dill, 1986). Indeed, threats may be perceived earlier by individuals in a large group, but because individuals feel safer in such groups an overt response, which may be costly in terms of time spent away from feeding, may be delayed. Finding exactly when a threat is detected represents an empirical challenge for studies of detection ability.

5.3. DECREASED VIGILANCE IN LARGER GROUPS

5.3.1. Meta-Analysis

The prediction that vigilance ought to decline in larger groups has been tested numerous times in a large range of species (Caro, 2005). All models of vigilance predict such a decline but not necessarily for the same reasons. For instance, Pulliam (1973) only considered detection effects, McNamara and Houston (1992) included detection and dilution effects while a more recent model also included the encounter/dilution effect (Ale and Brown, 2007). All of these mechanisms allow individuals to reduce their vigilance as group size increases (see Chapter 3).

Few researchers have tested the quantitative predictions of vigilance models, that is, exactly which level of vigilance should occur in groups of different sizes. Quantitative tests require detailed knowledge about predator speed, distance between predator and prey, scan duration, and scan frequency among other factors (Elgar and Catterall, 1982; Pulliam et al., 1982). Such knowledge is rather scant, and researchers have focused instead on the qualitative prediction that vigilance should decrease with group size.

Early reviews of the group-size effect on vigilance found overwhelming evidence in support of this prediction (Elgar, 1989; Quenette, 1990) (Fig. 5.2). However, failures to document the expected decline started to accumulate, and a more sophisticated appraisal of the evidence was needed (Barbosa, 2002; Treves, 2000).

In these early reviews, the evidence in support of the group-size effect on vigilance was simply based on the statistical p-value of the relationship between proxies of vigilance, such as the proportion of time spent vigilant or the number of vigilance acts per unit time, and group size. A study was deemed to provide support for the hypothesis when the p-value was statistically significant. P-values are notoriously unreliable in evaluating the evidence for a biological phenomenon. Different statistical tests of the same true relationship between two variables may yield different p-values depending on sample size. P-values also tell us little about the magnitude of the relationship between the two variables. Hence, p-values

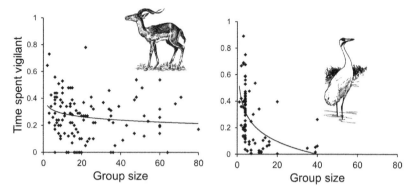

FIGURE 5.2   Vigilance and prey group size: vigilance decreases with group size in impalas (Périquet et al., 2012) and red-crowned cranes (Wang et al., 2011), illustrating the group-size effect on vigilance, which is known in many species of birds and mammals.

cannot really help us to estimate the proportion of the total variance in vigilance associated with changes in group size, which should be the main criteria to assess whether the results agree with the prediction. As an illustration of the pitfalls of using p-values, a recent study found that group size explained very little variation in vigilance in golden marmots despite the fact that the relationship between the two variables was statistically significant (Blumstein, 1996). The magnitude of the group-size effect is best quantified using standardized estimates, which allow us to directly compare results across a number of studies (Box 5.1).

I conducted a meta-analysis to examine the magnitude of the group-size effect on vigilance in birds (Beauchamp, 2008c). Studies on vigilance and group size typically report an estimate of their correlation. I pooled such standardized estimates across a large number of studies to provide an overall assessment of the hypothesis that vigilance ought to decline with group size (Fig. 5.3).

The meta-analysis revealed that time spent vigilant by an individual does decrease significantly with group size (Fig. 5.4). Nevertheless, the magnitude of the effect varied widely among studies and many confidence intervals overlapped the value of 0, which means that variation in vigilance is not always significantly associated with group size. Heterogeneity in the magnitude of the effect among studies was not related to sample size, suggesting that publication bias was probably not an issue here (Box 5.1).

The magnitude of the effect was estimated at −0.4 for time spent vigilant and scan frequency. This correlation coefficient corresponds to a medium effect, which is considerably higher than many effect sizes published in ecology and evolution (Møller and Jennions, 2002). The magnitude of the effect depends in part on the proxy used to measure vigilance. Indeed, scan duration was the proxy least correlated with group size (Fig. 5.4). Finding a correlation between scan duration and group size was not expected in the first place since most models assume fixed scan duration (see Chapter 4). Obviously, this assumption should

BOX 5.1 Meta-Analysis in a Nutshell

Meta-analysis is a statistical tool that pools results from a number of studies all testing the same hypothesis. The concept of a meta-analysis is a familiar one. Authors of a scientific paper always provide a review of the existing literature in which the evidence for or against a hypothesis is typically summarized verbally. Essentially, such narrative assessments provide a qualitative meta-analysis. A formal meta-analysis aims at providing quantitative estimates of the evidence for a hypothesis (Lipsey and Wilson, 2001). Meta-analysis has become a popular tool in animal behaviour research, and practical reviews have been published recently (Nakagawa and Santos, 2012).

Briefly, meta-analysis uses standardized estimates of a statistical effect gathered from a number of studies all testing the same hypothesis to determine whether the overall effect is significantly different from 0. Estimates from different studies are typically weighted using factors thought to be related to the reliability of the estimates, such as sample size. For instance, an estimate from a large study weighs more in the meta-analysis than the same estimate obtained from a smaller study. Commonly used standardized estimates include the difference between means from two experimental treatments or the correlation between two quantitative variables, such as vigilance and group size. If the magnitude of the effect varies substantially among studies, different sets of studies can be grouped to determine how their specific characteristics influence magnitude. For the relationship between vigilance and group size, one may wish, for example, to contrast experimental and observational studies or those carried out with birds and mammals.

The control of type II error rate, the probability of failing to reject a false null hypothesis, represents one of the main advantages of meta-analysis. In fields like animal behaviour, where achieving a large sample size is often difficult, statistically borderline results are quite frequent. However, our confidence in the value of a hypothesis should increase when many such borderline cases are reported along with other less equivocal results. By pooling results from many studies, meta-analysis provides a better handle on type II error rate. Meta-analyses are not without problems. Oft-discussed weaknesses include pooling results from studies that are fundamentally different in methodology, or using simple characteristics like sample size to assess reliability.

Obviously, the value of a meta-analysis, as in any review of the literature, depends on the studies available, which are typically culled from the published literature. However, published studies may only represent a subset of all studies that have tested the same hypothesis, and may be biased towards those that produced a statistically significant outcome. This is known as the file-drawer effect. If the null hypothesis were true for a particular hypothesis, 5% of the studies investigating this hypothesis may end up published on the basis that they rejected the null hypothesis with an alpha value of 5%, thus producing a totally biased sample of the reality.

Publication bias may be assessed using a funnel plot. In a funnel plot, standardized estimates are plotted against a measure of the validity of a study, typically sample size. If published studies represent an unbiased sample of the population of such

Continued

BOX 5.1 Meta-Analysis in a Nutshell—cont'd

studies, the plot should show a funnel shape with values of the estimates converging on the true effect size when sample size increases. Bites missing from the funnel shape are thought to indicate publication bias. For instance, studies with small effect sizes may only be published when the sample size is large enough to yield statistically significant results. Funnels plots have been criticized because small and large studies may be investigating different aspects of the same problem, which may be why they produce different results (Lau et al., 2006). Nevertheless, assessment of publication bias remains an important issue. Publication bias was assessed recently in animal behaviour research focusing on sexual selection. The researchers had access to published and unpublished results, and found little evidence for the file-drawer effect (Møller et al., 2005). Whether this is the case more generally is not clear.

As a final issue, different studies in a meta-analysis are assumed to provide independent estimates of the real effect size. Although this may be true in medical clinical trials run by different hospitals in far-ranging countries, it may not be so in a smaller field of research. In vigilance studies, for example, many researchers concentrate for convenience on a handful of species like the house sparrow, a tendency which will tend to underestimate the variability expected from truly independent studies. In addition, different species may be closely related and respond similarly to the same ecological challenges due to common ancestry. Therefore, effect sizes for related species in a meta-analysis may be expected to cluster around similar values. Lack of independence among estimates culled from different studies biases confidence intervals in a meta-analysis. Although effect sizes may increase or decrease after controlling for common ancestry, it is often the case that the overall effect is no longer statistically significant (Chamberlain et al., 2012). Recent advances in meta-analysis provide tools to deal with the lack of independence issue (Lajeunesse, 2009).

be relaxed in future models. Nevertheless, the results suggest that adjustments in vigilance are more likely to involve changes in scan frequency than in scan duration. As I pointed out in the previous chapter, it is still not clear how best to combine scan frequency and duration to achieve a given level of vigilance.

5.3.2. Why Vigilance Fails to Decline with Group Size

5.3.2.1. Low Statistical Power

What can explain why vigilance sometimes fails to decline with group size? In a statistical sense, studies with rather small sample sizes may lack the power to detect the expected medium effect size. Even in such cases, the magnitude of the effect should be reported, allowing a comparison with other studies. At least in birds, there is no evidence that the magnitude of the effect varies in a systematic fashion with sample size (Beauchamp, 2008c), suggesting that statistical power should be sufficient in all but the smallest studies.

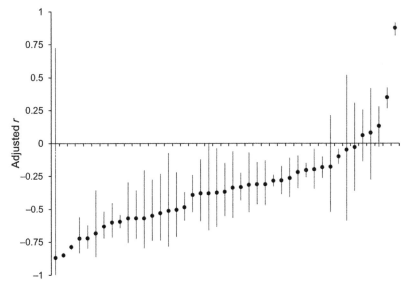

FIGURE 5.3   Meta-analysis of the group-size effect on vigilance in birds: many studies in birds report a negative correlation between time spent vigilant and group size ($n = 43$). Bars show the 95% confidence intervals around the correlation coefficient (shown as a dot). A non-significant relationship occurs when the confidence interval includes the value of 0. *Adapted from Beauchamp (2008c).*

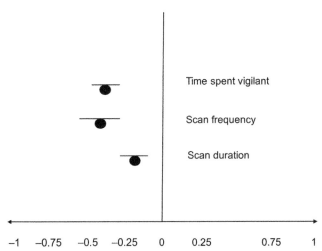

FIGURE 5.4   Meta-analysis of vigilance measurements in birds: overall magnitude of the group-size effect on vigilance in avian studies reporting time spent vigilant ($n = 43$), scan frequency ($n = 29$), or scan duration ($n = 20$). The pooled correlation coefficient is shown along with the 95% confidence intervals for each type of measurement. *Data source: Beauchamp (2008c).*

5.3.2.2. Other Targets of Vigilance

An alternative hypothesis for the lack of effect of group size on vigilance emphasizes the targets of vigilance rather than statistical power. Vigilance is usually thought to be aimed at predation threats, but rival sources of attention certainly exist. For instance, individuals may use vigilance to monitor threatening individuals within or outside their groups (Gaynor and Cords, 2012; Steenbeek et al., 1999). The need to monitor group members may actually increase with group size. Such an increase is predicted in two different contexts. First, if individuals use one another to discover food sources, the need to monitor companions should increase with group size (Chapter 2). In this case, overall vigilance may fail to decrease with group size because the decrease in vigilance aimed at predators in larger groups is compensated by an increase in vigilance aimed at companions to locate food sources (Beauchamp, 2001b). Second, if individuals steal food or displace one another from feeding sites, the need to monitor such threats should also increase in larger groups where such threats are more common. Indeed, increased aggression has often been invoked to explain why vigilance fails to decrease in larger groups (Barbosa, 2002; Blumstein and Daniel, 2002; Cameron and Du Toit, 2005; Teichroeb and Sicotte, 2012).

Distinguishing between the targets of vigilance may help us to determine whether social monitoring interferes with the group-size effect on antipredator vigilance. Unfortunately, it is no easy task to determine what animals are looking at. In chickens, for instance, directly pointing the beak at an object implies that the bird is looking at it, but only if the object is not too far away (Dawkins, 2002). Birds also make substantial head movements when looking at an object, viewing the same object with the two eyes in quick succession. Such head movements have been noted in vigilance studies (Fernández-Juricic et al., 2011), suggesting a potential marker to determine the target of vigilance.

Two studies directly assessed the targets of vigilance. In nutmeg mannikins, stationary bouts of vigilance only correlated with factors known to influence predation risk, such as distance to cover. By contrast, vigilance during movements was associated instead with scrounging success, suggesting that the first type of vigilance was aimed at predators while the second type was aimed at companions to detect food sources (Coolen et al., 2001). In eastern grey kangaroos, vigilance aimed at predation threats was thought to occur when animals scanned their surroundings facing away from companions while social vigilance involved looking directly at a companion (Favreau et al., 2010). Overall, vigilance in kangaroos did not vary as a function of group size, but closer inspection of the results showed that social vigilance increased with group size while antipredator vigilance declined, thus cancelling their effects (Fig. 5.5). As mentioned earlier, a detailed knowledge of the field of view of an animal may be needed to determine what individuals are looking at. The evidence reviewed here certainly suggests that the need for social monitoring is a reason why vigilance may fail to decrease in larger groups.

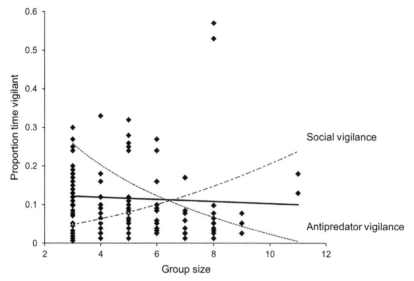

FIGURE 5.5   Targets of vigilance: the expected decrease in antipredator vigilance with group size may be compensated by an increase in social monitoring in larger groups, resulting in an overall lack of effect of group size on vigilance (thick line). *Trend lines inspired by Favreau et al. (2009).*

5.3.3. Interaction with Other Factors

The preceding analysis pointed out that the magnitude of the group-size effect on vigilance varies substantially among species. The magnitude of the effect may also vary within a species. In particular, the group-size effect on vigilance may interact with various factors, including sex, predation risk, and personality traits. Interactions of this type mean that different patterns of vigilance may be expected within the same species depending on the situation.

5.3.3.1. Sex

Sexual differences in overall vigilance have been discussed extensively in the literature (Caro, 2005), but less attention has been paid to sexual differences in the magnitude of the group-size effect on vigilance. Such differences in magnitude have been noted in many studies in birds and mammals (Cameron and Du Toit, 2005; Childress and Lung, 2003; Li et al., 2012; Rieucau et al., 2012), but no clear pattern has emerged as to which sex shows the largest magnitude.

Sexual differences in the magnitude of the group-size effect suggest that males and females may aim their vigilance at different targets. In elk, for example, the group-size effect on vigilance occurs in females without calves, but not in females with calves or in males (Childress and Lung, 2003; Lung and Childress, 2007). Females without calves presumably use vigilance to detect predation threats, and the need to monitor such threats decreases in larger groups. By contrast, females with calves may devote most of their vigilance to monitoring their offspring while males probably monitor rivals extensively. In other ungulates, it has also been shown that males in large groups can devote a large amount of time to monitoring rivals, which limits the scope of adjustments in vigilance with group size (Li et al., 2012). In giraffe, males monitor rivals and females monitor threatening males, and in both cases vigilance failed to vary as a function of group size (Cameron and Du Toit, 2005). The occurrence of sexual differences in the magnitude of the group-size effect makes it important to sample the sexes separately to document vigilance patterns.

5.3.3.2. Predation Risk

Predation risk can also interact with group size to determine the magnitude of the group-size effect on vigilance. Predation risk has been related to several ecological variables, including distance to cover, position in the group, and predator density. Depending on the species, the largest magnitude occurs in high-risk (Frid, 1997; Lendrem, 1984; Lima, 1987a) or in low-risk (Lima et al., 1999; Manor and Saltz, 2003; Martella et al., 1995) settings.

Why should the magnitude of the group-size effect on vigilance change at all with predation risk? Frid (1997) proposed that the effect of group size on vigilance should be more pronounced when predation risk is high. When predation risk is low because, say, individuals are closer to cover, a simple decrease in vigilance in groups of all sizes will overemphasize safety. Individuals in smaller groups should be selected to decrease their vigilance to a greater extent than those in large groups to reap extra benefits from foraging at low predation risk, dampening the magnitude of the group-size effect on vigilance. Manor and Saltz (2003) also proposed various ways to adjust vigilance with group size when predation risk varies. In the empirical test of their predictions in gazelles, the largest magnitude occurred in the low-risk setting instead. I point out that without explicitly modelling the costs and benefits of vigilance as a function of group size under different levels of predation risk, it is difficult to make actual predictions regarding the scope of interactive effects.

Extending earlier, more quantitative analyses (Bohlin and Johnsson, 2004; Grand and Dill, 1999), Bednekoff and Lima (2004) used just such a cost-benefit analysis to examine the optimal allocation of effort to foraging in groups of different sizes under different levels of predation risk. An increase in foraging effort was hypothesized to reduce the ability to survive attacks, by reducing the investment in antipredator vigilance, but decrease the probability of starving. In their first model, the authors examined the effect of predation risk on the magnitude of the group-size effect assuming no competition for resources in the group. Their model suggests that the magnitude of the group-size effect should be larger when predation risk is low rather than high, and that it is the larger groups that adjust vigilance to the greatest extent (Bednekoff and Lima, 2004). This is consistent with the pattern of change reported by Manor and Saltz (2003) but not with the one reported by Frid (1997). In conclusion, the magnitude of the group-size effect should not be expected to be a fixed trait under different levels of predation risk.

5.3.3.3. Personality Traits

An interesting new development suggests a potential interaction between personality traits and vigilance. It has long been recognized that vigilance levels may vary consistently among individuals in relation to phenotypic traits like age, sex, or dominance status (Caro, 2005). As I discussed earlier, males and females within a species may show consistently different levels of vigilance, and the magnitude of the group-size effect on vigilance may also vary between the sexes. Within the same class of individuals, further differences among individuals have been related to personality traits, such as shyness or boldness (Wilson et al., 1994). It is of note that models of vigilance assume that all individuals within a given phenotypic class should respond equally to changes in group size.

To assess the relevance of personality traits, vigilance for the same individuals must be documented across a range of conditions. Only a handful of studies have reported such individual vigilance profiles. In a laboratory study with nutmeg mannikins, the magnitude of the group-size effect on vigilance was found to be similar for each individual, but some birds were consistently more vigilant than others (Rieucau et al., 2010) (Fig. 5.6). In a field study of redshanks, a small shorebird, different individuals adopted consistently different vigilance levels across two different locations over many weeks (Couchoux and Cresswell, 2012), suggesting again the existence of stable inter-individual differences in vigilance. Another study with shorebirds,

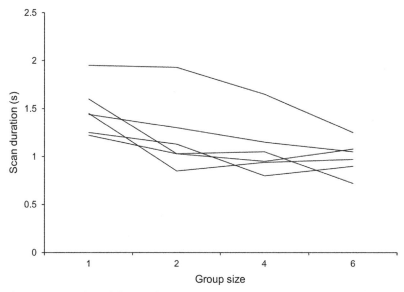

FIGURE 5.6   Individual differences in vigilance: individual nutmeg mannikins show distinct patterns of vigilance (scan duration during food handling) over a range of group sizes. *Adapted from Rieucau et al. (2010).*

conducted in the laboratory this time, documented consistent inter-individual differences in vigilance across three different levels of predation risk (Mathot et al., 2011). The two shorebird studies, however, did not examine plasticity in vigilance as a function of group size. In eastern grey kangaroos, the magnitude of the group-size effect this time varied from one individual to another (Carter et al., 2009). The lack of group-size effect for some individuals was related to their greater need for social monitoring, which, as argued earlier, can mask the expected decrease of vigilance with group size.

The consequences of stable inter-individual differences in vigilance are far-reaching. For instance, it means that an individual that consistently invests little in vigilance will fare better in a group composed of vigilant rather than nonchalant companions, suggesting that individual fitness may vary depending on group composition. Inter-individual differences may also help to explain the wide scatter in plots of the relationship between vigilance and group size in many species.

To summarize the evidence regarding interactive effects in general, the magnitude of the group-size effect on vigilance can vary among individuals and in relation to several ecological factors such as predation risk and sex. In a scatter plot of the relationship between vigilance and group size, interactive effects are likely to create an uneven scatter with wider variance at one end of the plot. Indeed, an analysis of published relationships between vigilance and group size in birds and mammals revealed widespread variance heterogeneity (Beauchamp, 2013). In addition to raising statistical issues regarding the proper analysis of vigilance data, this finding highlights the need to consider interactive effects in models of vigilance.

5.3.4. Alternative Hypotheses to Explain the Group-Size Effect on Vigilance

Vigilance models, based on a trade-off between safety and foraging, predict that vigilance should decrease with group size. However, it became readily apparent that alternative hypotheses, based on different mechanisms, can also explain why vigilance may be adjusted downward in larger groups. Any discussion of the group-size effect on vigilance should examine the relevance of these alternative explanations.

5.3.4.1. Edge Effect

Vigilance has long been known to be higher at the edges than at the centre of a group (see review by Caro, 2005). Because the proportion of individuals at the edges of a group decreases in larger groups, vigilance at the group level should decrease with group size even though vigilance at the individual level may remain unchanged in groups of different sizes (Inglis and Lazarus, 1981). To counter this argument, it is necessary to document whether vigilance decreases with group size after controlling for position in the group. For instance, vigilance in European starlings is indeed higher at the edges than at the centre of the group. However, all individuals, whether at the centre or at the edges of the group, reduce their vigilance as group size increases (Jennings and Evans, 1980), suggesting that it is group size rather than geometry that explains lower vigilance in larger groups.

5.3.4.2. Targets of Vigilance

An earlier explanation for vigilance in animals emphasized the detection of foraging opportunities (Krebs, 1974). Vigilance would allow foragers in poor food patches to obtain clues from companions about the location of richer feeding areas. Vigilance should thus decrease with group size as the need to obtain food-related information becomes less relevant in larger groups, which tend to occupy rich food patches. This explanation has been largely ignored, save for a few studies (Coolen and Giraldeau, 2003; Robinette and Ha, 2001). A case may be made that individuals using vigilance to detect foraging clues from companions in a nearby group should actually be considered part of that group, in which case vigilance to detect foraging opportunities may simply be called scrounging. The validity of this hypothesis thus hinges on where a group starts and ends.

5.3.4.3. Food Competition

Food competition represents the alternative explanation for the group-size effect on vigilance that has received the most attention. Because most vigilance models only consider the influence of predation on scanning patterns, the actual process of foraging for food has been largely ignored. Nevertheless, individuals must obtain food to survive, and patterns of vigilance may be expected to vary in response to the distribution of resources. For instance, when foraging for scarce resources, a unilateral decrease in vigilance may allow an individual to obtain a disproportionate share of the food items and increase its relative success (Clark and Mangel, 1986). The best response by other group members to such a defection should be to decrease their own vigilance to exploit resources just as quickly, bringing vigilance down in the whole group. Exploiting resources more quickly, however, increases the energy cost of obtaining resources and predation risk as well. At some point, the extra foraging benefits from a decrease in vigilance will not compensate for the loss of safety and the rising exploitation costs, and vigilance will settle at a new, lower equilibrium. The general point remains that a decrease in vigilance in larger groups may be a response to competition for limited resources.

The model by Bednekoff and Lima (2004), presented in the preceding section, can be used to predict the effect of competition for resources on vigilance. Their models include cases where group size only influences the probability of surviving an attack, or only the level of competition for food, or both. Without competition, the group-size effect on vigilance is driven by changes in the probability of surviving an attack. As I explained earlier, the model predicts that the magnitude of the group-size effect on vigilance should be larger in a low-risk than in a high-predation risk setting (Fig. 5.7), inducing an interaction between

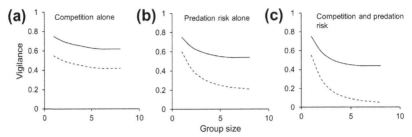

FIGURE 5.7 Prey group size and willingness to take risk: willingness to take risk varies with group size and interacts with overall predation risk (high risk: solid line; dashed line: low risk). In this particular example, low vigilance indicates a greater willingness to take risk. An increase in group size can increase competition (a), decrease perceived predation risk (b), or change both (c).

group size and predation risk. When group size only influences the level of competition for resources in a group, and not the probability of surviving an attack, changes in vigilance are driven by competition. In this particular case, the model also predicts a decrease in vigilance with group size, supporting the unilateral defection argument presented above. Interestingly, the magnitude of the group-size effect is remarkably similar under the two predation risk regimes, suggesting little interaction between predation risk and group size. When group size influences both competition and predation, the group-size effect on vigilance becomes steeper, corroborating results from an earlier model (Beauchamp and Ruxton, 2003). Here again, the largest magnitude occurs under the low-risk setting. These various interaction patterns between group size and predation risk could thus be used to determine whether the group-size effect on vigilance is driven by competition, by predation, or both.

The crux of the model lies in documenting an interaction between group size and risk level. However, to show that competition acts alone in driving changes in vigilance with group size requires demonstrating a lack of interaction. Lack of effect can be very tricky to demonstrate statistically. Documenting an interaction can only tell us that competition did not act alone because an interaction is predicted when predation risk acts alone or in combination with competition. Only quantitative predictions can tell us whether competition acted together with predation. Although this is not an impossible task, I note that very few quantitative tests of vigilance models have been performed thus far, partly because quantitative information about most model parameter values is lacking.

An alternative strategy to demonstrate an effect of competition on the group-size effect consists in documenting vigilance when competition can be effectively ruled out. In dark-eyed juncos, vigilance decreased with group size even when birds fed in an environment with a large supply of food, which presumably ruled out competition (Lima et al., 1999). Since vigilance was measured when birds handled food head up, the researchers hypothesized that food handling speed, and thus indirectly vigilance, would only be adjusted when food is scarce in response to competition. The authors thus concluded that competition did not drive changes in vigilance with group size in this species. However, I note that the magnitude of the group-size effect on vigilance in this study did not differ between the high- and low-risk settings. According to the above model, the lack of interaction between predation risk and group size tells us instead that competition acted alone, a rather different conclusion. Because it is difficult to reject the hypothesis that competition acts alone, it is perhaps better to discount competition by actually manipulating the level of competition.

Manipulation of food competition has been performed in a handful of studies. In a study with tammar wallabies, the number of food bins available to a group was reduced to induce competition, but vigilance failed to decrease as predicted (Blumstein et al., 2002). In a field study with coots, competition was induced by providing a highly valued but scarce food (Randler, 2005a). Vigilance decreased when birds fed on the scarcer food, controlling for group size, suggesting that the decrease in vigilance aided birds in obtaining a greater share of the limited food. A more recent approach simulated competition by presenting virtual companions on a screen to nutmeg mannikins foraging in a cage (Rieucau and Giraldeau, 2009a). Previous work with this species had shown that virtual companions on a screen are just as effective as real birds in inducing a decrease in vigilance in larger groups (Rieucau and Giraldeau, 2009b). In the simulated competition study, competition intensity was manipulated by varying the amount of time virtual companions spent feeding. The authors hypothesized that an increase in time spent feeding would indicate to focal subjects that competition for resources is more intense. Focal birds watching virtual companions feeding intensively head down decreased their vigilance to a greater extent than when watching the same number of individuals feeding less intensely, suggesting that perceived competition induced the change in vigilance (Fig. 5.8). However, vigilance copying, which appears to be quite common in animals, as I discussed in the preceding chapter, can also explain why vigilance would decrease when virtual companions spend more time feeding. In addition, the amount of food available to the focal birds did not vary in response to the simulated competition by virtual companions, making it difficult to ascribe the changes in vigilance to real competition.

It is clear that competition for food is not the only factor influencing vigilance in view of the overwhelming evidence that variation in predation risk influences scanning patterns in birds and mammals (Caro, 2005). In addition, vigilance also decreases when animals are resting or preening, in which case competition for food can be safely ruled out (Dominguez and Vidal, 2007; Randler, 2005b). Competition may also play a limited role when animals feed on plentiful resources or can overlap foraging and vigilance. Nevertheless, at least on a theoretical basis, a role for competition cannot be neglected and more empirical evidence should help us assess its importance.

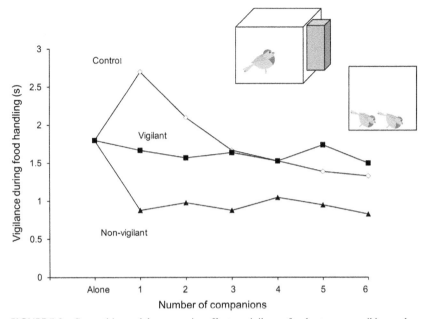

FIGURE 5.8   Competition and the group-size effect on vigilance: focal nutmeg mannikins are less vigilant when companions on a video screen feed more intensely, suggesting that competition drives changes in vigilance. The control group consisted of companions alternating between vigilance and feeding. Errors bars are not shown for clarity. The inset illustrates the apparatus with one focal subject in a cage watching a video screen showing two feeding companions. *Adapted from Rieucau and Giraldeau (2009b).* (For colour version of this figure, the reader is referred to the online version of this book.)

5.3.4.4. Confounding Variables

I now turn to the effect of confounding variables on the relationship between vigilance and group size. As most studies of vigilance are observational in nature, the possibility arises that part of if not all the variation in vigilance may be explained not by group size directly but indirectly by variables correlated with group size. Essentially, variation in vigilance may be caused by unmeasured variables, which seriously impairs our inference about the effect of group size. Assessing the role of confounding factors is important when discussing causality in the relationship between vigilance and group size.

To illustrate the role of confounding factors, consider the potential effect of food density on vigilance. Rich food patches are known to attract foragers. If feeding interferes with vigilance, vigilance may be lower in larger groups not because individuals occur in bigger groups but rather because they spend more time feeding in these rich food patches (Barnard, 1980; Elgar, 1989). An increase in time spent feeding in rich food patches may simply reflect the greater availability of food. However, recent models also suggest that foragers should spend proportionately more time feeding in rich food patches, thus reducing vigilance in such patches even more. In rich patches, the balancing act between achieving greater food gains and maintaining safety is expected to shift to the foraging side (Ale and Brown, 2007; McNamara and Houston, 1992). However, this only holds true when foragers face time constraints and starvation is a real possibility. With time constraints, a reduction in vigilance will yield a larger decrease in the probability of starving when food intake rate is high rather than low. When food intake rate is low, a reduction in vigilance may in fact produce negligible effects on starvation risk but greatly increase predation risk. In general, sacrifices in safety are worthwhile when foraging gains, and ultimately fitness, increase rapidly with a decrease in vigilance. For these various reasons, it is conceivable that food density can confound the relationship between vigilance and group size.

Few studies have directly tested the relationship between vigilance and food density. As predicted by the above argument, vigilance decreased as the density of food available to caged sparrows increased (Repasky, 1996). However, vigilance failed to vary with food density in the two other studies that manipulated food density (Blumstein et al., 2002; Slotow and Coumi, 2000). In observational studies, the effect of food density on vigilance has been equivocal (Beauchamp, 2009a). These early findings suggest that food density should be taken into account when reporting the effect of group size on vigilance.

Although I limited my discussion on confounding factors to the effect of food density, any factor correlated with group size, such as distance to cover and temperature, may act as a confounding factor. In view of the various factors that are not controlled in observational studies of vigilance, and because confounding factors, such as food density, can explain why vigilance decreases in larger groups, caution should be exercised when concluding that variation in vigilance is caused by changes in group size.

5.4. VIGILANCE WHEN PREDATION RISK IS NEGLIGIBLE

Predation risk represents one of the main engines of change in vigilance. Indeed, a reduction in perceived predation risk allows individuals to decrease their investment in vigilance in larger groups. If predation risk drives adjustments in vigilance, it follows that vigilance should drop to very low levels in the absence of predation. As predation risk becomes negligible, any change in vigilance implies that other factors, such as competition, must be involved, enabling us to tease apart their relative contribution to vigilance.

A number of studies have examined whether vigilance varies with group size when predators are rare or absent. Most of these studies have focused on islands long isolated from predators. On such islands, costly antipredator adaptations like vigilance are expected to be eliminated by natural selection over evolutionary time if they no longer provide benefits (Lahti et al., 2009). In line with this expectation, vigilance did not vary with group size in one bird species foraging on a predator-free island located in the Great Barrier Reef off Australia

(Catterall et al., 1992). Our confidence that vigilance no longer co-varies with group size when predation risk is negligible would be stronger if such results are the rule rather than the exception. To this end, Blumstein et al. compiled the results of several studies investigating the relationship between vigilance and group size in marsupial populations exposed to varying levels of predation. The meta-analysis revealed that the group-size effect on vigilance was indeed less common in populations located on predator-free islands than on the predator-rich mainland (Blumstein and Daniel, 2005). In one particular species, disappearance of the group-size effect on vigilance took place in less than 130 years following isolation from predators (Blumstein et al., 2004a) (Fig. 5.9).

Findings with other mammalian species on predator-free islands have been equivocal. On Lundy Island off the United Kingdom, free-ranging soay sheep maintained low levels of vigilance unrelated to group size (Hopewell et al., 2005). However, on Rum Island more to the north, feral goats decreased their vigilance in larger groups (Shi et al., 2010).

Although the loss of antipredator behaviour following isolation from predators can become evident after only a few generations (Blumstein et al., 2004a; Stankowich and Coss, 2007), ancestral traits may be retained despite thousands of years of isolation (Blumstein et al., 2000; Zheng et al., 2013). Persistence of traits under relaxed selection has been related to the cost of maintaining traits that are no longer useful, but may also imply that more than one type

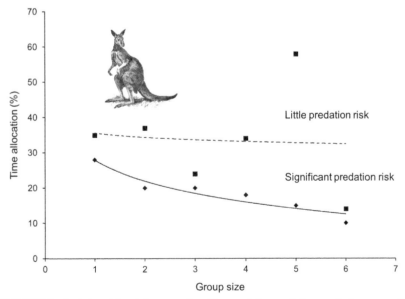

FIGURE 5.9 Predation risk and the group-size effect on vigilance: vigilance only decreases with group size in tammar wallabies at locations with significant predation risk. *Adapted from Blumstein et al. (2004a).*

of predator shaped antipredator responses in the past (Blumstein and Daniel, 2005; Coss, 1999). Antipredator vigilance may also persist because it is useful in other contexts, such as competition (Shi et al., 2010).

Taken as a whole, the results that I reported support the hypothesis that relaxed predation pressure can considerably weaken the group-size effect on vigilance. Studies of vigilance when predation risk is negligible still face important challenges. Showing that vigilance is not associated with group size, which is the equivalent of accepting the null hypothesis, is always tricky statistically. In addition, it is important to remember that the lack of effect of group size on vigilance need not imply that animals have lost altogether the ability to respond to predation threats; they may simply show plasticity in their responses to negligible predation risk.

5.5. CONCLUDING REMARKS

I draw here general conclusions from the two chapters on antipredator vigilance. The study of vigilance in animal behaviour represents a success story on many fronts. In particular, vigilance research has managed to strike a fine balance between theoretical and empirical work. Numerous empirical studies have tested predictions from vigilance models and uncovered deficiencies that have stimulated the formulation of alternative models, whose predictions have been tested once again in a process of ever-increasing refinements. Empirical research on vigilance also encompasses a broad range of species, in both birds and mammals, and effectively combines observational as well as experimental approaches. Meta-analyses of the vast literature on vigilance, which has grown tremendously over the last four decades, have delineated general tendencies and uncovered irksome exceptions that are sure to stimulate further research. Support for the qualitative predictions of vigilance models, with respect to the detection of predation threats and the effect of group size, has for the most part been overwhelming. It is now certainly the time to move on to more quantitative tests of these predictions.

I think the greatest weakness in vigilance studies has been the neglect of non-evolutionary issues. Proximate mechanisms and developmental issues have been set aside by the intense focus on the adaptive value of vigilance. Consider the simple question of how animals adjust vigilance to group size. Models of vigilance implicitly assume that animals are somehow able to estimate group size to adjust vigilance accordingly (Bekoff, 1996; Elgar et al., 1984). How this assessment is done is not known. Work on this question may reveal interesting constraints on the ability to adjust vigilance to group size. For instance, the ever-slower decline in vigilance as group size increases may reflect, in part, the fact that animals are unable to accurately estimate group size in larger groups.

Which cues could animals use to assess group size? That vigilance varies with group size need not imply that individuals estimate group size directly by counting. Indirect cues to adjust vigilance may include factors such as food availability, which is often correlated with group size, as I showed earlier. Foraging individuals also produce visual and non-visual cues that are not enumerable. Animals produce scents and noises and occupy a certain area in space, all of which can be estimated on a qualitative scale (e.g. less or more). In terms of auditory cues, animals can adjust their behaviour to the amount of noise produced by rivals (McComb et al., 1994; Radford and Du Plessis, 2004) or in response to the production of calls by companions (Radford and Ridley, 2007), suggesting possible auditory cues to estimate group size. In terms of visual cues, the area occupied by a group and contour length must be closely correlated with group size. Therefore, an adjustment of vigilance to group size may indicate a response to the area occupied by a group or contour length rather than the number of individuals involved.

Where to draw the line that defines group membership is another thorny proximate issue. Empirical researchers typically count the number of animals within an arbitrarily defined area and assume that individuals adjust their vigilance according to the number of animals therein. But just how far individuals can interact with one another is not very clear. Results from many studies show that near neighbours have more influence on vigilance (Elgar et al., 1984; Fernández-Juricic and Kowalski, 2011; Lima and Zollner, 1996) (Fig. 5.10), suggesting that group membership may not extend very far. Therefore, it appears primordial to consider the perceptual abilities of species when defining group size. In general, sensory ecology information, namely, what animals can see and hear when foraging and vigilant, is typically lacking, which makes it difficult to determine the validity of many assumptions in vigilance models (Fernández-Juricic, 2012).

How individuals achieve the stable, predicted level of vigilance represents yet another proximate issue that has attracted little attention. Game-theory models of vigilance tell us little about the actual process of reaching the stable solution. Vigilance, ultimately, represents an attempt to adjust behaviour to perceived predation risk. Perception of predation risk is typically viewed as a dynamic process, with individuals continually acquiring information from the environment to adjust their behaviour (Bouskila and Blumstein, 1992). Instead of assuming that animals know the level of predation risk, the perception of predation risk itself could be modelled as a function of time (Sirot and Pays, 2011), which may allow us to determine how vigilance eventually reaches the predicted value. I also mentioned earlier the possibility that individuals may monitor each other as the group progresses towards the predicted stable level. Recent developments in the study of vigilance copying also suggest that a more dynamic view of vigilance is needed, one that focuses on information acquired by individuals in the process of adjusting vigilance.

With respect to developmental issues, it is not clear whether animals know instinctively how to adjust vigilance to group size. Indeed, animals may need to learn that large groups are safer. Very few studies have focused on the development of vigilance during the lifetime of an individual (Loughry, 1992; Sullivan, 1988). Such studies may also shed light on how individuals achieve the stable level of vigilance.

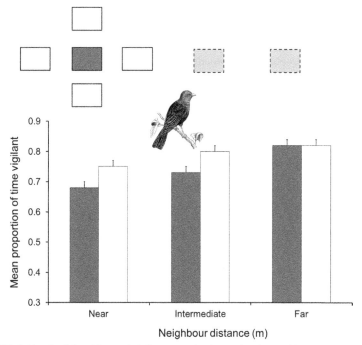

FIGURE 5.10 Spatial position and vigilance: captive brown-headed cowbirds are more vigilant at the edges than in the centre of a group when neighbours are nearby. The spatial arrangement of cages holding edge and central birds is shown above. To increase neighbour distances, all edge birds were moved farther away from the centre cage. Error bars show one standard error. *Adapted from Fernández-Juricic and Beauchamp (2008).*

The focus on the current costs and benefits of vigilance also means that little is known about the evolutionary history of vigilance. Vigilance varies tremendously from species to species, and yet we know little about the factors that have driven this variation over evolutionary time. In mammalian grazers, for instance, the smallest species are thought to be more vigilant because of their greater exposure to predation threats and also, perhaps, due to their more restricted field of vision (Berger and Cunningham, 1988; Burger and Gochfeld, 1994; Treves, 1997). However, body mass has emerged as a poor predictor of vigilance in other species (Blumstein and Daniel, 2003; Du Toit and Yetman, 2005; Scheel, 1993; Treves, 1998). In birds, differences in vigilance between species have been mostly related to foraging ecology (Barbosa, 1995; Guillemain et al., 2002).

In view of the large amount of variation in vigilance within a species, even when controlling for group size, one may wonder whether vigilance is not too labile a trait to allow a comparison across species. As a first step in this direction, I performed a comparative analysis of vigilance in a large number of avian species, and found that vigilance was lower in smaller species and in those foraging in larger groups (Beauchamp, 2010a). A similar analysis may be useful in mammals to determine whether the evolutionary correlates of vigilance are similar.

In the end, I propose that a broader perspective on vigilance, encompassing proximate and ultimate questions, will ensure that vigilance studies remain at the forefront of animal behaviour research for years to come.

The Selfish Herd

6.1. INTRODUCTION

This is a story that made the rounds on the Internet a few years ago although few appreciated its biological significance. Two intrepid campers (or lawyers, in other versions) are chased by a bear and take refuge in a tree. As the bear refuses to leave, the campers decide to go down and make a run for it. Seeing his companion fitting his brand-new shoes, one camper remarks that there is probably little point in outrunning a bear. What they really need to do, of course, is to outrun each other, not the bear. While drawing a morally objectionable conclusion, this story illustrates the concept that in a group, an individual may use companions as a shelter from predation. Sheltering from predators in a group adds a new dimension to antipredator ploys. The ploys reviewed in the preceding three chapters typically decrease absolute predation risk for a group member. This mechanism allows, instead, a diminution in relative predation risk.

The idea that group members can serve as layers of protection in the face of predation was first formalized by Hamilton in 1971. He coined the term 'selfish herd' for such aggregation behaviour and developed a number of key testable ideas (Hamilton, 1971). The selfish-herd hypothesis, as it became known, spawned a lot of research over the years, but perusing the vast literature citing the work, one has the distinct impression that the hypothesis was more successful as a suggestion that selfish mechanisms can lead to group formation rather than as a theory that could be tested empirically in the field.

Social Predation. http://dx.doi.org/10.1016/B978-0-12-407228-2.00006-8

It is important to remember the context in which this hypothesis was first proposed to appreciate the novelty of the work. Early animal behaviour research was concerned with explanations that emphasized benefits for the group, or the species more generally, rather than for individuals. Birds, for example, were expected to restrain the number of eggs they produce in times of food scarcity to ensure population viability over several generations (Wynne-Edwards, 1962). Such group selection arguments have gradually been abandoned since explanations based on benefits that individuals can achieve in the struggle to pass their genes can capture the essence of most behaviour in nature (Williams, 1966). The spirit of Hamilton's paper was therefore to provide an explanation of group formation based on the selfish interests of group members. The paper thus works best as an example of how a behaviour that seems to work for the benefit of the group actually results from individual impulses to reduce predation risk at the expense of others, which is the crucial point.

Hamilton used the simple scenario that follows to illustrate this concept. Consider a circular pond where frogs are exposed at random times to predation attempts by a water snake. The snake can appear anywhere in the pond and catch the nearest frog. In response to this threat, the frogs arrange themselves around the edge of the pond. A particular frog will be captured on each strike by the snake, and the best solution is thus to minimize the chances of being selected by the snake. As the snake appears randomly in the pond and captures the nearest frog, the odds of capture for a frog are proportional to the amount of space in which this particular frog is the one nearest to the predator. If a snake appears in this area, the frog will be attacked. This area has been called the domain of danger (DOD), and represents the area in which all points from the point of view of the predator are closer to a particular group member than to any other companions. In the simple case of frogs arranged on the perimeter of a circle, the odds of capture are calculated as the ratio between the DOD of a particular frog and the circumference of the pond. Relative predation risk thus increases proportionately to the size of the DOD.

All frogs, therefore, have a DOD, and this domain can vary from very small to very large depending on the spatial distribution of frogs around the pond. Although it is certain that a frog will be captured, a frog that has a larger DOD than other companions may do well by reconsidering its position in the pond so as to reduce its relative risk of attack. Each frog should thus consider the size of the gaps between neighbouring frogs and jump in the smallest gap (Fig. 6.1). By doing so, the DOD of that frog will decrease at the expense of at least one other. The end result will be the formation of a much tighter group than was originally the case, thus showing how selfish interest can lead to the formation of groups that do not benefit the group as a whole but rather benefit individuals, some of which have, at least temporarily, decreased their relative risk of attack. The key point is not that predation is unavoidable, but that individuals may reduce their relative risk by positioning themselves differently. Any decrease in predation risk in the group is at the expense of others.

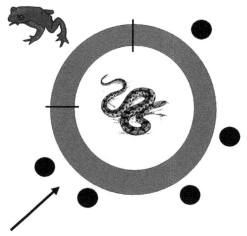

FIGURE 6.1   Domain of danger: the domain of danger of a green frog on the edge of a pond is shown as the shaded area between the two thick lines. Anywhere in the domain of danger of the green frog, the water snake will be closer to the green frog than to any of its companions, here shown as black dots. The best move by the green frog to reduce relative predation risk is to hop into the smallest gap between two adjacent frogs (shown by the arrow). (For interpretation of the references to colour in this figure legend, the reader is referred to the online version of this book.)

This thought experiment is for the simple case of prey arranged in a one-dimensional plane. What happens when prey occupy two or three dimensions? For instance, herds of ungulates move along two dimensions and flying flocks of birds and schools of fish travel in three dimensions. The concept of DOD can, in fact, be extended to these cases. Domains of danger in a two-dimensional space are calculated by Voronoi tessellation (Okabe et al., 1992). These spaces form a polygon inside of which all points are closer to a particular individual than to any other. Obviously, such polygons extend to infinity at the edges of a group and are smaller inside the group as long as the predator can attack anywhere within the group. The concept of risk is still the same in two dimensions and, for the case of a predator that can attack anywhere in the group, risk represents the ratio between the area of a polygon and the total area occupied by the group.

In the frog example presented earlier, the rule of movement to reduce relative predation risk, namely, jumping into the smallest gap between any two neighbours, was quite simple. However, this is not the case in a two-dimensional space, where many individuals have more than two neighbours. Hamilton considered the rule of moving to the nearest neighbour (NN) but soon realized that such a rule often results in the formation of many small clusters rather than one large aggregation. Such smaller aggregations may then coalesce into one larger unit although this explanation, ironically, involves benefits to the subgroup rather than to the individual (Viscido, 2003). The types of rules that individuals may possibly follow when under attack has been the subject of renewed interest lately, and I will review these results and present new research below.

6.2. NEW THEORETICAL DEVELOPMENTS

Several extensions of the selfish-herd hypothesis have been proposed since Hamilton's paper was published. In many cases, these developments have high-lighted some of the weaknesses in the original hypothesis.

6.2.1. Predators Attack from the Outside

At about the same time as Hamilton's paper, Vine (1971) asked whether a preda-tor that attacks from the outside, rather than from anywhere in the group as in Hamilton's model with the frogs, would still favour aggregation (Vine, 1971). Vine's model requires more information about predator behaviour and because it addresses many of the same issues raised by Hamilton's paper, it did not quite achieve the same level of success. Vine's model works best when showing that grouped prey may evade predation by reducing their odds of detection. Vine also articulated the principle of risk dilution, which I considered in Chapter 3. The model also suggests that individuals in the middle of the group may be less at risk even when predators strike from the outside. I will focus here on this geometrical argument.

In Vine's model, the predator visually patrols an area along the edges of a circle. Prey animals are distributed along the edges of that circle. The preda-tor can only scan the perimeter with a finite arc of view, which may encom-pass from one to several members of the group. Once a prey animal has been detected within the arc of view, the predator attacks one of the prey animals inside. The key point is that while a prey at the edge of the group may be the sole individual detected by the predator, should the arc of view only encompass the edge of the group, an individual inside the group will always be detected at the same time as the companions closer to the outside. In such a case, Vine assumed that only one of these individuals will be attacked. In this case, individuals at the centre of the group are on average less likely to be attacked because they always share predation risk with at least one other companion. Vine did not address how individuals achieve their position in the group, but the end result is greater protection inside the group even when a predator attacks from the outside.

Vine addressed what is indeed an important issue: in many cases it cannot be assumed that the predator can appear anywhere inside the group (Viscido et al., 2001). The emphasis on predator sensory information is a welcome addi-tion since it suggests that differences in sensory ecology can lead to differ-ent outcomes. However, the attack scenario appears contrived. Why would a predator that detects one prey in the group not take an extra second to scan the rest of the group and then choose to attack one prey animal in particular? Is it possible for a predator to maintain its target as the group flees upon attack? One could imagine that an animal at the edge may try to quickly take shelter behind others or may be lost in the chase, thus making the whole argument by Vine moot.

6.2.2. Defining Domain of Danger

A DOD at the edge of a group extends by definition to infinity. This is not very practical when it comes to making empirical measurements. In addition, a peripheral animal that moves toward the group but still remains at the edge would have again an infinite DOD even though its relative predation risk is probably much lower closer to the group. The limited domain of danger (LDOD) puts boundaries on peripheral domains based on the predator's detection and capture abilities (James et al., 2004). Although it is possible to use natural boundaries to limit the DOD, like a change in habitat type or the presence of cover (Viscido and Wethey, 2002), natural boundaries are difficult to establish in habitats with no strict demarcation. The LDOD has a radius like the conventional DOD but the radius is now determined by predator behaviour. For instance, the predator may be unable to detect the prey beyond this radius, or must approach to within this distance to launch a successful attack. The LDOD of a peripheral animal is still larger than that of centrally located companions. For centrally located animals, the LDOD is the same as the traditional DOD.

Beyond the detection range of the predator, prey animals that cluster together following any movement rule may gain an advantage by reducing the odds that the predator locates the group, as was the case in Vine's model. However, once the predator has located the group, movements by prey animals are made in the selfish-herd fashion so that now any decrease in predation risk by one individual necessarily increases that of at least one companion. One of the interesting consequences of the LDOD concept is that different rules of movement to reduce LDOD can be investigated because the size of each domain can be calculated exactly based on the biology of the system.

6.2.3. Movement Rules

Hamilton recognized the difficulty in finding a rule that governs movement to reduce the DOD in a two-dimensional space. Computer simulations have been developed to examine the consequences of using different rules for aggregation. Simulations put rules to the test and determine which ones may be plausible in the field. However, it should be clear that finding a rule that approximates aggregation behaviour is no guarantee that this particular rule is actually used by animals in nature. Simulations are useful in guiding empirical research by pinpointing possible rules and their consequences.

Hamilton's original rule was to move to the NN, but as pointed out earlier this often leads to the formation of a large number of small clusters. Integrating spatial information from more than one companion has been considered in one simulation model (Morton et al., 1994). For one rule, an individual could move toward the average position of two or more neighbours (up to nine). For another rule, the foraging area was divided into four quadrants starting from the centre, and individuals could move toward the middle of the most crowded quadrant. These more complex rules of movement resulted in a more compact DOD, but

the end result was often a number of clusters rather than one single large aggregation. In a different simulation study, animals could scan the whole horizon
and move toward the most crowded part (Fig. 6.2). This local-crowded-horizon
rule produced movements toward a central point and larger aggregations, which
is more in line with the expected bunching response (Viscido et al., 2002).

Using the LDOD concept, James et al. (2004) considered two simple rules
of movement: (1) move to the nearest neighbour in space, or (2) move to the
nearest neighbour in time. The distinction between nearest neighbour in space
or in time is important because a neighbour may be closer in space than another
but more distant in time if the animal that moves must change direction to reach
the new position and lose time doing so (Fig. 6.2). Their simulations looked at
the behaviour of 20 agents located in a two-dimensional space, one of which
was allowed to move using one of the above rules while the others remained stationary. A large number of simulations were run using the same rules but different initial positioning for the 20 agents. The LDOD of each moving agent was
shown to decrease with time to a greater extent when agents used the nearest-
neighbour-in-time rule than when agents used the nearest-neighbour-in-space
rule. The decrease in LDOD with time is consistent with the prediction derived
from the selfish-herd hypothesis.

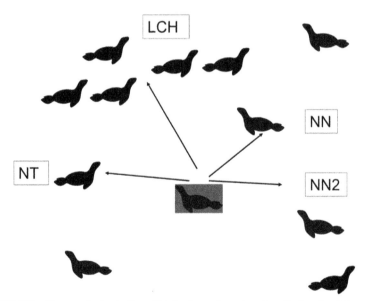

FIGURE 6.2   Movement rules in the selfish-herd paradigm: to reduce its domain of danger, the
individual in the middle may move to the nearest neighbour in space (NN), to the nearest neighbour
in time (NT; here this target may be further away than the nearest neighbour in space but because
of the cost in changing direction, this position may be nearer in time), to the average position of
the two nearest neighbours (NN2), or to the area with densest aggregation of companions, the
local crowded horizon (LCH). (For colour version of this figure, the reader is referred to the online
version of this book.)

Although it is interesting to see the consequences of using different rules for aggregation, in evolutionary terms, it is equally important to determine whether an individual that uses a different rule from the majority can achieve greater success, in this case, reduced LDOD size. A rule, when used by all individuals, may be quite successful; however, it may not be able to compete with an alternative rule, in which case it may be replaced by the alternative after several generations. To show this, James et al. used the same agents described above and allowed each individual the opportunity to move using the same rule. One of the agents, however, used an alternative rule. Results indicate that an agent using the time rule in a population of agents using the spatial rule was more successful than the others. Crucially, a single agent using the spatial rule could not achieve greater success when invading a population using the time rule, suggesting that there is an evolutionary advantage in using the time rule. The time cost of changing direction is thus an important consideration in modelling rules of movement. The point here is to establish an evolutionary framework to evaluate different rules using plausible biological arguments for defining both the LDOD and movement rules. An evolutionary framework has also been used to establish whether the local crowded-horizon (LCH) rule could be evolutionarily stable. Results indicate that the evolved solution can lead to strong aggregation and involve gathering information from many companions in the group (Reluga and Viscido, 2005).

Models that investigate movement rules examine their success for all group members regardless of position in the group. However, an individual at the edge of a group when a predator attacks from the outside may not experience the same returns from a rule as individuals situated more inside the group, who may simply want to retain their beneficial positions. The benefits of using different rules based on position within the group have been investigated in a recent simulation study (Morrell et al., 2011a). The authors found that only complex rules, such as those involving a consideration of the position of multiple neighbours, allowed individuals at the periphery to move to more centrally located positions and individuals at the centre to retain their positions. Therefore, it is important to take into account the spatial position of a forager when assessing the value of a movement rule.

6.2.4. Selfish-Herd Effect in Time

If all individuals in a group follow an aggregating strategy, such as the frogs in Hamilton's example or the agents in the above simulation studies, one would expect that by the end all individuals would be very close to one another and predation risk would again be equal for each group member. Selfish-herd benefits may be transitory and more evident early on in the aggregation phase. Yet most models assess the value of a movement rule when the group has reached its final configuration. A new simulation model, along the lines described above, has focused on the build-up phase of aggregation (Morrell et al., 2011b). The model allowed simulated predators to target individuals at different time periods during the aggregation process. The authors considered the three rules described earlier: moving toward the NN, moving toward the average position of three nearest neighbours (NN3), and moving to the area with the densest concentration of companions (LCH). Like before, the success of each rule was measured by the size of the LDOD. In the first set of simulations, all three rules were equally represented in the population. The results show that if the predator attacks very early on after detection and the initiation of movement, the LCH rule performs worst and the two alternatives perform equally well. The LCH rule performs better than the others later in the sequence and proved evolutionarily stable against invasion by the other rules (Fig. 6.3). Results from this simulation study show that different rules may perform better than others, but it is critical to identify when in the sequence the predator is most likely to attack. This model makes predictions about movement rules in biologically plausible scenarios that involve short or long attack times, which may correspond to different types of habitats (e.g. open versus closed) or different types of predators (e.g. ambush versus pursuers).

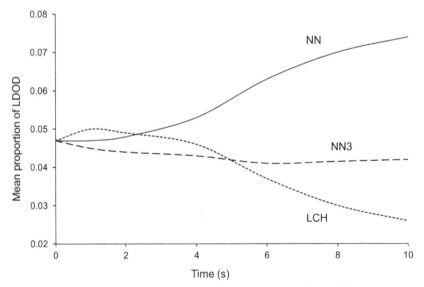

FIGURE 6.3 Success of different movement rules in the selfish herd: changes in the mean proportion of total LDOD occupied by individuals are shown as a function of time since the beginning of an attack. Three different rules were equally represented at the onset of the simulation: (1) move to the nearest neighbour (NN), (2) move to the average position of the three nearest neighbours (NN3), and (3) move to the area with the densest concentration of companions (LCH). *Adapted from Morrell et al. (2011).*

6.2.5. Vigilance and the Selfish Herd

For a model that purports to determine behaviour in a group, the selfish-herd hypothesis is relatively devoid of information about the individuals moving about. The frogs in Hamilton's model just sit on the edge of the pond managing their predation risk but doing little else. In the simulations described above, individuals are only concerned with achieving a better position. Animals in the wild are no doubt also concerned about finding food and keeping an eye on predators. Models of antipredator vigilance, which have been described in detail in the previous two chapters, rarely consider spatial issues, such as those arising in a selfish herd, while selfish-herd models typically ignore all activities except spatial displacements.

To address this issue, I used a genetic algorithm model to allow individuals to search for food and detect predation threats (Beauchamp, 2007). Genetic algorithm models are increasingly used in animal behaviour research and provide useful solutions for less tractable problems (Box 6.1). In my model, two individuals lie at opposite ends of an imaginary line and allocate time to vigilance and feeding. The dilemma faced by the two foragers is that investing time in vigilance increases the chances of detecting the predator but decreases feeding rate, both of which are important to increase survival (Pulliam et al., 1982). The solution to maximize survival is to allocate an intermediate proportion of time to vigilance, the exact value of which depends on predation risk and feeding rates. The predator appears at random times on either side of the imaginary line and starts chasing the nearest prey. Upon detection of the predator, the targeted prey moves toward the companion, which may be overtaken should it fail to be vigilant during the chase. The predator eventually captures the nearest forager with a fixed probability. I showed that the level of vigilance increases when the two foragers are closer together. There is therefore a cost in achieving the sorts of benefits predicted by the selfish-herd effect, and the instant reaction assumed in most models may be most relevant when individuals are close to one another. If such costs vary among individuals and perhaps also according to spatial location in the group, then not all individuals are expected to react in the same way to a sudden predation threat. It would be interesting to see consequences for aggregation in the above simulation studies when a delay in responding is imposed by investment in other fitness-enhancing activities, such as vigilance and feeding.

6.2.6. Evolving the Selfish Herd

Much theoretical effort has focused on the movement rules adopted by individuals under threat. Would selfish-herd effects evolve in a population that has never been exposed to predation? This is an important issue; after all, the driving theme of Hamilton's paper was that gregariousness could evolve as a result of predation selection pressure. The evolution of gregariousness has been examined in

BOX 6.1 Genetic Algorithms

Mathematical modelling has a long tradition in animal behaviour research and has been a driving force in fields such as optimal foraging theory (Stephens and Krebs, 1986). Initially, simple models, with a solution derived analytically in the form of an equation, were able to capture the essence of many problems, such as optimal prey choice. However, as more and more variables are integrated into the modelling framework, finding solutions becomes a real challenge, especially for mathematically challenged biologists. Computer simulation studies can provide a solution to problems that are intractable using simple analytical tools. Hence, a trade-off emerges in the very tools that one uses to model a biological problem: elegant but simple analytical solutions versus messier but more realistic models.

Genetic algorithms (GAs) involve the use of simulations to systematically search for the solution of a problem in the space defined by the possible values assigned to model variables. GAs were first developed in the mid-1970s (Holland, 1975) and are part of a family of heuristic tools referred to as evolutionary computation. Such models have been widely used in fields like biology, chemistry, and economics (Haupt, 2004). GAs use biological evolution as their framework. A simple genetic algorithm in animal behaviour research would consider a population of individuals, each with one chromosome. A chromosome contains one or more elements associated with the different strategies or choices that are involved in the optimization process. Differential fitness associated with each strategy or choice is translated into differential reproduction, thereby creating a new population. After several generations, the solution emerges in the form of a stable proportion of strategies or choices in the population (Sumida et al., 1990).

Here, I describe an example of a genetic algorithm dealing with antipredator vigilance (Ruxton and Beauchamp, 2008). Individuals in a group must choose how much time to allocate to two mutually exclusive activities: detecting predation threats and finding food. An increase in time spent vigilant reduces predation risk but increases starvation risk. In addition to the trade-off between time spent vigilant and time spent foraging, animals in the group must also take into account the behaviour of companions. Vigilance is a frequency-dependent problem in the sense that it pays an individual to reduce its own vigilance when the rest of the group is more vigilant and can detect predators for everyone. Analytical solutions to this problem exist (Ale and Brown, 2007; McNamara and Houston, 1992), but Ruxton and I wanted to compare one such analytical solution with the solution from a genetic algorithm to illustrate the value of such models in animal behaviour research concerned with frequency-dependent strategies.

The chromosome in this vigilance problem contains one element, the proportion of time allocated to vigilance by an individual. Initially, each individual has a value set randomly between 0 and 1. The first part of the genetic algorithm is to simulate the fate of each individual in the population when using the vigilance strategy encoded on their chromosome. Thus, an individual searches for food and interrupts feeding to scan its surroundings, during which time it cannot obtain any food but can detect a predator sooner and escape with a greater probability. At the end of a set period, the success of each individual is tallied and those that died from predation or from starvation are removed from the pool of individuals. Each

---

BOX 6.1 Genetic Algorithms—cont'd

member of the next generation picks a value of vigilance from a random survivor to which a slight random perturbation is added. Such perturbations are similar to mutation: mutation can allow slight changes in the available strategies in the population. More complex reproduction rules can also be considered, such as crossover, when parts of the chromosome of two parents are swapped to produce the genotype of the next generation. The next generation is tested again in the same environment and with the same rules. Reproduction and mutation are performed over thousands of generations, thereby simulating evolution. The solution, here in terms of the optimal allocation of time to vigilance, emerges from the model.

We found a close match between the analytical solution and the solution derived from the genetic algorithm. We also found a number of issues unique to genetic algorithms. It is important, in particular, to determine whether the solution achieved by the model is independent of the initial set of conditions. This is not a trivial issue since many suboptimal solutions may exist in the search space while the goal is to find the best solution. Other issues include allowing enough time to reach a stable solution and using a proper mutation rate.

Genetic algorithms are particularly helpful for finding solutions to complex problems, but because they involve many modelling choices—for example, mutation rate, length of a generation, number of simulations—it is important to assess their impact on the solution that emerges.

---

a model where individuals were exposed to predators (Wood and Ackland, 2007). The authors used a genetic algorithm approach to evolve parameters that control the movements of individuals. For instance, speed of movement and the radius of interaction with other individuals were not set initially but evolved as the simulation progressed. This approach is strikingly different from the one used in the simulation studies presented earlier in which movement rules are immutable and arbitrary. The key point is that all sorts of configurations could evolve, from individuals moving independently of each other to the formation of a tight aggregation. With a genetic algorithm, the attributes of individuals can be passed on to a new generation, as we saw above. Obviously, those that possess attributes that contribute to a greater risk of predation are less likely to propagate. The predator moves into the habitat and, when detecting a prey in its arc of view, attacks the closest individual. The predator cannot be confused by the prey with this simple rule. Individuals that detect the predator move away but independently of one another. This is in sharp contrast to the simulation studies described earlier, where movement is dictated by the location of companions.

After several generations, the model reached an equilibrium, which included the formation of a tight aggregation. Given that individuals cannot purposefully hide in the group, what is the benefit of forming aggregations in the first place? The encounter principle does not apply here given that similar aggregations formed regardless of the viewing angle of the predator. Following

Vine's argument that I presented earlier, a smaller arc of view for the predator should have selected for tighter aggregation by prey so as to reduce the rate of encounter of the whole group with the predator. Since this was not the case here, I suspect that once a prey was targeted, it benefited from being in a group because the predator could select an alternative target when another individual became equally close. However, this is difficult to assess since the authors did not describe what happened when the predator faced a choice between two equally distant targets. Another possibility is that the targeted individual could hide in the group and pass along the risk to another individual that happened to move closer to the predator. Despite these caveats, this model is instructive in describing how predation pressure can lead to the evolution of gregariousness. More work is needed to establish exactly how individuals benefited from forming groups, and therefore whether a selfish-herd effect was at work.

6.3. EMPIRICAL EVIDENCE

The various models presented in the preceding section lead to testable predictions, which can be organized into two main categories: (1) animal groups should become more tightly bunched when under predation threat as individuals struggle to reduce their LDOD relative to one another, and (2) individuals with a smaller LDOD should experience proportionately less predation. This second prediction is crucial, because showing a decrease in LDOD will be meaningless if no relationship with predation risk can be demonstrated.

Some studies have shown that as predation risk increases, individuals are more likely to seek cover within the group and that more central positions are coveted. For example, individual fish exposed to a fright substance attempt to move inside a nearby school (Krause, 1993), as would be expected if central positions are more protected. A particularly intriguing study investigated the positioning of gray-breasted martins roosting on electrical wires in Mexico (Watt and Mock, 1987). Interestingly, the situation on a wire is very similar to that faced by frogs on the periphery of a pond. Assuming that an aerial predator can attack from any direction but obviously from outside the group, the LDOD of central individuals will certainly be smaller than that of birds at the edges of the group. It turns out that central positions were more contested than those at the edges and were less likely to be abandoned spontaneously. Food-related hypotheses, which could also explain position preferences, are unlikely to explain this result since the birds were roosting. A follow-up study of predation risk as a function of position on the wire would provide the crucial evidence that edge birds with a larger LDOD are more likely to be attacked.

The tendency to bunch up upon attack or threat has clearly been shown in simulation studies where it is easy to keep track of individual movements. The empirical literature contains much evidence for clumping in the face of predation threats (Mooring and Hart, 1992). Some examples include clustering upon attack in many species like schooling fishes and ungulates in herds. In amphibians, introducing a predator or the scent of a predator clearly increases cohesion in tadpoles of toad species (Spieler and Linsenmair, 1999; Watt et al., 1997). The arrival of potentially threatening bachelor gelada baboon males triggers a clustering of individuals in the group (Pappano et al., 2012). In the shorebird species that I study, loose flocks of semipalmated sandpipers flying across the water will quickly cluster in a tight formation when attacked by a peregrine falcon.

The next step for these studies is to track individual movement. This is essential to uncover the rules that animals use to alter their LDOD. With sandpiper flocks, often including thousands of birds in a three-dimensional space, the task is nearly impossible. A British study has looked at a much smaller system in a two-dimensional space: flocks of sheep herded by a trained sheepdog (King et al., 2012). The sheep were fitted with GPS devices allowing researchers to follow individual trajectories through time and space as the flock was herded. The prediction from the selfish-herd hypothesis would be centripetal movement in the herded flock, a tendency to move toward the centroid of the group as the dog comes closer. Such a tendency was documented in three separate trials, with individuals taking generally less than 1 min to move about 20–30 m to the centroid. The next step with such a system would be to investigate the possible rules that can account for individual trajectories in space and time as sheep move toward the centroid.

Is there any evidence that predation risk is lower for centrally located foragers, that is, those with a smaller LDOD? A review of the literature has been undertaken to determine whether rates of attack or odds of capture differ for peripheral and central animals (Stankowich, 2003). This review included a total of 29 studies, covering a vast array of species from invertebrates to fish, birds and mammals. The majority (21/29) reported greater predation pressure at the edges, and only three showed greater pressure at the centre of a group. While the overall impression is certainly that predation pressure increases at the edges of groups, we need to rule out alternative hypotheses to explain such differential effects.

In particular, many of the studies used in the meta-analysis dealt with breeding colonies. It has been argued that greater predation pressure at the edges of colonies may simply reflect the fact that predators have more difficulty reaching the centre of a colony due to more efficient mobbing. In addition, the level of parental involvement in nest defence may vary between peripheral and central individuals, making peripheral nests more likely targets. Phenotypic differences related to spatial position is an issue that also arises in non-breeding groups. For instance, individuals at the edge of a group may be slower to escape (Quinn and Cresswell, 2006), and may thus be preferred by predators. Stankowich also noted that different studies use different definitions to determine which individuals are on the edge a group, and that depending on the definition used, the same individuals may be classified as peripheral or central (Box 6.2). In view of these issues, the relationship between spatial position and predation pressure requires more empirical scrutiny.

BOX 6.2 Defining Central vs Peripheral Position with a Group

Determining who in a group occupies a central or a peripheral position is crucial to the selfish-herd hypothesis. This issue is also relevant to a number of other behavioural patterns that have been linked to spatial position, such as vigilance and foraging efficiency (Krause, 1994). Ambiguities in defining spatial position arise when groups are large. In a small group, composed of, say, two or three individuals, all group members are literally at the edge of the group. Stankowich (2003) reviewed different ways of defining position in a group. Other methods have been considered elsewhere (Hirsch and Morrell, 2011).

With the polygon vertices method, an individual is declared peripheral if it occurs at the vertex of the smallest closed convex polygon that encloses all group members. Essentially, one has to join by a straight line all individuals that are the farthest away from the centre. In a larger group, the polygon may have many faces but the smallest polygon will have three faces. Another similar conceptualization is that a peripheral animal has no neighbours within a semicircle of 180° on one side.

In the layer method, central individuals are those that are entirely enclosed by more than one layer of other group members. The circle method consists of drawing a circle around the entire group, and starting from the centre, drawing one centre circle, and then a further circle, each with the same area. The animals located outside the central circle are declared peripheral. For the distance method, the distance between each individual and the centre of the group is measured and the mean distance is calculated with all measurements. A circle is then drawn with the mean distance as radius. All those individuals outside the circle thus drawn are declared peripheral.

Notice that all these methods are based on geometry and fail to take into account how animals themselves may perceive their position as peripheral or central. A consideration of how quickly individuals may move to attain central position or how much information can be obtained from nearby companions may be relevant in assessing how peripheral an individual really is. Nevertheless, the key finding from the review by Stankowich is that the same individuals may be classified as central or peripheral depending on which of the above definitions is used. In particular, definitions which are based on layers of protection or the mean distance to centre generally classify more individuals as peripheral.

Many studies still seem to find an association between spatial position, however defined, and fitness correlates, suggesting that methodological issues related to the definition of spatial position are not masking biologically important phenomena. In view of these difficulties, the minimum that needs to be done is to clearly specify which definition is used and perhaps carry out a sensitivity analysis to determine the robustness of the findings.

Further to this point, Hirsch and Morell (2011) examined predation risk as a function of spatial position in simulations of groups under attack. The predator moved in a straight line across the group and captured the nearest prey located fatally close. Spatial position was determined according to various methods, including some of the definitions presented earlier. The polygon vertices method was found to predict predation risk quite well, as opposed to other methods such as the mean distance to centre. Similar work needs to be carried out with real biological systems in order to develop biologically accurate ways of defining spatial position in a group.

Very few studies have actually measured LDOD in the field owing to the technical challenges of measuring spatial location from images. In simulation studies, the LDOD is shown from above and all distances between individuals are known. In the field, animals are rarely viewed in this way and measuring distances at the viewing angle used is quite challenging due to a distortion effect. Animals also move frequently and their LDOD may thus change rather rapidly. It is also important to establish boundaries for DOD at the edge of the group, and also to determine whether a predator can attack individuals regardless of position in the group. The challenges faced when attempting to measure LDOD in the field are illustrated well by a study of fiddler crabs (Viscido and Wethey, 2002). They taped fiddler crabs foraging on sandy beaches and attempted to determine LDOD for many groups. Crabs foraged on a two-dimensional space bounded on one side by the edge of a creek, which was used to put a limit on the DOD of crabs at the edges of the group. This is important since otherwise the DOD of animals at the edge extends to infinity and cannot be calculated. Crabs are preyed upon by a variety of shorebirds but the authors did not specify whether attacks occurred from the outside of the group or could target any animal within the group. Their drawings of an LDOD assume that the predator can target any individual in the group. Using natural and simulated attacks, the authors showed that the bounded DOD decreased in size after panic set in. Interestingly, individuals that were on the other side of the group when panic set in actually moved toward the source of disturbance to reach inner positions in the group, indicating that individuals were not simply moving away from the threat but actually trying to reduce their relative LDOD. Unfortunately, the authors could not determine whether predation pressure was related to size of LDOD.

In perhaps the most convincing study investigating the relationship between LDOD and predation risk, Quinn and Cresswell (2006) studied the odds of attack for individual redshanks, a small shorebird species, foraging in a tidal marsh. Redshanks that were attacked by birds of prey were more widely spaced than their nearest non-targeted neighbours. The appeal of this study was the ability to rule out alternative explanations for targeting edge birds. The authors could determine which individuals were targeted and which were not because the flocks were rather loose and individuals were typically slow to react to an approaching predator (Quinn and Cresswell, 2006). Comparing targeted and non-targeted close neighbours (that is, birds occurring in the same part of the flock) is important because it avoids confounding factors related to potential phenotypic differences associated with different spatial position in the group.

A simple explanation for selective predation in the redshank system is that birds of prey target the nearest redshank after launching an attack. However, the distance between predator and targeted and non-targeted redshanks did not differ significantly. In addition, the effect of spacing on the odds of being targeted persisted in the statistical model after controlling for predator proximity. Similarly, targeted redshanks could simply be the ones who escaped more slowly. Although it turned out that targeted redshanks took longer than their neighbours to initiate escape, the effect of spacing still was the only statistically significant factor remaining when escape delay was incorporated in the model. Interestingly, individuals on the riskiest side of the flock (that is, the side of the flock closest to the area where birds of prey launched their attacks) were more tightly packed the closer to the source of danger, as one would expect if birds were trying to reduce their LDOD. It is not clear whether such preferences for widely spaced prey are common in other species. In a laboratory experiment with fish, where a predator could attack all positions within the group, the predator did not selectively target individuals with the largest nearest-neighbour distance but, rather, preferred those at the edge (Romey et al., 2008). It is possible that the fish predator was more easily confused than the bird of prey attacking redshanks. Experiments or observational studies with varying densities of prey may be able to settle this issue.

A strong test of the selfish-herd hypothesis is one that controls for potential differences among individuals with a different LDOD, as in the redshank study. Ideally, all individuals should have the same attributes except for their LDOD. A recent, particularly inventive, study investigated rates of attacks of Cape fur seal decoys by white sharks in the waters off South Africa (De Vos and O'Riain, 2010). Potential confounding factors, such as body size or level of vigilance, which may vary in relation to position in the group, were controlled with the use of inanimate decoys. Decoys, simulating the shape and size of seals, were deployed behind a boat and towed across a danger zone where seals are often attacked by sharks. The LDOD of each decoy seal was measured and several independent trials were conducted over many days. The results strongly supported the selfish-herd hypothesis: the LDOD of attacked decoys was significantly larger than those of non-targeted decoys (Fig. 6.4). Confusion of the predator, which may arise when prey panic and escape in tight groups, was unlikely because decoys could not change position during a trial.

In a further study, the authors investigated possible movement rules used by real fur seals when swimming in these dangerous waters (De Vos and O'Riain, 2012). Simple rules, such as moving to the NN in time or in between the two nearest neighbours, accounted for more variation in the choices made by seals than more complex rules like the LCH rule. In addition, after such movements, the DOD of seals decreased, as would be expected by the selfish-herd mechanism. By contrast, when fur seals swan in safer waters, individuals typically moved straight ahead regardless of the position of neighbours. The lack of success of more complex rules may not necessarily indicate that such rules are too complex to implement cognitively. Seals may, in fact, have limited information about the position of many neighbours due to turbidity.

6.4. CONCLUDING REMARKS

The selfish-herd hypothesis has generated a large amount of theoretical work, most of which aimed at uncovering the movement rules used by individuals to achieve greater protection. These rules vary in complexity and can involve

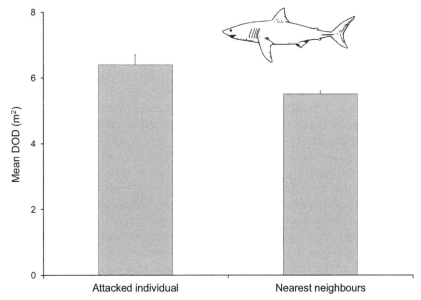

FIGURE 6.4   Selfish herd in the field: Cape fur seal decoys attacked by white sharks have larger mean DOD than those of non-attacked nearest neighbours. Error bars represent 95% confidence intervals ($n = 36$). *Adapted from De Vos and O'Riain (2010).*

gathering spatial information from one to several companions in the group. Only two studies have documented the sorts of rules individuals use when clustering under threat. In one fish species, there is some evidence that individuals move toward the NN in time when under attack (Krause and Tegeder, 1994). In fur seals, individuals appear to use simple rules, such as moving to the NN in time or in between the two nearest neighbours (De Vos and O'Riain, 2012). An obvious direction of future research on the selfish herd is to document the rules used by different animal species. Simulation studies can only go so far in helping us understand aggregation processes. Empirical feedback is needed to guide further theoretical work and pinpoint the actual decision rules used by animals under predation threat.

Tracking the spatial position of group members using GPS devices is certainly a very promising avenue for future empirical research. The use of decoys or other types of inanimate individuals, such as robots (Fernández-Juricic et al., 2006; Halloy et al., 2007), is also a promising avenue for studying differential predation pressure in relation to spatial position. Such model animals can be deployed to manipulate LDOD and their behaviour can be precisely controlled. This may prove a particularly valuable approach for studying prey animals that move along one or two dimensions.

Relating LDOD to predation risk is no easy empirical task. It is crucial that the LDOD of both targeted and non-targeted individuals be calculated but it may prove difficult to determine which individual was initially targeted. The targeted individual can be obvious in a small group or in a group where individuals are widely spaced, like in the redshank study discussed above. However, if the predator attacks from afar, all individuals are likely to flee, making it difficult to determine which individual was initially targeted (Lima and Bednekoff, 2011). In addition, it must be possible to follow the fate of the targeted individual, which may be difficult in a large group where individuals cross paths as they escape. It is also important to determine whether the predator switches targets during the attack.

A challenge for the selfish-herd hypothesis is to rule out alternative explanation for the formation of aggregations. There are a number of reasons why aggregations may form, including pressure to reduce the rate of encounter with predators (Turner and Pitcher, 1986). In such a case, animals may be expected to cluster but such a mechanism does not involve any specific rule for aggregating as per the selfish-herd explanation. Animals in these aggregations may also benefit from the dilution effect that arises from the presence of several alternative targets in the group. Encounter-dilution effects can arise regardless of the spatial position of individuals in the group, and the jostling for position predicted by the selfish-herd hypothesis should be absent. Documenting specific rules of movement and ensuing changes in LDOD during the process of group formation should enable us to rule out the encounter-dilution effect acting alone.

Perhaps a more important challenge is to rule out the confusion effect. Such a mechanism has been discussed by several researchers in their study of predation of peripheral members of the group. Predators may prefer to attack peripheral individuals to avoid the denser part of the group where the presence of many individuals can reduce the probability of capturing any targeted prey. The confusion effect is a cognitive limitation that can be exploited by grouped prey (Curio, 1976; Landeau and Terborgh, 1986; Neill and Cullen, 1974). It has long been noticed that predators that can easily tackle a single prey are often less successful when chasing prey in cohesive groups (Kenward, 1978; Schaller, 1972). Criss-crossing of paths by fleeing animals may make it more difficult for the predator to select a specific target. Predators may thus attack at the edges of the group and select peripheral individuals irrespective of their LDOD. This can be an insidious problem for the biologist trying to rule out alternative explanations. Even with the use of decoys, as in the shark study I presented earlier, where confusion was probably limited since the prey never changed their relative position and group size was small, one could argue that perhaps the sharks have learned to avoid the denser part of a seal group because of lower attack success. One can assess the relevance of the confusion effect by looking at attack success rate when prey density changes. If attack success rate decreases in denser groups, controlling for confounding factors like group size, escape speed and vigilance levels, then there would be evidence that confusion is probably involved. Confused predators may change targets, and it may be possible to document this empirically, as the attack proceeds. By contrast, non-confused predators should retain their original targets and be less influenced by prey spacing (Quinn and Cresswell, 2006).

Can the selfish-herd mechanism be responsible for the evolution of gregariousness as originally proposed by Hamilton? In his review of the selfish-herd hypothesis, Viscido (2003) concluded that there was no evidence for this assertion. In an evolutionary scenario where two solitary individuals come closer to one another to form a proto-group, the size of their relative LDOD cannot be changed by altering position in the group. However, if individuals can change their position once an attack is under way, one individual may be able to hide behind the other, and thus achieve a reduction in predation risk. Unless individuals are inherently different in their ability to achieve a better position during an attack, the benefits of grouping should, on average, be the same over several attacks, making this temporary advantage moot in evolutionary terms. Many other benefits of grouping are more likely to work in such a proto-group, such as the encounter-dilution effect. In addition, each individual may be able to use the detection abilities of others to escape more quickly. It thus appears doubtful that selfish-herd effects alone can trigger aggregations. However, once groups have evolved, this mechanism may provide benefits for individuals.

As a final comment, the selfish-herd hypothesis has usually been tested in the context of predation. However, grouping may be expected to provide benefits in other contexts. One example is the case of parasites, such as biting flies, harassing animals. Grouping may be helpful not only in reducing encounter rate with parasites and diluting harassment among more individuals, but also in gaining protection through seeking more sheltered positions in the centre of the group (Fauchald et al., 2007; Mooring and Hart, 1992). Similarly, the presence of threatening conspecifics may also trigger clustering-like behaviour, as exemplified by gelada baboon responses to approaching bachelor males that can harass females and challenge males already in the group (Pappano et al., 2012). It would be fruitful to extend the selfish-herd hypothesis to a broader range of problems where protection is required.

Part C

General Considerations

Group Size and Composition

7.1. INTRODUCTION

The lodgepole pine is a conifer species found in the western parts of North America. The thin, straight trunk of this tree made it especially well suited for building tepees, the traditional houses of some native North Americans, whence its name. The tree grows in dense stands whose life cycle is controlled by recurring fires. Its thin bark provides little protection against fire and also makes the tree more vulnerable to attack by the mountain pine beetle. The size of a grain of rice, this beetle usually attacks old and vulnerable trees, thereby favouring the development of young ones. However, the unusually dry summers and mild winters of recent years have led to an epidemic of biblical proportions, felling large stands of trees in many parts of western Canada and the United States. The mountain pine beetle infestation is now the largest insect blight ever seen in North America (Safranyk and Wilson, 2007).

The pine beetle lays its eggs under the bark and introduces a fungus that prevents the tree from repelling and killing the attacking beetles. Pine beetles emit a pheromone that attracts conspecifics to the attacked tree. They can also emit other signals to prevent the arrival of further beetles, suggesting that the number of beetles attacking a tree can be manipulated by the beetles already present on the tree. Larval feeding and fungal colonization girdle the tree and can kill a mature tree in a matter of weeks. The predator–prey relationship between the pine beetle and the lodgepole pine provides a remarkable example of the consequences of group size in a system in which the costs and benefits of group foraging are reasonably well understood.

Are group attacks beneficial for mountain pine beetles? When pine beetles attack a pine tree, the probability of killing a tree increases slowly with group size at first, and then very rapidly as the combined larval feeding and fungal accumulation becomes more and more effective in overcoming tree defences. The rate of increase in the probability of killing a tree eventually slows down when the number of pine beetles exceeds 100 per square metre of tree (Berryman et al., 1985). On the cost side, individuals in larger groups must share the resources from the tree, which naturally sets a limit on the amount each individual can obtain. In addition, pine beetles compete more directly with one another when in large numbers, which can slow down their food intake. Food intake provides energy to the beetle, which eventually fuels its reproductive output. As group size on a tree increases, the expected reproductive output of an individual decreases in an exponential-like fashion. Combining the probability of killing a tree and reproductive output provides an estimate of expected reproductive success for individual beetles in groups of different sizes. This particular fitness function first increases and then decreases with group size, peaking at about 75 beetles per square metre of tree (Fig. 7.1). Average beetle density should thus cluster around this value in nature. Indeed, variation in infestation

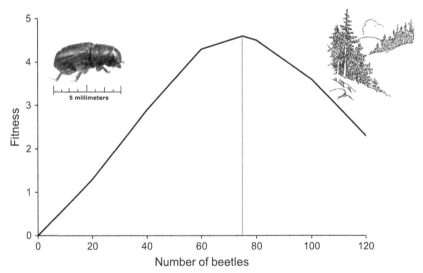

FIGURE 7.1   Optimal group size in pine beetles: individual fitness first increases and then decreases when the number of mountain pine beetles per square metre of lodgepole pine increases. The vertical red line in the middle indicates the optimal number of beetles for maximum individual fitness. The probability of killing a tree multiplied by the reproductive output of a beetle determines fitness. *Adapted from Berryman et al. (1985).* (For interpretation of the references to colour in this figure legend, the reader is referred to the online version of this book.)

from one tree stand to another in western North America typically ranges from 60 to 90, encompassing the optimal value (Berryman et al., 1985). Despite simplifying assumptions, this example nicely illustrates how a careful consideration of the costs and benefits of foraging in groups can lead to insights into the aggregation pattern of animals in nature.

Earlier chapters in this book covered the costs and benefits of foraging in groups for predators and prey, treating the different effects, such as risk dilution and encounter rate with prey, separately. In this chapter, I consider the combined effects of such influences on the expected size of animal groups. Of course, which particular factors come into play varies between species. For predator species that face few threats of predation, like orcas and wolves, the costs and benefits of foraging in groups will be related solely to the acquisition of resources. For predator species exposed to predation threats, and for prey species in general, the costs and benefits are related not only to the acquisition of resources but also to the ability to escape predation. The principles that I detail in this chapter apply to all of these situations.

The costs and benefits of group living are often measured in very different currencies, which compounds the difficulties of predicting group size in animal groups. For instance, the benefits of group foraging may be expressed as units of food intake and the costs as probability of surviving an attack. To develop a predictive model of group size, these costs and benefits must be expressed in the same units. Wilson (1975) realized this early on; he expressed the costs and benefits of foraging in groups in terms of energetic payoffs, and predicted that groups form when the energy gains exceed the energy maintenance costs (Wilson, 1975). A few years later, Caraco (1979) used a similar approach but focused instead on costs and benefits expressed in time units rather than in energetic payoffs (Caraco, 1979b). Ultimately, the costs and benefits of foraging in groups must be expressed in units of reproductive success or, at the very least, in comparable units related as directly as possible to fitness.

In this chapter, I focus exclusively on situations where foraging in groups provides higher fitness than foraging alone over at least a certain range of group sizes. This leaves out situations where individuals obtain no benefits from aggregating, which Giraldeau and Caraco (2000) refer to as a dispersion economy. For example, oystercatchers in a mudflat compete over burrowing prey and experience a decrease in food intake rate when the density of nearby birds increases (Goss-Custard, 1996). In this situation, the aggregation of birds in a food patch reflects the paucity of better alternatives rather than any attraction between companions. By contrast, in an aggregation economy, individuals benefit from the presence of companions and are expected to form groups.

In the first part of this chapter, I examine the concept of optimal group size. I review the relevant theory and empirical evidence supporting the notion that predators and prey forage preferentially in groups of specific sizes. Perhaps just as important as the notion of optimal size is the effect of group composition. This is the idea that groups may be composed of subsets of individuals that have similar phenotypic traits, such as body size. The assumption is that individuals are not identical and so may experience the costs and benefits of foraging in groups differently. Different individuals may, therefore, prefer to forage in groups of different sizes or composition. Homogeneity in group composition is explored in the second part of this chapter.

7.2. OPTIMAL GROUP SIZE

7.2.1. Optimal and Equilibrium Group Size

Evolutionary theory predicts that animals select courses of action that maximize their reproductive success or proxies of reproductive success like food intake rate or survival. With respect to choosing a foraging group size, individuals are expected to join a group when doing so increases their payoffs, and to abandon a group when foraging alone becomes more profitable. Group size, in this evolutionary framework, will reflect individual choices shaped by natural selection to maximize fitness (Pulliam and Caraco, 1984). Returning to the mountain pine beetle example, individuals in groups that are too small to inflict much damage on a tree may turn on the pheromone signal to attract companions. However, when the group becomes too large and competition intensifies, the pheromone signal may be turned off to avoid attracting more competitors, and some individuals may actually leave the group. This produces an optimal group size. Reflecting the balance of costs and benefits, which may vary in different ways in different species, the optimal group size may vary across and even within species. For instance, when pine beetles tackle healthier trees, the optimal group size shifts to higher values (Berryman et al., 1985).

The notion of optimal group size implies free entry and exit from any group, with individuals moving in and out of groups to maximize their own fitness. Entry or exit from a group may, however, be tightly regulated. As an example, damselflies settling together for the night in a sheltered location may be unwilling to leave the group because good sites are rare and dangerous to locate (Grether and Switzer, 2000). Similarly, colonies of spiders may grow quite large because individuals leaving the colony have very low survival prospects (Avilés and Tufiño, 1998). With respect to entry into a group, individuals may prevent others from joining the group. Groups of lionesses, for example, can evict intruders trying to join their groups (Heinsohn and Packer, 1995). Thus, constraints on entering and leaving groups have to be considered when assessing optimal group size.

In a system with free entry, the optimal group size may, in fact, be unstable. This counterintuitive expectation appears to have been reached independently by different researchers in the early 1980s (Clark and Mangel, 1984; Sibly, 1983). The optimal group size may well represent the ideal size for individuals already in a group, but unless group members can prevent entry, other individuals may join the group to increase their own fitness (Fig. 7.2). As long as the

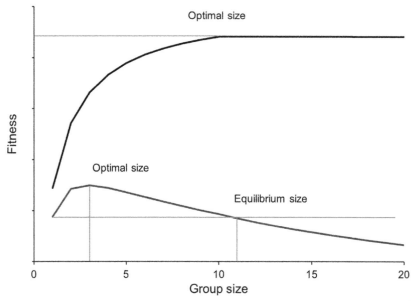

FIGURE 7.2   Optimal group size theory: how fitness varies with group size determines the occurrence and stability of the optimal group size. In the upper curve, fitness increases rapidly with group size and reaches a plateau. The optimal group size, 10 in this case, is the group size that maximizes fitness. Larger groups may occur but these groups will show no tendency to rejoin when split. In the lower curve, fitness first increases and then decreases with group size, producing an optimal group size of 3. However, the equilibrium group size is 11 because the fitness of an individual joining a larger group will be no greater than the fitness of an individual foraging alone. (For colour version of this figure, the reader is referred to the online version of this book.)

fitness from joining a group exceeds the fitness from remaining alone, a solitary individual will benefit from joining the group, and group size will increase beyond the optimal value. This process continues until the group becomes so large that joining brings the fitness of the last joiner below that expected from foraging alone. At the stable, or equilibrium, group size, which is larger than the optimal size, the fitness of a group member equals the fitness of a solitary forager.

Splitting a group of optimal size should entice individuals to reform a group of the same size. The same is not true for a group at the equilibrium group size. Given the opportunity, individuals in such groups should try to reform smaller groups near the optimal size (Kramer, 1985). However, additional splitting of these smaller groups will eventually force individuals to join larger groups, pushing group size again beyond the optimal value. Optimal group size in a free-entry system is, therefore, unstable, and groups larger than the optimal size should be expected in nature. It is ironic that individuals aiming to increase their fitness by living in groups end up with no more success than when foraging alone. This constitutes the group-size paradox.

It is important to bear in mind that changes in the shape of the function relating fitness to group size can reduce the gap between fitnesses at the equilibrium and optimal group sizes. For example, if fitness just beyond the optimal group size drops below that expected when foraging alone, the optimal group size then becomes evolutionarily stable (Giraldeau and Gillis, 1985), allowing foragers to get the maximum payoffs. In one example of this type of function, the daily net per capita food intake reaches a peak value in pairs of wolves hunting moose (Vucetich et al., 2004). Trios, on the other hand, experience a lower per capita food intake than solitary individuals, which implies that size two represents the optimal as well as the equilibrium group size. The group-size paradox applies to cases where fitness at the equilibrium group size is much lower than at the optimal size. How far group size can stray from the optimal value depends on a host of factors, which are explored below.

7.2.2. Solving the Group-Size Paradox

7.2.2.1. Competition between Groups

The fitness of a solitary forager is key to understanding why individuals join groups larger than the optimal size. Individuals leaving groups not only face the burden of finding food and surviving alone, but may also compete with other larger groups (Zemel and Lubin, 1995). I explained earlier how larger groups may be able to displace smaller groups from preferred feeding areas or defend their prey more successfully against food stealers (see Chapter 1). Larger groups may also be more efficient in managing their resources. In this context, individuals foraging alone may be quite disadvantaged when competing against larger groups, which may explain their tendency to join groups larger than the optimal size. In a free-entry system, competition between groups is thus expected to move the expected group size away from the optimal size toward the equilibrium size. Other factors may, by contrast, entice foragers to forage in groups closer to the optimal size.

7.2.2.2. Effect of Relatedness

Thus far, my discussion of optimal group size has assumed individuals aim to maximize their personal fitness, which, as shown earlier, often takes place at the expense of others in the group. However, many groups are composed of relatives. For example, insect larvae feeding communally on a plant part often have the same mother (Costa, 1997). Although relatedness in social vertebrate species never approaches the levels seen in invertebrate societies, it still may be quite high. This is true, for example, of lions (VanderWaal et al., 2009). In groups composed of relatives, the choice to join or leave a group must be evaluated by considering the fitness consequences for all related individuals (Giraldeau and Caraco, 1993; Higashi and Yamamura, 1993). In such cases, the consequences of group-size choices are measured in terms of what is called inclusive fitness. This measure is based on the number of offspring produced by an individual and the number produced by its relatives, weighted by the degree of relatedness, which are expected as a result of a particular choice or action (Hamilton, 1964).

Consider a population where the average coefficient of relatedness among individuals is r, with higher values representing more closely related individuals. In a free-entry system, a solitary individual that joins a group of relatives near the optimal size expects to increase its own fitness at the expense of all individuals already in the group. Both fitness consequences matter when the individual joining is related to the group members. Specifically, for joining to occur, the losses in fitness incurred by all individuals already in the group, weighted by the factor r, must be smaller than the gain in fitness enjoyed by the solitary forager when joining the group. Calculations reveal that in a free-entry system, the expected group size should shift from the equilibrium value predicted for when individuals are not related toward the optimal group size as the coefficient of relatedness increases (Giraldeau and Caraco, 1993). Surprisingly, empirical support for this rather intuitive prediction is scant. Fish species can recognize kin from non-kin in the laboratory (Ward and Hart, 2003), but there is little evidence that relatedness plays a role in structuring groups in the field (Croft et al., 2012). In perhaps the best supporting evidence for a free-entry system, salmon appear to avoid aggregating with more related companions when resting because larger groups attract more predators, which reduces the fitness of close relatives (Griffiths et al., 2003).

In a group that controls entry, by contrast, it is now the group members that choose whether to expel or accept related joiners. Accepting a joiner adds one individual to the group and reduces fitness for individuals already in the group near the optimal size. The cost of repelling represents the difference in fitness experienced by the potential joiner when foraging in the larger group versus foraging alone, which is weighted by the relatedness coefficient and then divided among all existing group members. The benefit represents the sum of all gains obtained from preventing an increase in group size to beyond the optimal size. In this case, the expected group size should shift toward larger values as the degree of relatedness increases in the population (Giraldeau and Caraco, 1993). Such a mechanism has been proposed to explain why lions forage in groups larger than the optimal size (Rodman, 1981).

Earlier, I described how individuals in groups often exploit the food discoveries of others without investing their own time and energy in locating resources (see Chapter 2). If the producer of a patch controls the extent to which scroungers can join, then the situation becomes quite similar to the problem of group size when the group controls entry. Therefore, scrounging may be more prevalent in groups with related individuals. For producers, the loss in fitness caused by sharing resources with additional companions may be recouped by increasing the foraging success of related scroungers. In support of this prediction, scrounging in zebra finches was found to be more prevalent in groups with more closely related individuals (Mathot and Giraldeau, 2010b). However, it is not always easy to determine whether producers or scroungers control patch sharing. This is an important issue because scrounging may be expected to decrease with relatedness when scroungers control patch sharing. In a similar study, house sparrow scroungers joined fewer patches and modulated their aggression when they joined groups with relatives (Toth et al., 2009), again suggesting that relatedness matters when joining patches exploited by companions. Nevertheless, in both studies, familiarity, rather than relatedness, cannot be ruled out as an explanation for the scrounging patterns, because relatives were reared together. Although related individuals tend to be familiar with one another, familiar individuals need not be related, suggesting that familiarity may operate independently from relatedness. For instance, aggression may be toned down in groups composed of individuals already familiar with one another. More work on the effect of relatedness on scrounging is needed to settle these issues. Another interesting research question is whether groups of related individuals should allow non-related companions to join in order to obtain the benefits associated with an increase in group size (Aviles et al., 2004).

7.2.2.3. Unequal Competition

In systems with group-controlled entry, members of a group can expel potential joiners. In groups with free entry, competition within the group, rather than the threat of eviction, modulates the expected group size. In most groups, individuals typically vary in their ability to compete for resources. This may arise through differences in the speed at which individuals locate or exploit resources, or through differences in the ability to monopolize resources using aggression. The major consequence of these inherent differences among individuals is that not all individuals will achieve the same fitness in the same group. Competitors of varying quality may thus maximize their fitness by living in groups of different sizes (Pulliam and Caraco, 1984).

To benefit from foraging in a group, individuals that monopolize resources must balance the benefits of getting a larger share of resources with the benefits of allowing other companions to remain in the group. Indeed, individuals that obtain too little a share of the resources in a group may simply decide to forage alone, leading to a decrease in group size. More dominant group members should therefore be expected to provide enough resources to ensure that other companions remain in the group. Give and take in groups with unequal access to resources has mostly been studied in the context of reproduction (Vehrencamp, 1983), but the idea has recently been applied to foraging groups.

How unequal competitors in a foraging group should optimally partition resources has been investigated theoretically (Hamilton, 2000). The model includes recruiters, who locate resources, and joiners, who may join the recruiters to share resources. Recruiters control the division of resources, and thus have the means to repel or retain potential joiners. The optimal allocation of resources will thus be the result of a transaction between group members

(Johnstone, 2000). The amount of resources that an individual can obtain by foraging alone represents the key parameter in the model. When the benefits of grouping are low with respect to the benefits when foraging alone, join-ers will decide to stay only if they are offered a comparatively larger share of the resources. By contrast, recruiters can monopolize a greater share of the resources when individuals foraging alone expect much less success.

The balancing act between recruiters and joiners also has important implica-tions for the optimal group size. In a system with free entry and equal foragers, the expected group size shifts from the optimal value to the much larger equi-librium value. However, when recruiters can control the division of resources, the expected group size should remain closer to the optimal value. In essence, the recruiters have the final say on group size by altering the benefits of joining for potential joiners. Unequal sharing of resources then becomes another reason why expected group sizes may not stray away too far from the optimal value.

Recruiter-joiner models share some similarities with producer-scrounger models but differ in one key assumption. Models of producing and scrounging also consider two types of foragers, the producers of resources and those that scrounge their food discoveries (see Chapter 2). In producer-scrounger games, it is typically the scroungers that control the division of resources by forceful joining of already exploited patches. However, such models do not incorporate transactions between producers and scroungers, because neither can leave the group. In an open system, in which individuals can join or abandon existing groups, repeated losses to scroungers may entice producers to leave the group. Using the logic of transaction models, the threat of leaving the group may be sufficient to lower scrounging in foraging groups to levels more acceptable to producers, a prediction that awaits empirical scrutiny.

The role of unequal competition in transactional models of optimal group size assumes that more dominant group members can effectively control the share of resources in the group. Groups may become larger than expected if dominant individuals cannot achieve such control. In a recent study of coral reef fish, for example, subordinate individuals in larger territories were able to escape more often from the influence of dominant group members, leading to an increase in group size (Ang and Manica, 2010). However, these larger groups were less stable through time and often reverted to smaller sizes. This study nicely illustrates the notion that adequate control over resources is key to determining group size in societies with unequal competitors. It also highlights the point that group size increasing to beyond the optimal value can be a short-lived phenomenon.

7.2.2.4. Learning

So far, we have seen that the paradox that animal group size may exceed the optimal value can be solved by invoking relatedness among group members or unequal division of resources in the group. A further possibility has been inves-tigated recently considering learning about patch quality. Fernández-Juricic and I modelled group size as the adaptive outcome of individual foraging decisions (Beauchamp and Fernández-Juricic, 2005). By contrast to the recruiter-joiner model discussed above, we assumed that all individuals share resources in a patch equally. Foragers learn about food availability in their habitat as they obtain resources while searching for food. In the spirit of optimal foraging theory, we speculated that a forager should leave a patch when the current food intake rate falls below what it expects from the habitat (Stephens and Krebs, 1986). In a patch, food intake rate varies depending on the number of companions present, first increasing and then decreasing beyond an optimal value, reflecting the various costs and benefits associated with group size.

The dynamics of joining or leaving groups are therefore expected to reflect individual choices based on the experiences of each forager. For instance, a forager that estimates a high food intake rate from the habitat may be unwilling to join or stay in a large group with too much competition. Similarly, joining a small group may not be very advantageous for such individuals since food intake rate may be restricted by the need to be more vigilant against predators. Joining choices are ultimately based on the relative value of foraging alone versus foraging in the group in the habitat.

The results of our simulation models indicate that the distribution of group sizes at the level of the habitat peaks at the value predicted to maximize food intake rate in a patch, with some variation reflecting individual variability in the estimate of habitat quality (Fig. 7.3). Without the need to invoke relatedness or unequal competition, the results indicate that simple foraging processes at the individual level, inspired by optimal foraging theory, can lead to group sizes that are quite close to the optimal value. Whether the foraging rules that we implemented are optimal or evolutionarily stable in the face of challenges from alternative rules remains to be explored. Nevertheless, the results suggest that a greater understanding of the processes that lead foragers to join and leave groups may prove useful to solving the group-size paradox.

7.2.3. Empirical Evidence

To assess whether animals occur in groups of optimal sizes remains a challenging empirical issue. Documenting the costs and benefits of foraging in groups of different sizes represents the first challenge. Ideally, individuals should be allocated randomly to groups of different sizes and the groups themselves should be allocated randomly across the habitat. Both conditions are necessary to ensure that groups of different sizes are composed of phenotypically similar individuals that all experience similar conditions in terms of resource availability and predation risk. Experimental approaches that use such a random allocation procedure have been conducted with relatively immobile prey, such as insect larvae, which can be easily manipulated and placed on different plants as needed (Fordyce and Agrawal, 2001; Wise et al., 2006). In species such as birds

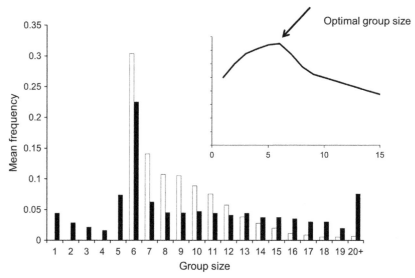

FIGURE 7.3   Learning and the optimal group size: the distribution of group sizes experienced by an individual peaks at the optimal value in a simulation model in which foragers learn about habitat quality and abandon food patches when their current food intake rate falls below the value expected for the habitat. Bars show the mean from 100 simulations. The scatter in the distribution of group sizes increases when food patches in the habitat vary in quality (white bars: no variation; black bars: variable patch quality). The inset illustrates how food intake rate in a food patch varies with the number of companions present, and peaks at the optimal group size of 6. *Adapted from Beauchamp and Fernández-Juricic (2005).*

and mammals, it is more difficult to randomly allocate individuals to close-knit groups because such groups can often be intolerant of intruders. The second challenge is finding a suitable currency to express the costs and benefits of foraging in groups of different sizes. Many currencies are simple proxies and may not be closely related to fitness. These issues are explored in the following case studies drawn from a broad range of taxa.

7.2.3.1. Lions

The controversy over optimal group size in large mammals like lions illustrates the pitfalls of using currencies that may not be ideally suited to estimate fitness. Using data collected by Schaller (1972) on hunting lions in the Serengeti, Caraco and Wolf (1975) determined the amount of meat obtained on a daily basis by individual lions foraging in groups of different sizes. Mean daily gross per capita food intake peaked in groups of two and was smaller in groups of three than in solitary foragers (Caraco and Wolf, 1975). Especially when hunting larger prey, individual lions could easily meet their daily food requirements, and tended to forage in groups larger than the optimal value (Fig. 7.4). Such large groups may provide lions with other types of benefits, such as the ability to deter scavenging hyaenas.

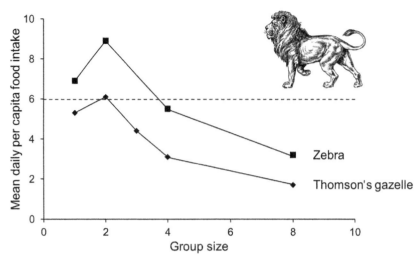

FIGURE 7.4   Optimal group size in lions: mean daily gross per capita food intake (in kg of meat) peaks in groups of two when lions hunt Thomson's gazelles, a small prey, or zebras, a larger prey. The food requirement for an individual lion was estimated at 6 kg per day (dashed line). Hunting larger prey in groups larger than the optimal size would still provide sufficient food to fulfil daily requirements. *Adapted from Caraco and Wolf (1975).*

However, there may well be other explanations for the seemingly greater than optimal group sizes in lion prides. Subsequent research pointed out that lions in prides tend to be closely related, and that maximizing daily per capita food intake, an individual measure, may not be the proper currency to evaluate fitness (Giraldeau and Gillis, 1988; Rodman, 1981). Lions in a pride may thus accept to increase group size to avoid paying the costs of letting a closely related companion forage alone, with low payoffs, rather than with the group. A further consideration is that using daily per capita food intake focuses on the mean value experienced by individuals and fails to consider variance in food intake, which typically decreases in larger groups (see Chapter 1). In this context, minimizing the probability of an energy shortfall may make more sense, and may be achieved in groups larger than the optimal size (Clark, 1987). Yet another issue is that daily per capita food intake represents a gross measure of intake that fails to take into account the energetic costs associated with foraging in groups of different sizes. Selection would presumably maximize the daily net, rather than gross per capita food intake (Packer and Caro, 1997). Foraging costs have tended to be ignored in earlier, influential studies of optimal group size (Caro, 1994; Creel and Creel, 1995; Fanshawe and FitzGibbon, 1993; Schmidt and Mech, 1997). Finally, the larger than expected group sizes in lions have been attributed to an increased ability to repel food stealers, and to prevent risk of infanticide by intruding males (Packer et al., 1990).

Teasing apart the contribution of all these factors remains challenging. The lion story, in which at least five different factors may come into play, illustrates

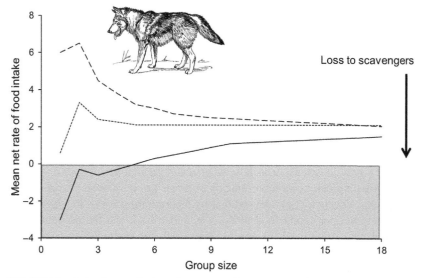

FIGURE 7.5   Optimal group size in wolves: mean daily net per capita food intake (in kg of meat) peaks in groups of two when wolves hunt moose. When food losses to ravens are substantial, however, a larger group size may provide more benefits. *Adapted from Vucetich et al. (2004).*

the challenges in explaining observed group sizes even in a relatively well-known species.

7.2.3.2. Wolves

A study focusing on wolves hunting moose on Isle Royale in Lake Superior, United States, addressed many of these issues (Vucetich et al., 2004). Although the authors did not manipulate group size, they concluded that differential access to resources in groups of different sizes was not a confounding factor. Daily net, rather than gross per capita food intake, was calculated by taking into account the energetic costs associated with daily travel within the territory and chasing prey. This is an important consideration because travel costs tend to increase in larger packs. Mean daily net food intake peaked in groups of two in this population (Fig. 7.5).

As many wolf groups are much larger than the optimal size, the authors considered factors that could be shifting group sizes to larger values. First of all, they considered the possibility that instead of maximizing mean per capita food intake, wolves may be minimizing the risk of starvation. Indeed, variance in food intake tends to decrease in larger groups, thereby perhaps allowing such groups to increase their chances of survival despite a lower mean food intake. However, it turned out that the probability of failing to acquire sufficient resources over a given period of time was lowest in groups of two, the group size that also maximized mean foraging efficiency.

Relatedness may also be an issue because wolf packs often include close relatives. Wolf packs control group size by the threat of eviction. Given the sharp drop in food intake in packs larger than the optimal size, the calculations revealed that individuals in small packs should evict close relatives only when they have acquired mature foraging skills. In this case, the costs of letting go of a close relative is more than compensated by the higher foraging success of the remaining individuals in the pack. Lack of eviction of non-mature individuals may thus push the size of the groups beyond the optimal value.

Losses to scavenging also play a role in determining the size of wolf groups, just as they did for lions as we saw earlier. Wolves can lose a substantial amount from their prey to ravens, especially when a carcass remains edible for several days. By eating carcasses more quickly, a larger group of wolves can reduce such losses. Vucetich et al. showed that by incorporating losses to scavengers, mean daily net per capita food intake was maximized, and the probability of starvation was minimized, in groups larger than the optimal size (Fig. 7.5). Such losses to scavengers are trivial for smaller prey, which can be consumed more rapidly or defended more effectively. Expected group sizes should thus cluster around the optimal value only when prey are relatively small. It would be interesting to estimate the losses to scavengers experienced by other large mammalian predators, like lions and wild dogs, to see if scavenging represents a common selection pressure favouring foraging in groups of greater than optimal size.

7.2.3.3. Northern Bobwhite

Flocks of birds typically fluctuate in size during the day, making it difficult to estimate daily food intake in groups of a fixed size. In addition, smaller species of birds, in contrast to the larger mammalian carnivores I considered above, also face the threat of predation. Survival in flocks of birds may thus depend not only on foraging efficiency but also on predation avoidance, making it even more difficult to estimate optimal group size. Birds living in relatively permanent groups, however, offer an opportunity to examine group size when the costs and benefits of group foraging vary along more than just one dimension.

Quails in North America form coveys during the non-breeding season that tend to be relatively stable in size and occupy a definite home range, making it possible to examine the fitness consequences of foraging in groups of different sizes. Williams et al. (2003) used a combination of aviary experiments and covey size manipulations in the field to determine whether northern bobwhites, a small quail species, forage in groups of optimal sizes. Aviary experiments, in which individuals were randomly allocated to coveys of different sizes, indicated that individuals in larger coveys spent more time in the foraging area and, when foraging, spent less time vigilant against predators. Birds in these larger coveys thus foraged more efficiently. Predator detection also occurred more rapidly in the larger coveys due to the presence of more eyes (Williams et al., 2003). These results show that the prospects of survival should increase with covey size, at least when food availability is not limited and predators do not target larger coveys preferentially.

In the field, the researchers manipulated covey size by removing individuals from some coveys in order to obtain a greater range of covey sizes and to break down possible associations between covey size and habitat variables, such as food density or vegetation height, which are known to influence survival in related species (Watson et al., 2007). They followed marked individuals in manipulated and unmanipulated coveys over the winter to determine survival rate. The results show that small and large coveys moved more than coveys of intermediate sizes, and that individuals lost mass over time to a greater extent as covey size increased. Overall, daily survival first increased and then decreased with covey size, peaking at intermediate values of about 10 to 15. Surveys in the area revealed that the usual size of coveys was about 11, suggesting that this value is optimal (Fig. 7.6).

Individuals in small coveys may feed less efficiently, as judged from the aviary experiment. An increase in time spent travelling also represents a threat to survival in this species, perhaps explaining why daily survival was rather low in the small coveys. Individuals in larger coveys may be better protected against predators, but lose more weight and also travel more, perhaps in response to greater food competition. Therefore, survival is maximized at intermediate covey sizes. The role of relatedness could not be assessed in this study but

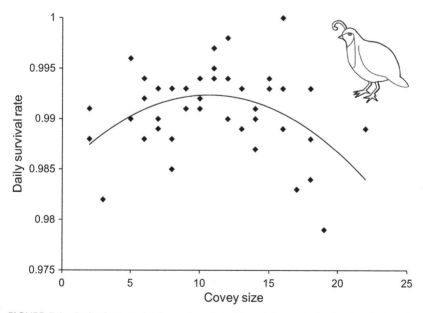

FIGURE 7.6   Optimal group size in northern bobwhites: daily survival rate of marked northern bobwhites in a controlled field experiment peaks at intermediate covey sizes. *Adapted from Williams et al. (2003).*

groups control entry in this system, suggesting that groups larger than the optimal size may occur occasionally.

7.2.3.4. Flatworms

Although studies of large mammalian predators, and to some extent birds, have garnered the most attention, studies involving invertebrate species can be particularly helpful for studying issues related to optimal group size. In relatively sedentary species, it is possible to randomly allocate individuals to different groups and to provide similar environments to all groups. This means observed effects on fitness can be directly related to group size. The relatively short life-span of such species also makes it possible to document changes in a number of relevant life history variables, such as growth, survival, and even reproductive success. Ultimately, the fitness consequences of group size variation should be related to reproductive success, but very few studies with vertebrate species can achieve this goal.

In a particularly impressive study, researchers used a freshwater flatworm species that naturally feeds in groups of varying sizes on prey captured by a mucous trap built collectively by group members (Cash et al., 1993). Facing few predation threats of their own, the consequences of group size choices for flatworms can be related to growth in the short term, which is a function of daily food intake, and to reproduction in the long term. In laboratory studies that simulate the freshwater pond habitat of these predators, individuals were randomly allocated to groups of different sizes, and food availability and other environmental variables were effectively homogenized among groups.

Mean daily gross per capita food intake first increased and then decreased with group size, peaking at intermediate values around two to four, depending on prey type and density (Fig. 7.7(a)). When maintained in groups of one or four for several weeks, survival rate did not vary over the experimental period,

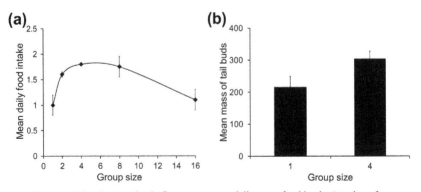

FIGURE 7.7 Optimal group size in flatworms: mean daily gross food intake (number of prey per day) peaked at intermediate group sizes (left panel), and reproductive performance, as expressed by the mean mass of bud tails (which will become offspring), was significantly higher in groups than in solitary foragers (right panel). Error bars show one standard error. *Adapted from Cash et al. (1993).*

although reproductive performance was significantly higher for individuals in larger groups (Fig. 7.7(b)). The ability to produce more mucous in larger groups provides an advantage for flatworms, but the results suggest that when groups become too large, other factors contribute to a decline in foraging efficiency. This is one of the few studies in which gains from group foraging are expressed in fitness components such as reproduction.

7.2.4. Measuring Group Size

To determine whether animals occur in groups that maximize mean foraging efficiency, or minimize the probability of starvation, implies a comparison between observed and expected group sizes. But what represents the best empirical measure of group size is controversial. Indeed, mean group size, the most common metric for measuring group size, may not represent the most frequently occurring group size or even the size experienced by most individuals in the population. Mean group size represents a group-centric view of sociality, and it is only useful for characterizing the most common group size when the distribution of group sizes is symmetrical. However, many distributions of group sizes are heavily skewed to the right, with many small groups and few very large groups, a pattern that tends to move the mean away from the modal or median group size (Jovani and Mavor, 2011; Reiczigel et al., 2008). Perhaps more problematic is the fact that mean group size may tell us very little about the size of a group experienced by an individual in a population, which, in the end, is the most relevant metric for assessing whether or not individuals occur in groups of optimal sizes.

Consider the following simple situation where ten individuals are split into one group of two and one group of eight. Mean size for the two groups is five and yet the majority of individuals in the population live in a group of eight. As pointed out early on in the group size literature, an individual typically lives in a group far larger than the mean group size (Jarman, 1974; Lloyd, 1967). A new measure, typical group size, following Jarman's terminology (or mean crowding, following Lloyd's terminology), attempts to capture group size from the point of view of an individual in the population. In the above example, the typical group size is 6.8, a value larger than the mean and closer to the size experienced by most individuals in the population. In the Northern bobwhite example presented earlier, the mean covey size is 11 while the covey typical group size is closer to 13. The mismatch between the two values increases in more skewed distributions.

For species that I examined above, the conclusion that group size exceeds the optimal value was based on empirical measurements at the group level. As the individual group size would be even more extreme, we have an even more compelling argument that animals often forage in groups that are larger than the optimal size. Statistical techniques to calculate group size and confidence intervals at the individual level, and to compare sample values are detailed elsewhere

(Neuhäuser et al., 2010; Reiczigel et al., 2008). As many processes underlying evolution occur at the level of the individual rather than the group, which is certainly the case for the choice of group size, individual-centric measurements of group size may be better suited to uncovering relationships with ecological variables.

7.3. GROUP COMPOSITION

Optimal group size theory typically focuses on groups with interchangeable individuals, individuals who are equal in their ability to obtain and exploit resources. Introducing individual variation may have important consequences for optimal group size. For instance, the optimal group size in a population may be shifted to larger values if dominant group members can recruit subordinates into their groups. Therefore, the notion of optimal group size may be intimately linked to group composition.

7.3.1. Predictions from Models

The consequences of individual differences on group size and composition have attracted much attention in a dispersion economy. In a dispersion economy, adding one forager to a group decreases individual fitness, and this is true for all group sizes. Faced with limited foraging opportunities, individuals must distribute themselves among food patches to reduce their losses, leading to the so-called ideal free distribution (Fretwell and Lucas, 1970). Typically, the relative number of foragers in a patch will match the quality of that patch relative to all available patches. However, a model that introduced individual differences in competitive ability showed that various combinations of unequal competitors across patches can be stable, ranging from patches with only one type of competitor to patches with a mixture of competitor types (Parker and Sutherland, 1986).

There have been fewer attempts to predict optimal group composition in an aggregation economy. Ranta et al. (1994) developed a model where individuals are free to enter and leave any group, which corresponds to a free-entry rather than a group-controlled entry system. Larger individuals are assumed to find resources more quickly and also obtain a greater share of resources from each patch discovered (Ranta et al., 1994). Results from the model showed that smaller individuals fare poorly in groups with larger companions. The larger number of food patches they can exploit over time, due to the presence of better food finders, fails to compensate for the smaller share of resources they obtain from each patch. Larger individuals, on the other hand, fare better with large companions because the overall increase in the rate of food patch finding compensates for the greater competition. The model thus predicts size segregation among groups, such that there may well be more than one optimal group size in a population with competitively unequal individuals.

The possibility that phenotypically matched groups compete with one another in the same habitat was not considered in Ranta's model. The higher rate of food patch finding by the larger competitors may, for instance, translate into more visits to empty patches by the smaller individuals in the other groups, making it therefore more advantageous for them to stay in groups with larger companions. In the spirit of the recruiter-joiner model I examined earlier, allowing transactions between unequal competitors within the same group may also lead to groups with such a mixture of individuals. It would be interesting to see whether flexible patch sharing would yield more benefits to the larger competitors. This model, caveats aside, points the way to further theoretical developments on optimal group composition in terms of individual competitive ability.

7.3.2. Empirical Evidence

Individuals can obtain clear benefits from joining others, but these benefits can also be enhanced by associating with individuals of a particular phenotype. In fact, choosing a particular group composition may even trump group size considerations. In the laboratory, banded killifish, a common freshwater species, prefer to join the larger of two groups when threatened by a predator, which makes adaptive sense in terms of reducing predation risk (Krause and Godin, 1994). However, when given the choice of joining similar-sized or larger companions, individuals prefer to join groups with similar-sized companions irrespective of group size (Fig. 7.8), which was interpreted in the light of the oddity effect, given that individuals did not actually compete for food in the experiment.

The expectation that group composition may have important fitness consequences has led to numerous studies of phenotypic assortment in groups. However, it is important to bear in mind that several factors other than competition may lead to phenotypically assorted groups. Earlier, in Chapter 3, I examined how minimization of phenotypic oddity can reduce individual predation risk. Differences among individuals in activity level (Conradt, 1998), locomotion speed (Gueron et al., 1996; Watkins et al., 1992), or habitat preferences (Bremset and Berg, 1999), which are more passive mechanisms, can also lead to phenotypic segregation among groups.

Much of the research on such questions uses small fish species that can be brought to the laboratory. In cafeteria-style experiments, individuals are able to choose between groups of various sizes and composition set by the researcher (Krause et al., 2000b). Fortunately, many of the key findings from the laboratory have been validated in the field (Krause et al., 2000a). A study with Trinidadian guppies, for example, found that a combination of active and passive mechanisms appears to produce shoals of similar-sized guppies (Croft et al., 2003). Shoals readily split while foraging, providing individuals with a chance to relocate with similar-sized companions when nearby shoals come into view. Nevertheless, individuals of different sizes occupy different areas of the

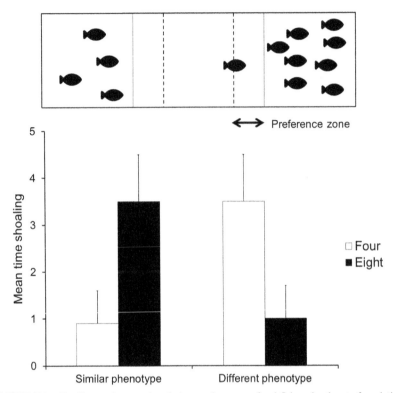

FIGURE 7.8 Shoaling preferences in relation to phenotype: focal fish under threat of predation preferentially chose to be near the larger of two shoals when companions were the same size (similar phenotype) but preferred the smaller shoal when the larger shoal contained larger companions (different phenotype). The inset illustrates the experimental setup. Time focal subjects spent in close proximity (preference zone) to either stimulus shoal was noted. Error bars show one standard deviation. *Adapted from Krause and Godin (1994).*

habitat presumably matched to their foraging skills and susceptibility to predation. Active sorting thus probably takes place within the confines of a particular habitat that passively selects for a limited range of phenotypes. Although this study did not examine whether guppies in size-assorted groups benefit from a reduction in predation risk or an increase in foraging efficiency, it is clear that active sorting provides a mechanism by which individuals select groups that increase their fitness. A further study with this species has revealed that size segregation occurs in populations exposed to high predation risk, just as the oddity mechanism would predict (Croft et al., 2009). However, size segregation was also documented in some populations experiencing low predation, suggesting there may be other factors like competition or differential locomotion speed involved.

By contrast, size segregation in prawns has a somewhat different explanation. Individual prawns prefer to join larger groups rather than single companions, but laboratory experiments failed to document a preference for associating with similar-sized companions (Evans et al., 2007). Therefore the size segregation found in the field probably reflects the action of a passive mechanism like differential habitat selection.

Recent studies point to another mechanism behind phenotypic assortment: parasite prevalence and load, with larger groups composed of less parasitized individuals (Hoare et al., 2000; Jovani and Blanco, 2000). Such assortment in response to parasitism may arise because more parasitized individuals persist less often in groups or are avoided by others. Parasitized individuals may show signs of infection (conspicuous black spots in fish, for instance), which may be used as a cue for active avoidance (Krause and Godin, 1996b). If infection causes noticeable signs, individuals can avoid parasitized companions and thereby reduce the likelihood of infection or oddity (Barber et al., 2000).

Individuals may also obtain fitness benefits from preferential association with familiar or related companions. Several species of fish have been shown to discriminate individuals on the basis of kinship and to associate preferentially with kin (Brown and Brown, 1996). Associating with kin has been shown to reduce aggression levels in groups while also providing opportunities to gain indirect fitness benefits. It is also clear that familiarity alone, independent of relatedness, may drive assortative behaviour. For instance, association with familiar individuals can increase shoal cohesion and antipredator responses in fish (Chivers et al., 1995) and reduce levels of aggression in goats (Stanley and Dunbar, 2013). In addition, associating with familiar individuals can increase survival (Seppä et al., 2001) and food intake rate (Höjesjö et al., 1998). Nevertheless, several researchers have warned that association patterns between individuals may not reflect active preferences, but rather more passive mechanisms such as similar habitat choice (Carter et al., 2013; Wey et al., 2008). Therefore, it is important to ascertain how much variation in observed associations can be explained by active preferences for kin or familiar individuals.

The ability of fish to discriminate among potential group mates not only relies on features like size and familiarity, but also on personality. In a recent study of mosquitofish, individuals preferred to shoal with the larger of two groups and, more importantly, with groups composed of more sociable individuals as reflected by the proximity of individuals to each other (Fig. 7.9) (Côté et al., 2012). Sociability was an individual trait that measured the tendency of an individual to be attracted to companions. Joining a group with more sociable individuals may provide benefits to the joiner since such groups are more cohesive and show less aggression. A tendency for social individuals to join groups may also promote segregation of shoals with respect to sociability. It would be interesting to see whether sociability represents a greater attractive force than kinship or familiarity, both of which have been shown to influence shoaling preferences in fish.

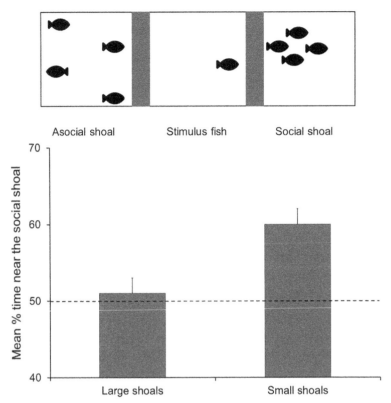

FIGURE 7.9   Shoaling preferences in relation to sociability: focal mosquitofish spent more time close to the shoal with more sociable companions when their shoal was small, but not when their shoal was large. The inset illustrates the experimental setup and shows the preference zone (grey area) and the two shoals differing in sociability. Error bars show one standard error. *Adapted from Côté et al. (2012).*

7.4. PROXIMATE MECHANISMS

Research on group size at the ultimate level has made remarkable strides. Yet little is known about the mechanisms underlying variation both within and among species in group size. Such work may allow us to get a different perspective on sociality and to understand potential constraints on the evolution of group size and composition.

The tendency to form groups involves motivational factors including social arousal, approach/avoidance, anxiety, and dominance, whose combination probably dictates whether animals seek others rather than simply fight or flee upon contact. It has long been known that gonadal hormones like testosterone can modulate aggression in a species and thus influence the expression of sociality (Emlen, 1952). More recent research has focused on neurotransmitters that modulate behavioural patterns related to sociality. This research links variation

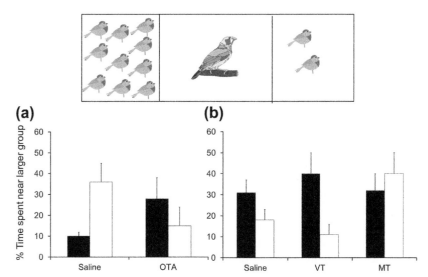

FIGURE 7.10   Hormones and flocking preferences: focal zebra finches (females: white bars; males: black bars) spent less time on average with the larger of two companion flocks, which is indicative of less sociality, when injected with an antagonist of oxytocin (OTA), but only in females (a). In a separate experiment, infusion of vasotocin (VT) but not mesotocin (MT) produced in females a decrease in time spent with the larger of two flocks (b). The inset shows the experimental apparatus in which focal subjects in the middle of a cage faced a choice between spending time near a large or a small group of conspecifics. Error bars show one standard error. *Adapted from Goodson et al. (2009)*. (For colour version of this figure, the reader is referred to the online version of this book.)

in grouping behaviour to the anatomy and function of hormone-secretion systems in the brain.

Estrildid finches, which include species like the zebra finch and the nutmeg mannikin, have become a key group to investigate these issues (Goodson and Kingsbury, 2011). These species have also been instrumental in research on scrounging behaviour, as we saw earlier in Chapter 2. This family of ground-foraging birds, originating from Southeast Asia, exhibits a large variation in the tendency to form groups, from territorial species to others, like the zebra finch, which forage in very large groups much of the year. Experiments with these species of birds have uncovered a close relationship between hormone-secretion systems and grouping behaviour. By contrast to territorial species, gregarious species have more binding sites in the brain for hormones such as oxytocin and vasotocin. Exposure to a companion increases markers of neural activity in these regions for gregarious species but not for territorial species (Goodson and Wang, 2006). These findings suggest that some brain circuitry is involved in gregariousness. As expected if hormones play an important role in grouping behaviour, injection of an antagonist of oxytocin in the brain of birds reduced their tendency to associate with companions while injection of meso-tocin restored the natural flocking tendency (Fig. 7.10) (Goodson et al., 2009).

It is interesting to note that many of these hormones have evolved millions of years ago and occur in one form or another in almost all vertebrates, suggesting a common foundation to modulate sociality in this large group of species.

Variation in grouping behaviour also occurs in some species from season to season, which provides us with another opportunity to look at the proximate correlates of sociality, but this time within the same species. In line with the above results, annual variation in brain circuitry is apparent in avian species showing marked seasonal changes in sociality but not in species remaining territorial year-round (Goodson et al., 2012).

The transition from solitary to gregarious living in the desert locust surely represents the most spectacular case of plasticity in sociality in the animal world. Solitary locusts avoid one another but can transform into a form that lives in very large and coherent swarms, much to the alarm of farmers in equatorial Africa. When forced together and with appropriate sensory stimulation, the solitary locust will undergo a transition to the more gregarious form. Without serotonin, a neurotransmitter, the sensory stimuli alone are not sufficient to induce gregariousness. However, injection of serotonin without external stimulation can lead to more sociality. Serotonin released in the nerve circuitry, therefore, is both necessary and sufficient to explain the transition between these two forms (Anstey et al., 2009). Whether serotonin acts alone or allows other mechanisms to come into play to further maintain or develop gregariousness is not known.

Taken together, the results with birds and locusts certainly show how various hormone-secretion mechanisms are involved in the expression of sociality. It remains to be shown whether similar mechanisms act in other species. Although it is clear that such mechanisms can explain transitions from solitary or territorial forms to very gregarious forms, an all-or-none occurrence, it is not known whether they can also account for more quantitative variation in group size, say, from small to large.

7.5. CONCLUDING REMARKS

Although the tendency to form groups provides opportunities for increasing foraging efficiency and decreasing predation risk, it also entails costs such as increased competition (Alexander, 1974). The balance between the costs and benefits of foraging in groups should thus lead to the tendency to forage in groups of the particular size that maximizes individual fitness. However, the expected group size will vary depending on whether entry in the group is free or constrained. In a free-entry system, groups are expected to be larger than the optimal size, while groups that control entry may be closer to the optimal value. The seeming paradox that groups often tend to be larger than the optimal size in a free-entry system can be solved when considering factors such as relatedness, within-group competition, and learning about habitat quality. In general, testing the notion of optimal group size remains challenging. Controlling for phenotypic differences among individuals in groups of different sizes and ensuring that all groups experience similar constraints in terms of food availability and predation pressure, represent two important challenges that few empirical studies have successfully overcome.

Group composition has emerged as an important consideration when assessing the fitness consequences of group living. Segregation of groups along phenotypic traits, such as size, has been documented in many species, and this finding certainly suggests that individuals pay attention to not only the size but also the composition of groups. Perhaps the greatest impediment to our understanding of the factors that control joining preferences is the almost complete lack of data from animals other than fish. A few exceptions to this bias are one study on a bird species (Senar and Camerino, 1998) and another on a marine invertebrate (Evans et al., 2007).

Non-random assortment of individuals in groups, arising for instance from preferences to join particular companions, has important consequences for the notion of optimal group size. Optimal group size is based on the idea that group members are interchangeable. However, if individuals obtain benefits from associating with particular companions, such considerations may trump size considerations and explain deviation from the optimal group size. In this context, the ability to form and maintain partnerships with particular companions may set a limit on the size of a group, which may be quite different from the size predicted from simple considerations about food requirements and predation risk (Lehmann and Dunbar, 2009). The use of social network analysis, which can help identify group cohesiveness and particular bonds among individuals, may be extremely useful to examine issues about optimal group size and composition (Blumstein, 2013; Stanley and Dunbar, 2013).

I also note that much of the literature on phenotypic assortment focuses on free-entry systems and the joining preferences of single individuals. In group-controlled entry systems, preferences may be expressed by the forceful ejection of group members and a refusal to allow entry. It would be useful to examine why a group allows entry to certain joiners, rather than simply focusing on which group an individual prefers to join. Relatedness of the potential joiner will definitely be an important factor, but other considerations may also prevail. For instance, a group that is already large may be unwilling to accept a potential joiner that will increase competition and yet only provide a slight contribution to predation risk dilution. However, the group may accept individuals that are expected to contribute disproportionately to vigilance because, say, of their large energy reserves (Jaatinen and Öst, 2013). Observations on the outcome of encounters between groups in the field may provide a way to investigate such choices.

Phenotypic assortment among groups has been linked to parasite prevalence and load. The possibility that parasites can manipulate their hosts to increase grouping tendency is intriguing. Parasite-induced sociability would increase the transmission of parasites within the group. If large groups are more subject to predation, increased sociability may also favour transmission to host predators

(Rode et al., 2013). This may be especially relevant in species where parasitized individuals show little signs of infection. If grouping provides effective protection against predators—through the confusion effect, for instance—increased sociability may be a poor choice to increase transmission to a host predator. This leads to the intriguing possibility that manipulation of grouping tendency by parasites may be a function of predator behaviour.

Considerations about group size typically focus on optimal or equilibrium values. However, other approaches that I did not consider here can enable us to determine the overall distribution of group sizes. These approaches view group size as the stochastic realization of individual choices based on ecological factors (Caraco, 1980; Cohen, 1971). In such models, for example, the probability that an individual leaves a group can be made dependent on group size, as would be expected if aggression or competition increases in larger groups. The probability of joining a group can also be made dependent on group size, mimicking local enhancement, or independent of group size if individuals are attracted to a site regardless of how many foragers are already present. Combining different individual arrival and departure rules leads to quantitatively different distributions of group sizes, which can be fitted with different known statistical distributions, such as the negative binomial or the power law. Fitting probability laws to observed distribution of group sizes can be helpful in our attempts to understand underlying processes at the individual level (Beauchamp, 2011b; Bonabeau et al., 1999; Jovani et al., 2008). However, to be more than just an exercise in curve fitting, individual choices must be clearly linked to natural ecological processes.

Mixed-Species Groups

8.1. INTRODUCTION

Walking in the woods in a Canadian winter is not for the faint of heart. Ploughing through the snow is hard enough without the chilling wind freezing hair in your nostrils. The woods are eerily quiet as most birds have long gone to warmer climes. Year-round resident birds, like the black-capped chickadee and woodpeckers, are few and far between. Nevertheless, the patient trekker is eventually rewarded with the sight and sounds of a mixed-species flock passing through. First to be seen and heard, a noisy flock of black-capped chickadees may be followed by a pair of red-breasted nuthatches, brown creepers, and downy woodpeckers. Given the vastness of the woods, it is no coincidence these species are all found so close to one another. The mixed-species flock soon flies on to other areas continuing their search for food, and again one is left with the quiet desolation of a winter wood.

Such mixed-species groups occur in many animal communities across the world and involve a wide array of species including arachnids, fish, mammals, and birds (Lukoschek and McCormick, 2002; Sridhar et al., 2009; Stensland et al., 2003; Terborgh, 1990). Mixed-species groups have been the focus of attention since the early days of animal behaviour research. Indeed, more than 100 years ago, naturalists speculated on the reasons why species may join other species while foraging

Social Predation. http://dx.doi.org/10.1016/B978-0-12-407228-2.00008-1

(Bates, 1863; Belt, 1874). In birds, where the aggregation of many species takes its fullest expression, mixed-species flocks are found in all habitats and may involve permanent associations of dozens of species and hundreds of individuals (Greenberg, 2000). Near-permanent associations between species have also been reported in monkey species (Heymann and Buchanan-Smith, 2000). What benefits can be derived for joining and exploiting resources with other species? Crucially, are these alleged benefits any different than those that individuals would get by simply forming a group with conspecifics?

This question has been approached from two different angles. First, from a non-adaptive perspective, mixed-species groups may simply represent an aggregation of species that serves no specific purposes for any of the participating species. It may also be the case that participating in mixed-species groups allows a species to join a group when conspecifics are not available. Second, from an adaptive point of view, participating in a mixed-species group may provide benefits beyond those that can be achieved by joining a similar number of conspecifics. While there may also be unique costs associated with mixed-species groups, those benefits outweigh the costs and favour participation in a mixed-species group. Much research has therefore been allocated to document the costs and benefits of participating in mixed-species groups, especially in birds and mammals.

To reap any benefits from participating in a mixed-species group, the different species must gather together and maintain cohesion over an extended period of time while searching for resources. Another way to examine the adaptive value of mixed-species grouping is to determine whether some features have evolved to favour the formation and cohesion of mixed-species groups, such as acoustic or visual signals. If mixed-species groups have evolved to become functional units with clear advantages to participating species, the formation of mixed-species groups may represent an example of niche construction (Odling-Smee et al., 2003), with the intriguing possibility that participants in such groups have evolved traits in response to novel selection pressures that arose from joining other species (Harrison and Whitehouse, 2011; Laland and Boogert, 2008).

In this chapter, I will first define mixed-species groups and then examine the adaptive value of mixed-species grouping. I will review the various costs and benefits of participating in mixed-species groups and examine whether specific traits may have evolved to favour the formation and cohesion of such groups.

8.2. WHAT IS A MIXED-SPECIES GROUP?

Deciding what constitutes a mixed-species group is no easy task, but all definitions recognize the participation of at least two species. Most authors distinguish between mixed-species groups and mixed-species aggregations. In an aggregation, individuals of different species occur at the same location at the same time as a result of responding independently to the presence of clumped resources. Examples are the occurrence of many species of birds at a fruiting tree (Willis, 1966) or many species of birds and mammals at carcasses (Selva et al., 2005). However, the possibility that some of these species may use one another to locate food, and thus derive benefits from joining other species, certainly warrants more attention and may broaden the definition of mixed-species groups.

The best examples of mixed-species groups are cases where many species travel together in search of food. The spatial clustering of many species over an extended period of time makes it less likely that such groups are accidental assemblages. In addition, species may join and abandon such groups at will, making a stronger case for a functional unit. As in single-species groups, the same difficulties arise in defining who should count or not as members of a group. This is not an issue when individuals of different species are close to one another, but when individuals are spread out in the habitat an inclusion criterion must be applied such as the maximum distance between any two group members that can still allow interactions. In view of the different sensory capabilities of different species, the radius in space that limits interactions between group members may not be the same, thereby complicating matters.

Another issue is whether groups composed of two species should count as instances of mixed-species grouping. The association between Thomson's gazelle and Grant's gazelle is typically considered an example of a mixed-species group in which each species receives some benefits (FitzGibbon, 1990). However, should the pairing of a cattle egret and a cow be considered a mixed-species group (Fig. 8.1)? Close associations of this type between two species are very common in nature (Sazima et al., 2012) and typically represent cases of commensalism where one species derives benefits without affecting the other. Reviews of mixed-species groups often do not include instances involving two species to exclude cases of commensalism and obviously cases of parasitism (Goodale and Beauchamp, 2010; Sridhar et al., 2009) but see Stensland et al. (2003).

Perhaps a criterion to decide when to call an association between species a mixed-species group is whether individuals of each species appear to function as a group, which means implicitly that each species has a reason to join and stay in the group. While a cattle egret clearly seeks a cow to obtain the insects that it flushes, the cow probably does not count the egret as part of its own group and probably does not derive any benefits from the association. In fact the egret may follow anything that flushes food, such as a tractor, in which case one would clearly not call such an association a mixed-species group. Mutual benefits thus appear to be a minimum criterion to define a mixed-species group.

FIGURE 8.1   Mixed-species groups in nature: examples of mixed-species groups composed of two species: (a) A cattle egret and the cow that provides flushed insects (*photo credit: Sarah and Iain*) and (b) Thomson's and Grant's gazelles (the larger of the two gazelle species) (*photo credit: NH53*). Which one represents a mixed-species group? *This figure is reproduced in colour in the colour section.*

8.3. THE FORMATION OF MIXED-SPECIES GROUPS

8.3.1. Non-Adaptive Hypotheses

The first question that arises from the formation of mixed-species groups is whether such groups really are functional and coordinated units rather than a random assemblage of individuals from various species that just happen to be at the same place at the same time.

Just as individuals of a single species often encounter one another during their daily routine, it is conceivable that mixed-species groups may simply result from the chance encounter of individuals of different species (Waser, 1984). For instance, different groups of different species may all be attracted by the same resources at the same time, and each species would not derive any benefits from such aggregations. This can be considered a null hypothesis for the formation of mixed-species groups. To reject the null hypothesis, one can use data on group composition and habitat use to show that the formation of mixed-species groups is not random. Finding that species derive an advantage from joining other species can also be used as argument against the null hypothesis. The difficulty with any random model is to derive the proper movement rules for species in their habitats. The Waser model assumes random movement but generally species have a limited number of routes they can travel each day and also have specific activities to fulfil at particular times of day, such as resting in specific shelters or searching for food in specific locations. These restrictions should be taken into consideration to make null models more biologically accurate.

The random mixing hypothesis rarely has been addressed directly in the literature (Holenweg et al., 1996; Waser, 1984; Whitesides, 1989). An extensive knowledge of habitat use and group composition must be available for each species to test the hypothesis in a statistical sense. The probability of occurrence of each species in one habitat at a point in time must therefore be known to determine whether the joint occurrence of two or more species exceeds the joint independent probabilities. It may seem futile to test this hypothesis in a vast ocean to determine that two dolphin species are unlikely statistically to be together at the same place at the same time. Nevertheless, other habitats may not be so vast and such a test may prove worthwhile.

A recent study of mixed-species groups in two gazelle species illustrates some of the issues (Li et al., 2010). The western Chinese province of Qinghai is often referred to as the third pole of the world given the dryness and coldness that characterize the high Tibetan plateau. Two endemic gazelle species occur near the Qinghai Lake, the Przewalski's gazelle and the Tibetan gazelle. These two gazelle species are closely related to one another and readily form mixed-species groups in the summer and winter. Li et al. (2010) made regular censuses in the study area to measure group size in single- and mixed-species groups of the two species. Each species occurred alone in single-sex or mixed-sex groups, and mixed-species groups can have just about every combination of species and sex possible.

The total number of groups and the distribution of group size for each species were remarkably similar, suggesting that population size in the study area was about the same for the two species. Single-species groups typically ranged from 1 to 30 individuals (Fig. 8.2) while mixed-species groups contained up to about 60 individuals.

The breakdown of single-species groups in terms of sex was also similar. Given these findings, the random hypothesis of indiscriminate mingling would suggest that each species should be equally represented in mixed-species groups irrespective of group size and sex composition. However, the proportion of Przewalski's gazelles in mixed-species groups increased with group size and was higher in all-female groups than in all-male groups. The results thus suggest that mixing of the two species was not random. The authors speculated that there may be some foraging competition between females of the two species.

Another approach to determine whether mixed-species groups form functional units is to document their stability over time. It would be very unlikely that the same group composition would occur repeatedly over time if species simply aggregated randomly. Data gathered at the same location in French Guiana over a 20-year period made it possible to test this hypothesis in mixed-species avian flocks. The results indicate that species composition in these flocks remained remarkably similar over the years and well beyond what would be predicted from null models (Martínez and Gomez, 2013). Flock species composition can

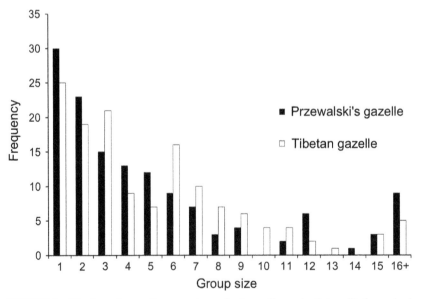

FIGURE 8.2   Mixed-species gazelle groups: two endemic gazelle species that readily form mixed-species groups in the Tibetan plateau have very similar intraspecific distributions of group sizes. *Adapted from Li et al. (2010).*

outlive the individuals that form mixed-species flocks, strongly suggesting that species composition is important to individual species.

As an additional non-adaptive hypothesis, it is possible to consider a best-of-a-bad-job argument to explain the formation of some mixed-species groups. Indeed, species may join a mixed-species group only when the formation of single-species groups is prevented. For instance, it was suggested that common dolphins join groups of other species only when their numbers are too low to allow the formation of single-species groups (Frantzis and Hertzing, 2002). In this case, there is no evolved tendency to participate in a mixed-species group. In fact, it may be more disadvantageous to forage with other species than with the same number of conspecifics, but joining a mixed-species group may be more advantageous than remaining alone. Evaluating this hypothesis requires detailed knowledge of the costs and benefits of foraging in single- and mixed-species groups of different sizes.

8.3.2. Adaptive Hypotheses

Showing that mixed-species groups are not formed randomly does not tell us the reasons why such groups may form in the first place. The key question when examining the adaptive value of mixed-species grouping is whether joining such groups provides unique advantages over joining a similar number of conspecifics.

Before considering the adaptive value of mixed-species grouping, it is important to realize that many factors may limit opportunities to form mixed-species groups. Typically, one would expect different species to occur together if they show the same habitat preferences and have similar feeding requirements (Heymann, 1997; Stensland et al., 2003). Niche differences between species should not be too slight, as it would increase interspecific competition, but not too large either as it would limit potential benefits of foraging together. In support of this, Heymann (1997) found that body size differences between associated species of tamarins ranged between 8% and 17% but only between 1% and 4% in non-associated species. The body size divergence in the two gazelle species mentioned above was about 20%, which is in line with the findings for the tamarins. If different species can join one another, the question then becomes what unique advantages and disadvantages may ensue from the formation of mixed-species groups.

8.3.2.1. Unique Benefits

Various adaptive hypotheses have been developed to explain the evolution of mixed-species grouping (Table 8.1). The first hypothesis is concerned with unique properties of information transfer within a mixed-species group. Other types of benefits from joining a mixed-species group are usually divided into three mutually non-exclusive categories: reproduction, foraging, and anti-predation.

TABLE 8.1 Summary of Benefits and Costs of Participating in Mixed-Species Groups

Types of Benefits and Costs	Description
Benefits	
Reproduction	
Breeding competition	Decrease the intensity of competition during the breeding season
Mating opportunities	Allows individuals to mate with other species (known in fish only)
Foraging	
Niche breadth	Increases the range of habitats and resources available
Encounter with resources	Increases the rate of encounter with resources through greater detection abilities, by exploiting the knowledge of other species about resource distribution, and by exploiting resources made available by other species
Resource exploitation	Avoiding costly visits to areas already exploited by other species
Predation	
Detection of predators	Better detection through differential sensory abilities
Group-size effects	In larger groups, dilution, detection, and confusion effects operate more forcefully to reduce predation risk *per capita*
Risk of attack	Attacks may be deflected to species that are more attractive to predators, more vulnerable, or simply odd-looking
Defence against predators	Joint defences among species can deter predators
Identification of predator	Naïve individuals can learn to identify predation threats from watching alarm reaction from other species
Costs	
Reproduction	
Mating	Increase the risk of breeding with another species
Foraging	
Competition	Food stealing or displacement from feeding sites by stronger species
Foraging speed	Sub-optimal foraging speed dictated by the needs of other species

TABLE 8.1 Summary of Benefits and Costs of Participating in
Mixed-Species Groups—cont'd

Types of Benefits and Costs	Description
Predation	
Vigilance	Increased monitoring of threatening species in groups
Risk of attack	Attacks may be deflected to a more vulnerable species
Encounter with predators	Mixed-species groups may be more conspicuous by virtue of their size and the sounds they produce

8.3.2.1.1. Information

One adaptive hypothesis for the formation of mixed-species groups argues that information provided by individuals of another species may be more valuable than information obtained from conspecifics (Seppänen et al., 2007). Information in this context may be related to foraging resources or predation threats. This information hypothesis specifies how species participating in a mixed-species group may obtain special benefits but does not address specifically the nature of these benefits.

The thrust of the hypothesis is based on the distance that can be maintained between conspecifics and between members of different species. The authors argue that in general, members of different species may occur closer to one another than different individuals of the same species. This is because competition for resources may be greater with conspecifics than with members of other species. Consequently, the distance that can be maintained between individuals of different species may be smaller than among conspecifics, allowing heterospecifics to provide more valuable information. The basis for this argument is the assumption that the value of information typically decreases with the distance between the sender and receiver of a signal. Generally, as the distance between sender and receiver increases, the message is less likely to be received and one would think that the message may also be less relevant as the distance increases. Evidence for this phenomenon with respect to antipredator ploys is described in Chapter 3.

At any given distance between sender and receiver, information provided by other species may also have special qualities. For instance, different species may have different sensory capabilities and each may be best able to detect specific types of stimuli. Forming a group with other species can thus increase the breadth of detection abilities in the group (Moynihan, 1962). Similarly, species may vary in the quantity and quality of the signals that they produce. If signals can be exploited by other species, there may be an advantage in joining species that have more developed signal production. A specific example would be the exploitation of conspicuous alarm calls produced by one species by other species lacking such developed calls (Goodale and Kotagama, 2005a). Beyond the issue of information quality, it is also certainly the case that phenotypic attributes may differ between species and some attributes may be exploited by others. An example is a timid species nesting within the territory of an aggressive species that can drive potential nest predators away. Many cases of less aggressive species nesting with more aggressive species have been documented (Quinn and Ueta, 2008).

The point of the previous arguments is to demonstrate that joining other species may be beneficial for reasons unique to this particular situation. Therefore, mixed-species groups may allow individuals to reap benefits related to aggregating, *per se*, the same sorts of benefits they could get from joining conspecifics, but in addition to obtaining services that are uniquely provided by other species.

8.3.2.1.2. Reproduction

Joining mixed-species groups in mammals is thought to reduce breeding competition and to allow individuals to gain social experience (Stensland et al., 2003). In some fish species, mixed-species groups are actually needed for mating opportunities (Schlupp et al., 1994). The formation of mixed-species groups during the reproductive season may be paradoxically costly if mating occurs between species, as was thought to be the case in the two Chinese gazelles described earlier (Li et al., 2010).

8.3.2.1.3. Foraging

Foraging advantages include cooperation in finding and exploiting resources. For instance, individuals in mixed-species groups may detect food resources more easily (Struhsaker, 1981), acquire food disturbed by the activity of other species as shown in many bird and mammal species (Munn, 1986; Terborgh, 1983), gain access to a broader range of resources (Gauthier-Hion et al., 1983; Hodge and Uetz, 1996; Morse, 1970), cue on other species to determine the location of food (Barnard and Stephens, 1983; Krebs, 1973), defend resources together (Garber, 1988), and avoid previously exploited areas (Cody, 1971; Terborgh, 1983). For instance, by associating with blue monkeys, red-tailed monkeys are thought to benefit by avoiding areas blue monkeys have already visited and presumably depleted (Cords, 1990a). Species of monkeys that associate in mixed-species groups are often frugivorous, and may benefit by joining each other to obtain precise information about the location of fruiting trees, which are widely scattered in the habitat (Norconk, 1990; Porter, 2001).

8.3.2.1.4. Predation

Antipredator benefits have been a main focus of research on mixed-species groups. The list of potential antipredator advantages for mixed-species groups overlaps the list presented earlier for single-species groups; it includes better detection of predators, reduced vigilance, and effects related to dilution and confusion. Here, I want to draw attention to unique advantages that arise from the formation of mixed-species groups.

At first sight, it may seem a rather easy task to pinpoint special advantages of mixed-species groups by simply comparing them with single-species groups. However, this can be problematic for species that only occur in mixed-species groups. Furthermore, many species only associate with others at certain times of year, meaning that single- and mixed-species groups may not form for the same reasons. For instance, mixed-species groups may only form when predation risk is high or when food is more patchily distributed. Level of risk and food availability may thus act as confounding factors when comparing the two types of groups. A further difficulty lies in the fact that the size of single- and mixed-species groups can differ, with mixed-species groups tending to be larger. Meaningful comparisons can only be made between the two types of groups by controlling group size.

With respect to detection of predators, mixed-species groups could be better at detecting predators by having a greater range of detection abilities than single-species groups (Altmann and Altmann, 1970; Thompson and Barnard, 1983). Different eye morphology, different scanning strategies, or different habitat use among species may result in differences in predator detection, which may increase the odds of detecting predators and also make certain species attractive as companion species (Moore et al., 2013). For example, vertical stratification is a common occurrence in primate mixed-species groups so that different species may be able to detect predators from different heights (Heymann and Buchanan-Smith, 2000). Some species may be better able to react to terrestrial predators and others to aerial predators, increasing the breadth of detection range.

Just like the example of less aggressive species nesting in the vicinity of more aggressive species, more alert species or those that respond to predator cues at a greater distance, in general, often tend to attract other species. For instance, the stonechat, a heathland bird that is quite vigilant and tends to respond to predators from farther away, attracts many other species from the same habitat (Greig-Smith, 1981). In shorebirds, the dunlin occurs so often in association with golden plovers that it has been nicknamed the plover's page. Dunlins have been hypothesized to take advantage of the greater wariness of plovers to escape sooner (Byrkjedal and Kålås, 1983; Thompson and Thompson, 1985). A similar argument has been made for species that join troops of Diana monkeys, which is a species that raises the alarm more frequently than others (Bshary and Noë, 1997), for dolphin species that form groups with the more alert spotted dolphin (Norris and Dohl, 1980), and for cowtail stingrays resting more often with reticulate stingrays because they provide earlier warning against predators (Semeniuk and Dill, 2006).

Reduced vigilance in mixed-species when compared to single-species groups has been documented in many species of birds (Sridhar et al., 2009, for a review). In mammals, consequences of participating in mixed-species groups for vigilance tend to be more variable than in birds. Although many studies have documented a decrease in vigilance in mixed-species groups (Cords, 1990b; Hardie and Buchanan-Smith, 1997; Rasa, 1983; Scheel, 1993; Wolters and Zuberbuhler, 2003), others have reported instead an increase in vigilance in mixed-species groups (Chapman and Chapman, 1996; Stanford, 1998) or no effect at all (Bednekoff and Ritter, 1994; Treves, 1999). An increase in vigilance may take place if some species within the mixed-species groups represent a potential threat and need monitoring (Barnard and Stephens, 1981; Chapman and Chapman, 1996; Popp, 1988).

One difficulty when comparing vigilance between the two types of groups, as mentioned earlier, is controlling for differences in group size. This is important because the decrease in vigilance may come about from an increase in group size and/or from joining species that provide extra safety. In mixed-species groups of African gazelles, for instance, the reduction in vigilance when group size increased was similar whether Thomson's gazelles joined the same number of conspecifics or heterospecifics (FitzGibbon, 1990), suggesting no extra benefit in terms of vigilance of joining a mixed-species group *per se*. Finding that vigilance decreases in mixed-species groups also does not also tell us which mechanisms are involved and, crucially, whether the reduced investment in time spent vigilant actually provides any benefits for other fitness-enhancing activities such as foraging.

Some species may also be more attractive to predators and deflect predation risk onto others (FitzGibbon, 1990). For instance, Grant's gazelles foraging in the Serengeti in Africa may benefit from an association with Thomson's gazelles, which is the preferred prey of cheetahs and would be more likely to be targeted in mixed-species groups. A similar observation was made for brook sticklebacks, which under high predation risk prefer to associate with fathead minnows, the preferred prey of their common predator (Mathis and Chivers, 2003).

Mixed-species groups are often much larger than conspecific groups. For instance, mixed-species groups of Chinese gazelles could be twice as large as single-species groups (Li et al., 2010). The increase in group size *per se*, rather than joining other species with special attributes, may be sufficient to explain some of the benefits that accrue from mixed-species grouping. Teasing apart the effect of group size from the effect of group composition has been done in a few studies. Coming back to the African gazelles, cheetahs tend to avoid hunting larger groups and, therefore, Thomson's gazelles may benefit by joining groups of other species to increase group size and reduce encounters with predators (FitzGibbon, 1990). Solitary midges in groups of aphids experienced a lower per capita capture rate as aphid group size increased because the predator was less likely to encounter the midge in larger groups (Lucas and Brodeur, 2001). Dilution would not work to the same extent if one species in the mixed-species group was preferred by the predator and attracted a disproportionate number of attacks. The key issue is demonstrating that joining a mixed-species group provides extra benefits over joining a group of conspecifics, controlling for group size.

The presence of a stronger species in a mixed-species group may drive away predators, thus providing weaker species with a reduction in predation risk (Herzing and Johnson, 1997; Struhsaker, 1981). As an example, mixed-species groups of monkeys in Africa were able to drive away an avian predator, but on their own the smaller species would have been much more vulnerable (Gauthier-Hion and Tutin, 1988). This may be one reason why smaller species may join mixed-species groups more readily.

Just as in single-species groups, a large number of individuals fleeing from an attack may confuse the predator and reduce the odds of capture for all concerned (Landeau and Terborgh, 1986). The confusion effect would appear to work best when individuals are fairly similar but this is less likely to be the case in mixed-species groups. An odd-looking species in a mixed-species group may provide an easy target for the predator and reduce confusion. In a field experiment with reef fishes, the less numerous species in a mixed-species group was more likely to leave the group to seek protection in the coral (Wolf, 1985). This is similar to the observation that odd individuals in single-species groups are more likely to be selected by predators (Milinski, 1977b; Mueller, 1975). The trade-off here is staying with a larger group or paying the price for standing out. Similar observations have also been made for other reef fishes (Quinn et al., 2012).

This mechanism should also work the other way around, in the sense that joining a mixed-species group will be less attractive when fewer individuals of the same species are present. For instance, chub, a small species of fish, showed more reluctance in joining a mixed-species group with European minnows when proportionately fewer conspecifics were present (Ward et al., 2002). The above observations suggest that joining a mixed-species group may be based on the number of conspecifics already present and that the derived benefits may be frequency-dependent. However, other studies have failed to document an oddity effect. For instance, in mixed-species flocks of herons, which contained mostly white birds and the odd dark bird, an avian predator showed no preference for odd birds (Caldwell, 1986). In addition, reluctance to join a group as a minority species may be driven by other needs such as avoiding competition for resources (Ward et al., 2002). More work is needed to determine the scope of the oddity effect in mixed-species groups.

In mixed-species schools of fish, naïve individuals can learn to recognize predation threats from watching fright reactions from other species (Mathis et al., 1996). Learning about predators was also documented in mixed-species groups of tadpoles (Ferrari and Chivers, 2008). In such cases, individuals that participate in mixed-species groups can benefit from a wider range of experiences related to predators than from joining conspecifics. Information transfer about predation risk in mixed-species groups remains to be documented in birds and mammals.

8.3.2.2. Unique Costs

Although associating with other species may offer unique benefits with respect to single-species groups, there may also be unique costs. Many studies have raised the issue of food stealing among species in mixed-species groups. Small species, for instance, may get supplanted at food sources by larger species (Peres, 1996; Terborgh, 1983) or more aggressive species can steal food from other species (Barnard et al., 1982). The deflection of predation risk to odd species or to more vulnerable species, as discussed in the preceding section, can be considered a cost for those species that are targeted more often by predators in such groups. As mixed-species groups are often larger than single-species groups, such groups may be more conspicuous to predators and the increased encounter rate with predators may offset the decreased risk of attack from the dilution effect. Joining a mixed-species group for any length of time means that some species may need to adjust to the pace and habitat choices of other species. It has been suggested that some species in mixed-species flocks may actually incur some costs related to a reduction in feeding efficiency as they forage in suboptimal parts of their habitats and also at a suboptimal rate (perhaps too fast) (Hutto, 1988).

8.4. LARGE-SCALE SYNTHESIS IN AVIAN FLOCKS

Numerous studies have examined how species benefit from joining mixed-species groups. Meta-analyses are particularly helpful to paint an overall picture when many disparate studies have been conducted in a wide range of species and habitats (Thiollay, 1999). Sridhar et al. (2009) have carried out a recent synthesis for avian mixed-species flocks.

The first part of their study looked at attributes common to the species that frequently join mixed-species flocks. Such attributes may tell us why species aggregate with other species. To draw general conclusions on why species join mixed-species groups, previous attempts had focused on a wide range of species but from single study sites. For instance, the tendency to join mixed-species flocks has been investigated in the forests of Costa Rica and French Guiana (Buskirk, 1976; Thiollay and Jullien, 1998). The results of such studies suggest that species that flock to a greater extent tend to be more vulnerable to predation. It is not known if such tendencies also occur in other flock systems. Perhaps a more pernicious problem in these analyses is the use of species as the independent unit of analysis. Many species in such flocks are closely related, and therefore it may not be surprising that these species show the same flocking tendencies. The lack of statistical independence between species is a recurring theme in evolutionary biology (Harvey and Pagel, 1991). Sridhar et al. (2009) used a statistical procedure that controlled for relatedness among species. They also derived a number of different hypotheses from factors thought to influence participation in mixed-species flocks.

1. If a species joins a flock for antipredator advantages, one would expect relatively small species, which probably face a broader array of predators, to join more frequently. This is similar to the argument made earlier for mixed-species groups of monkeys.

2. Intraspecific flocks of insectivorous species are often small due to their animal diet. Joining mixed-species flocks may allow insectivorous species the opportunity to forage in groups without incurring the costs of competing for resources with conspecifics.

3. Some foraging techniques and foraging locations are thought to reduce the ability to detect predators. For instance, species that use visually demanding foraging techniques or forage away from the ground, where they are more at risk of predation, should thus participate in mixed-species flocks more often.

4. Species that form their own groups to search for food may be less likely to join mixed-species flocks if this is the main benefit of foraging in groups.

By culling information from the vast literature on mixed-species flocking from around the world, Sridhar et al. (2009) tested the above hypotheses and found a greater propensity to participate in mixed-species flocks for insectivorous species, smaller species (Fig. 8.3), and to a lesser extent species foraging away from the ground. Overall, vulnerability to predation emerged as a key factor in the formation of mixed-species flocks. The greater occurrence of insectivorous species in mixed-species flocks does not appear to be related to the opportunity to forage in groups *per se*, because the group size for a particular species was not related to their propensity to join mixed-species flocks.

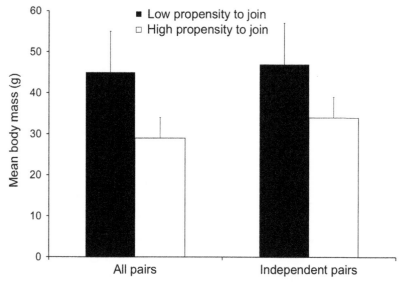

FIGURE 8.3   Meta-analysis of mixed-species flocking: smaller species show a higher propensity to join mixed-species groups in avian species. Mean body mass (g) between pairs of closely related species differing in flocking propensity (black bars: low propensity; white bars: high propensity). Data are shown for the full dataset and for the phylogenetically-independent dataset. Error bars show one standard error. *Adapted from Sridhar et al. (2009).*

Because many species in mixed-species flocks are insectivorous, it may simply be the case that it is easier for insectivorous species to join flocks with similar levels of activity and diet, a conclusion reminiscent of the observation that the type and size of a group may influence the formation of mixed-species groups (FitzGibbon, 1990).

Sridhar et al. (2009) also looked at whether species that join mixed-species flocks actually enjoy greater foraging efficiency and a reduced investment in antipredator vigilance. Using a meta-analysis technique that allows one to pool results from several studies to obtain an overall estimate of the effect, they found that foraging rates tend to increase and the amount of time allocated to vigilance tends to decrease in mixed-species flocks when compared to foraging alone or with conspecifics (Fig. 8.4). In contrast to species that typically follow other species in mixed-species flocks, the increase in foraging rate was not significant for leaders. This result suggests that benefits may vary among species in mixed-species flocks. One possibility to explain these results is that leaders in mixed-species flocks tend to forage in large intraspecific groups (see the following section) that are probably quite efficient in finding and exploiting resources on their own.

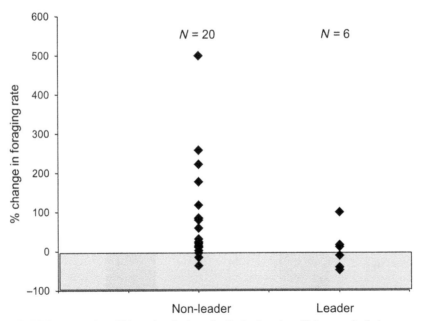

FIGURE 8.4   Foraging efficiency in mixed-species flocks: foraging efficiency typically increases when a species occurs in a mixed-species avian flock rather than alone or in single-species groups. Changes in foraging efficiency are documented for species that typically follow other species (nonleader) and those that lead (leader) mixed-species flocks. In the grey box, foraging efficiency in mixed-species flocks is lower than when foraging alone or in single-species flocks. *Data obtained from Sridhar et al. (2009).*

Obviously, the exact mechanisms that allow birds to forage more efficiently or to decrease their vigilance are not specified, but the overall finding is that species do benefit by joining mixed-species flocks. The caveat, as indicated earlier, is whether confounding factors, such as group size and food availability, have biased these estimates.

The different benefits of mixed-species flocking may have a significant impact on survival. A recent study found that the level of participation in mixed-species flocks was positively correlated with survival rate (Jullien and Clobert, 2000), which is just what one would expect if species enjoyed benefits by joining mixed-species flocks in terms of enhanced foraging efficiency and reduced predation risk.

8.5. EVOLUTION OF TRAITS ASSOCIATED WITH MIXED-SPECIES GROUPS

It has long been suggested that some species in mixed-species groups possess traits that favour the formation and cohesion of these groups (Moynihan, 1962). In birds for instance, large flock size, vivid plumage markings, or special calls may provide more stimuli to attract the attention of other species and thus facilitate the formation and cohesion of flocks (Diamond, 1981). Drab plumage has also been suggested as a trait that enhances flock formation; it is believed to tone down signals that may be perceived as aggressive by other species. I focus first on the possible role of a large flock size in species that lead flocks.

8.5.1. Leadership

The broader question is whether different species in flocks play different roles, one of which may be leading the flock. Obviously, these roles have not evolved to ensure the better functioning of the flock, which would be a group selection argument. Rather, some species may have evolved traits or possess traits that make them more suitable to play certain roles in flocks because these traits allow individuals to increase their success. The examination of roles in groups has mostly centred on bird species, although I would not be surprised if similar findings emerge in fish or mammals.

Moynihan (1962) defined nuclear species as those that through their behaviour contribute to stimulate the formation and enhance the cohesion of flocks. Most researchers have noted whether some species in their flock system tend to be nuclear. Common characteristics have often been noted in nuclear species: such species lead the flocks, have a high propensity to occur in such flocks, are gregarious, and tend to be conspicuously active and often very vocal (Hutto, 1994). Leadership is a characteristic often associated with nuclear species. Some aspects of leadership are difficult to document empirically, such as level of activity and vocal complexity. The relationship between leadership and gregariousness, on the other hand, is amenable to quantitative observations by simply measuring intraspecific group size. Gregariousness can simply increase the conspicuousness of a species, thus making it easier for others to join and follow the flock over a long distance. Other traits associated with gregarious species, such as conspicuous alarm calling, may also help the cohesion of flocks (Goodale and Kotagama, 2005b).

In support of this hypothesis of a relationship between leadership and gregariousness, leader species in neotropical flocks tended to have marginally larger group sizes than species that follow (Powell, 1985). This analysis was extended to a much broader range of flocks across many areas of the world (Goodale and Beauchamp, 2010). Focusing on mixed-species flocks that search for food together over long distances and that contained more than two species, we looked for evidence of leadership independent of flock size. For instance, species that tend to be located at the forefront of flocks, initiate movements, or are followed by other species were characterized as leaders. If leadership was random in the flock, one might expect a gregarious species to lead flocks more often simply because of their greater numbers. However, one would not expect to find such species consistently at the front of the flock; they should be found anywhere in the flock with equal probability.

The authors contrasted mean flock size in flock leaders, in associate species and in occasional species found within the same flocks. An associate species did not lead but was found regularly in the mixed-species flocks while an occasional species typically occurred in less than 50% of the flocks. In 45 different mixed-species flocks from around the world, mean group size was significantly larger in leader species than in associate and occasional species from the same flocks (Fig. 8.5).

These results support the hypothesis that gregariousness is associated with leadership. In another study, Sridhar et al. (2009) noted that leader species are more likely than follower species to be cooperative breeders. Cooperative breeders live in kin groups and are likely to have well-developed communication and alarm systems that can be used by other species to increase safety. Whether traits, such as gregariousness and cooperative breeding, are the cause or the consequence of different roles in mixed-species flocks remains to be established.

8.5.2. Vocal and Visual Mimicry

Other traits may have evolved in response to participation in mixed-species groups. Indeed, acoustic and visual signals produced by a species may be selected to converge within a mixed-species group.

The concept of mimicry, the ability to match phenotypic traits from other species (Ruxton et al., 2004), has been applied to mixed-species groups in a variety of contexts. Mimicry may involve vocalizations as well as the overall appearance of a species. With respect to vocalizations, vocal mimicry is common in the avian world, involving up to 25% of all species (Baylis, 1982). Nevertheless, an explanation for the function of vocal mimicry has remained elusive.

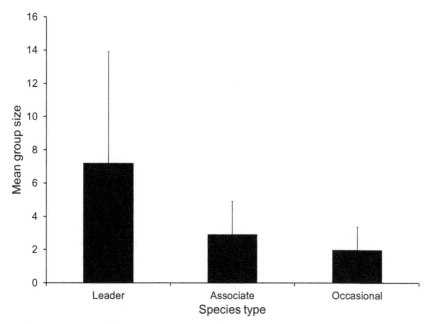

FIGURE 8.5   Leadership and mixed-species flocking: species that lead mixed-species avian flocks tend to occur in larger groups. The figure shows mean intraspecific flock size in mixed-species flock leaders, associate species, which were frequent followers, and occasional species, less frequent followers. Error bars show one standard deviation. *Adapted from Goodale and Beauchamp (2010).*

Mixed-species flocking, with the intermingling of many species, would appear to provide a context where vocal mimicry may be expected to provide benefits to species that can manipulate the behaviour of companion species (Goodale and Kotagama, 2005a). Vocal mimicry was investigated in Sri Lankan mixed-species flocks (Goodale and Kotagama, 2006). Vocal mimicry by the greater racket-tailed drongo in these flocks was already known, but its function was not clear. Drongos benefit from joining mixed-species flocks by capturing prey flushed by other species in the flocks and also by stealing their prey. By using playback experiments, the authors showed that the presence of mimicked elements in the calls of drongos manipulated other species in a way that benefited the drongos. They found that drongo calls with mimicked elements sounded like the sounds produced by a mixed-species flock and attracted nearby species. These other species would then form a flock, which could flush insects for the drongos.

This exploitative function of mimicry also resonates with the observation that some species use false alarm calls to startle other species into dropping their prey, which can then be captured easily by the false alarm caller (Flower, 2011; Munn, 1986). In the face of these exploitative ploys, why would species respond to these deceptive signals? While drongos do obtain benefits from attracting other species, they emit frequent and reliable alarm signals, which can benefit all other species. Such benefits to the other species can easily offset the cost of losing prey (Radford et al., 2011). Similarly, the cost of responding to false alarm signals may be small considering that alarm signals are generally reliable.

Visual mimicry has also been linked to mixed-species flocking in birds. Convergence in the appearance of birds has been documented in many mixed-species flocks (Cody, 1973; Diamond, 1982; Moynihan, 1962, 1968; Willis, 1963). As in many insect species, visual mimicry may involve signals of unpalatability. A potential example of Batesian mimicry, where a nontoxic species resembles a toxic species, was suspected to occur in mixed-species flocks in Papua New Guinea. Many species in the genus *Pitohui* are toxic (Dumbacher et al., 2008). Such species are prominent members of many mixed-species flocks and often act as leaders (Diamond, 1987). These authors noted that many species that join these flocks are very similar in appearance to toxic pitohuis, to the point that it took years for a researcher to tell that similar-looking birds in the flocks were actually different species. If nontoxic species participate in flocks in order to associate with pitohuis, one would expect the presence and calls of pitohuis to be very attractive to other species. However, it turned out that the presence and calls of another species were more attractive to other species (Goodale et al., 2012), providing little evidence that Batesian mimicry plays an important role in these flocks.

Factors other than the signalling of unpalatability may play a greater role in mixed-species flocks. Moynihan (1962) suggested that mimicry may have evolved to facilitate the formation and cohesion of mixed-species flocks because it reduces the number of signals needed to coordinate movements. The oddity effect, where odd-looking species attract a disproportionate amount of attacks in groups, may also favour the evolution of resemblance among species in a mixed-species group (Barnard, 1979; Landeau and Terborgh, 1986). Finally, subordinate species in mixed-species groups may benefit by resembling more dominant species because of reduced interspecific aggression (Diamond, 1982). In view of the potential advantages of mimicry in the context of mixed-species groups, one would expect a positive association between mimicry and flocking.

As it often the case with a simple hypothesis, the answer is bedevilled by methodological issues. Two alternative hypotheses must be ruled out to make a convincing case for mimicry. In mixed-species flocks, different species are necessarily found in exactly the same habitat. Therefore, one must rule out independent evolution of similar traits in the species in response to sharing the same habitat (Burtt and Gatz, 1982; Willis, 1976). As an example, consider the remarkable resemblance between semipalmated and western sandpipers (Fig. 8.6). These two species feed and flock together during the winter (Tripp and Collazo, 1997), and one needs a spotting scope to see the minute differences in plumage between the two species. Is this a case of mimicry associated with mixed-species flocking? One must consider that many species in the *Calidris* genus share the same open habitat with rocky shores, where a

Model: Semipalmated
sandpiper

| Putative mimic | Ancestry control | Habitat control |
| Western sandpiper | Little stint | Semipalmated plover |

Score: 1   1   3

FIGURE 8.6   Mimicry and mixed-species flocking: a case of putative mimicry in a mixed-species flock that is probably due to shared ancestry. The top picture represents the model species. The bottom row of pictures contains, from left to right, the putative mimic, the sister species of the putative mimic, and a species that participates in the mixed-species flock with the model and putative mimic but to a lesser extent. Scores represent the match judged by a panel of blinded raters, with 1 being a close match and 4 a poor match. *Adapted from Beauchamp and Goodale (2012). Photo credits: Semipalmated sandpiper (Guy Beauchamp); Western sandpiper (Winnu); Semipalmated plover (Dendroicacerulea); Little stint (Omarrun). This figure is reproduced in colour in the colour section.*

mainly dark plumage on the back and white plumage on the belly is the norm in many species. A stronger case for mimicry, in this case, would be to show that *Calidris* species that do not flock together resemble one another less than those that flock together.

The second alternative hypothesis that must be ruled out for putative cases of mimicry is simply that species that are closely related may have inherited the same plumage from a common ancestor. Shared ancestry may cause species to resemble one another even when they do not live in the same habitat. This may be the case for two sandpiper species mentioned above, because the two species are very closely related (Gibson and Baker, 2012). In addition to the two alternative mechanisms, a methodological hurdle lies in the fact that mimicry is often judged by only one person and may be specific to the samples that were judged.

With these caveats in mind, cases of putative mimicry were reassessed recently in mixed-species flocks (Beauchamp and Goodale, 2011). To address the methodological issue raised earlier, many different raters were asked to judge the resemblance between the model species in a mixed-species flock and the putative mimic, as determined by researchers in published papers. Photographs of the two species were used in one dataset, and drawings in the other, so as to have two independent samples. Two control species were added to each set: one to control for similarity caused by shared habitat requirements and the other to control for resemblance caused by shared ancestry. For the habitat control, another species from the same mixed-species flock was selected, and for the ancestry control a species closely related to the putative mimic species but not living in the same geographical area as the model was selected. Raters were asked to determine on a scale from 1 to 4, where 1 is a very strong match and 4 a weak match, which species most closely resemble the model species: the putative mimic, the habitat control species, or the shared ancestry control species. An illustrative example is shown in Fig. 8.6.

In the example shown in the figure, judges perceived a mismatch between the model species and the habitat control species. However, the match between the model and the putative mimic was judged similar to the match between the model and the sister species of the putative mimic as well. In this case, while the mimicry did not seem to arise from shared habitat requirements, the sister species, which does not flock with the model species, was very similar to the model, suggesting that the resemblance was probably caused by shared ancestry.

When the evidence was assessed across all potential cases of mimicry, it turns out that one or both alternative hypotheses for the occurrence of mimicry could not be ruled out in nearly half the cases. Support for mimicry in the other cases was stronger. However, there is a weakness in our analysis: only one among many potential habitat control species was evaluated. Perhaps, by coincidence, the selected species happened to be very different from the model species, and a broader search would have unearthed a species more alike.

This study raises important issues. Most importantly, more objective criteria are needed to evaluate resemblance, such as measurements of colour and luminance with spectral analysis (Cheney and Marshall, 2009; Endler and Mielke, 2005). Other taxa do not have the same visual systems as humans, and visual similarity (or the lack of it) to humans may not always match closely how species of more evolutionary relevance to the taxa concerned would judge similarity in appearance. However, sophisticated techniques may measure differences that the birds do not really use. From the point of view of a predator fast approaching a flock and only getting a glimpse at fleeing flock members, minute differences may not matter too much. A broad similarity among species may be just what is necessary to reduce oddity or increase confusion for the predator. Imperfect mimicry has been predicted and documented in many species (Penney et al., 2012). Fish predators, for instance, have been shown to respond quite similarly to a broad range of variation in the level of mimicry achieved by their prey (Caley and Schluter, 2003).

As a final comment on mimicry, even when the evidence suggests that mimicry did not evolve in association with mixed-species flocking, any resemblance between species in a group, regardless of whether or not it arose by chance, shared habitat requirements, or ancestry, may help individuals achieve some of the benefits that befall similar-looking species in a group. Looking like semi-palmated sandpipers may help western sandpipers by reducing oddity or by facilitating cohesion in the group. Species may prefer to participate in groups with other similar-looking species, an extension of the principle that single-species groups are often phenotypically matched (Hoare et al., 2000). Here, what has evolved is not mimicry *per se* but rather the ability to join species that share similar traits.

8.6. CONCLUDING REMARKS

The study of mixed-species groups has a long history but much of the evidence for their adaptive value has been anecdotal. While it is clear that there are unique advantages and disadvantages of joining a mixed-species group, it has proven remarkably difficult to establish how beneficial mixed-species grouping can be relative to simply living with members of the same species. As discussed earlier, this is because the two different types of groups may form for different reasons, and also because they also tend to vary greatly in size, making it difficult to disentangle the effect of confounding factors from the effect of participating in a mixed-species group.

Finding costs and benefits is only part of the story—combining them to get an estimate of fitness is quite another. Too often, discussion of the adaptive value of participating in a mixed-species group simply stops at arguing that benefits must offset the costs. Only one study has investigated survival consequences of forming mixed-species groups. The study contrasted species in mixed-species flocks with solitary species or species that flocked with other species occasionally (Jullien and Clobert, 2000). However, this study did not address the potential confounding factors listed above. Therefore, documenting fitness consequences of participating in mixed-species groups remains a challenge.

One of the research goals in the study of mixed-species groups has been to predict the species composition of such groups given the pool of species available in the habitat (Goodale et al., 2010). Large-scale studies, like the meta-analysis of Sridhar et al. (2009), are needed to establish whether there are some general principles for why species occur or not in mixed-species groups based on adaptive reasoning. More mechanistic studies of assembly rules for the formation of mixed-species groups can also shed some light on the constraints faced by species when forming such groups (Graves and Gotelli, 1993; Sridhar et al., 2012). It would be interesting to extend these approaches to other taxa that form mixed-species groups, such as fish and mammals, to test the generality of these tentative conclusions.

The evidence with respect to species roles typically relies on qualitative impressions by field researchers. For instance, nuclear species are defined according to the tendency to be followed by other species. More quantitative approaches based on the number of interactions between species, controlling for species availability, can provide more information about the centrality of a species in the social network of a mixed-species group and allow each species to be located on a continuous nuclearity scale rather than being rather arbitrarily defined as nuclear or not (Farine et al., 2012; Srinivasan et al., 2010). Phenotypic attributes of different species, such as intraspecific group size, can then be related to the location on this continuous nuclearity scale.

The niche construction hypothesis for mixed-species group, originally proposed by Moynihan (1962), suggests that mixed-species groups provide a fertile ground for the evolution of traits that can increase their formation and cohesion. Recent investigations have therefore tended to look at the association between leadership, group size and mimicry in mixed-species flocks. An obvious difficulty is establishing the chain of events in the evolution of such traits. For instance, did large group size in leader species evolve prior to the species participating in mixed-species flocks, or did it evolve after as the niche construction hypothesis would suggest? Novel phylogenetic techniques can be used to trace the sequence of events in the evolution of associated traits (see Sol et al., 2010, for a recent example).

In addition to the use of these more sophisticated inference tools, the niche construction hypothesis could be extended to other phenotypic traits and other taxonomic groups. Two phenotypic targets of evolution are suggested here as examples. First, Moynihan (1962) suggested that other plumage traits in birds may have been modified in mixed-species flocks, including drabness. Second, vocalizations can attract the attention of other species (Goodale and Kotagama, 2008; Suzuki, 2012) and thus be helpful in the formation of mixed-species groups. Such vocalizations are also known in primate species (Cords, 2000). Are these vocalizations simply the same as those used to attract conspecifics or do they possess special characteristics that would enhance communication across a broad range of species, each with their own specialized channels of communication? Finally, no doubt we need to broaden the taxonomic scope of the niche construction hypothesis by including other species, such as fish, many of which show mimicry (Moland et al., 2005). This could provide valuable opportunities for evaluating the association between mimicry and participation in mixed-species groups.

Evolutionary Issues

9.1. INTRODUCTION

Many species of spiders, scorpions, snakes, and lizards produce poisonous substances, known as venom, which can be directly delivered into the body of prey or competitors through bites or stings (King, 2011). Populous equatorial countries host a large number of venomous species, which leads to frequent, unfortunate encounters for human inhabitants. In Africa alone, half a million bites from venomous snakes are reported each year, causing approximately 20,000 deaths (Chippaux, 1998). Snake bites in humans are accidental because snakes typically attack much smaller prey like rodents or invertebrates. The fact that snake venom can kill much larger species indicates the potency of these substances, which have clearly evolved to produce debilitating and often lethal effects in their prey. At first sight, it would appear the evolution of venom in snakes must have been a very strong blow against prey species.

Phylogenetic studies of snakes and lizards pinpoint a single evolutionary origin of venom production, which is thought to have led to the tremendous diversification of advanced snake species (Fry et al., 2006). Despite the success associated with the evolution of venom production in snakes, not all species produce lethal varieties of venom, suggesting that there are physiological costs related to the production of these substances, and that toxicity may be driven in part by the type of diet (Barlow et al., 2009). In a telling example, snakes whose diet shifted over evolutionary time to immobile prey lost their venom production system. Venom resistance in prey species has emerged as a further factor in the evolution of venom toxicity. More toxic venom may have evolved in snake species whose prey have developed resistance to venom, suggesting an evolutionary arms race between predator snakes and their prey. In some sense, very toxic venom testifies to the ability of prey to evade their snake predators.

The snake story illustrates two key concepts in the study of predator–prey relationships. First, most studies of predator–prey relationships assume fixed traits in either the predator or their prey. However, this is often not the case, as seen in the above example of venomous snakes where venom toxicity is flexible and has evolved in relation to the type of responses exhibited by their prey. The same scenario may also hold for various aspects of social predation. For instance, models of the spatial distribution of predators typically predict that predators should concentrate their attacks in areas with higher prey density (Hassell, 1978). However, increased antipredator responses to predictable attacks may make prey in such areas particularly difficult to catch. The best response by predators may thus be to plant uncertainty in the mind of their prey by attacking areas of different prey densities in an unpredictable fashion (Roth and Lima, 2007). Such interplay in behaviour between predators and their prey often challenges traditional expectations of social predation models (Lima, 2002). In addition, recent models also suggest that interplay between predators and prey may have a stabilizing effect on population dynamics (Křivan, 2007) further extending the reach of behavioural games.

Second, tracing the evolution of traits through evolutionary time can be particularly helpful in identifying underlying selection pressures. In snakes, the relationship between venom toxicity and prey type across various lineages points to adaptive variation in venom toxicity in response to the costs and benefits of producing these substances. Adopting a historical perspective can thus shed light on the selective forces that have shaped the evolution of a trait in the past.

In this chapter, I deal with evolutionary issues related to social predation. Previous chapters have considered the current costs and benefits of social predation. I now adopt a more historical perspective that complements questions about the adaptive value of behavioural patterns. First, I want to illustrate the possible consequences of arms races between social predators and their prey, a specific force that may alter the expression of traits associated with social predation through evolutionary time. Second, I want to draw on phylogenetic studies in various animal species to assess patterns of covariation between traits related to social predation, such as group size and antipredator vigilance, and environmental and life history variables as a means to better understand the drivers of social predation in general.

9.2. CO-EVOLUTION BETWEEN PREDATORS AND PREY

The notion of co-evolution between predators and prey introduces a more dynamic view of their relationship. Instead of assuming fixed traits in predators or prey, this approach seeks to determine the best response of predators and their prey to potential adjustments in each other's behaviour within their lifetime or across generations. The best hunting strategy for a predator, then, should depend on what prey can do in response, leading to a stable solution from an evolutionary standpoint or evolutionarily stable strategy (ESS). The concept of a stable solution should be familiar to the readers by now. Earlier, I presented the ESS approach to producing and scrounging (Chapter 2) and to the level of antipredator vigilance maintained in a group (Chapter 4). Co-evolution extends the ESS approach to predator–prey relationships. The recent advances in our understanding of dynamic predator–prey relationships have clearly benefitted from the development of game theory, the basic approach to model evolutionary games (Sih, 1998). However, the more complex mathematics that underlies such models may explain why relatively few models have explored co-evolutionary issues in the context of social predation. I present some of these models in the following section to illustrate how intuitive expectations about predator and prey behaviour can be challenged when viewed in the light of an evolutionary arms race.

9.2.1. Predator–Prey Shell Games

Shell games involve an explicit arms race between predators and prey (Mitchell and Lima, 2002). The race, which is usually modelled over evolutionary time, occurs within a generation in this model since predators and prey are assumed to show flexibility in their behaviour within a generation. Adaptive phenotypic flexibility in predator and prey behaviour within a generation has been documented in many species (Agrawal, 2001), and so this assumption seems justified.

Instead of using a fixed predator attack rate across all food patches of a habitat, this model considers a predator that can learn to bias its attacks to particular patches based on prior hunting success. Prey can also show flexibility in their foraging patch choices. Stationary prey always forage in the best patch available to them while non-stationary prey move from patch to patch spending only a fraction of their time in the best patches available to them (Fig. 9.1). Predation is successful if the predator finds a prey during a visit to a particular patch. A learning predator is more likely to avoid revisiting a patch without prey or a patch where the prey have all been attacked and so now stands empty.

In the absence of predation (scenario 1), simulation results show that non-stationary prey, which spend time moving at the expense of feeding and occupy poor patches more often, are less likely to survive than stationary prey. Non-learning predators impose a fixed rate of predation on all patches, and not surprisingly, all prey types survive less but stationary prey still outperform moving prey (scenario 2). Learning predators, however, impose a high cost on stationary prey because they target their location more efficiently. Moving prey now survive better than stationary prey. To determine whether the moving strategy is evolutionarily stable, the authors allowed a few individuals using one strategy to

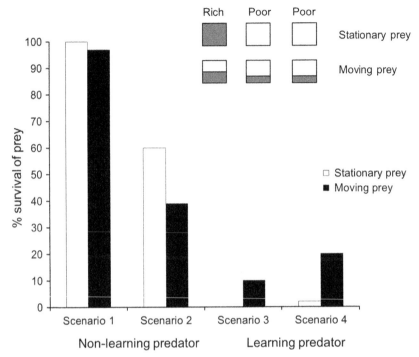

FIGURE 9.1 Predator–prey shell games: the probability that prey individuals survive over a fixed time horizon depends on the interaction between predator and prey behaviour. In the simulation model, each prey can exploit three types of patches, one of which is rich and two are poor. Stationary prey occupy the rich patch at all times while moving prey split their time between the three patches. In the absence of predation, stationary prey always accumulate enough food to survive while moving prey sometimes die of starvation because of their lower food intake (scenario 1). The presence of a predator that attacks patch locations randomly reduces the success of each type of prey, but stationary prey still survive better (scenario 2). A learning predator, which targets non-visited patches preferentially, eliminates all stationary prey, but movement by non-stationary prey allows them to escape predation more often. Moving prey perform better than stationary prey when the majority of the prey are stationary (scenario 3). Stationary prey, when rare, did not perform better than moving prey (scenario 4). Results are based on 1000 simulations. The inset illustrates the movement rules for the two types of prey. *Adapted from Mitchell and Lima (2002).*

invade a population where all members use the alternative strategy. Rare moving prey invaders outperform members of a population with stationary prey (scenario 3) but rare stationary prey invaders never do better than members of a population with moving prey (scenario 4). Moving thus represents an equilibrium solution in this game.

In the basic model, prey movement was the only solution to reduce the rate of encounter with predators. However, antipredator vigilance can reduce predation risk after an encounter by reducing the efficiency of an attack (Chapter 4). In a modified version of the basic model, Mitchell and Lima (2002) considered the possibility that prey may allocate time to antipredator vigilance while foraging. Such an allocation of time reduces the rate of food intake, which increases the chances of surviving an attack but also the chances of starvation. At low-to-moderate levels of vigilance, it turns out that prey movement still represents the equilibrium solution, suggesting that prey movement is a robust strategy when facing a learning predator.

The model thus predicts that prey should, counterintuitively, spend a fraction of their time in poorer food patches to reduce the risk of encounter with predators that are able to target their attacks to more productive areas. This approach could easily be used to evaluate other scenarios. For instance, can stationary prey benefit from occupying rich patches by foraging in larger groups that deter predators?

Such models could also benefit from adopting a framework where predator and prey behaviour emerges from individual interactions rather than being fixed at the onset. A genetic algorithm approach may be useful for examining stability issues in a more flexible framework (see Box 6.1 in Chapter 6) (Mitchell, 2009). Recent models along the same lines explore how predators and prey try to manipulate their rate of encounter with one another (Sih, 2005; Wolf and Mangel, 2007) or manipulate the outcome of an attack after an encounter (Hugie, 2003; Katz et al., 2013).

To illustrate the complexities that may arise when considering games between predators and prey, I return to the traditional models of antipredator vigilance (Chapter 4). These models assume that predators attack groups of all sizes to the same extent and that prey are essentially stationary since the rate of encounter with predators cannot be manipulated. By contrast, in the habitat selection model of Mitchell and Lima (2002), the rate of encounter between predators and prey may be manipulated by changes in prey movement tactics. Vigilance may thus be adjusted in part to reflect the spatial variation in expected predation risk that emerges from the habitat selection game between predators and prey. Furthermore, vulnerability to capture can be manipulated by the prey through factors such as aggregation (Cresswell and Quinn, 2004; Quinn and Cresswell, 2004), which may thus vary adaptively from habitat to habitat. In the end, if both predators and prey can show flexibility in their tactics and adapt their behaviour to one another, it may be much more difficult to predict how vigilance should vary across the habitat and in groups of different sizes.

9.2.2. Group Escape

An individual under attack by a predator should try to escape to safety as soon as possible. Vegetation cover near foraging areas, for example, provides shelter from predators. However, for many species, escapes take place in the open, and so escape speed and escape route difficulty determine the likelihood of capture. However, group-living prey escaping in the open can reduce their relative risk of capture by seeking more sheltered positions within the group (Chapter 6). While fleeing, prey may decide to remain in the group or leave the group to escape alone, and the pursuing predator may choose to attack the group or target the evader. A recent model explores this fleeing game for a predator going after prey in groups (Eshel et al., 2011).

When fleeing as a group, the fastest individual dictates the route of escape and the slowest individuals that follow are inevitably left at the rear, closer to the attacking predator. Should these more vulnerable individuals remain in the group? On the one hand, a group offers an effective means of protection for group members via effects like dilution and confusion. On the other, the laggards are directly exposed to predators and thus benefit relatively less from the dilution effect and certainly not from the selfish-herd effect. Leaving the group may be a viable option if the predator can actually miss their exit or if the prey can find shelter away from the group. Another consideration is that individuals in a fleeing group may hinder one another, and fleeing alone may increase the chances of evading the predator by increasing maneuverability. For example, in a sandpiper roosting group, a wave of alarm as individuals fly up spreads through the group more slowly when individuals are closer to one another, suggesting that the risk of collision forces individuals to slow down in denser groups (Beauchamp, 2012a).

If leaving the group represents a viable option, the choice to flee alone should always be made by the slowest group member first, which leaves the pursuing predator with the choice of attacking the group or a slower evader. An ESS analysis of this game indicates that the predator should always pursue the evader instead of the group. If the predator pursues the group, it will always pay for a group member to leave the group and pass along the risk to the remaining group members. In the end, the pursued group will slowly disintegrate and the predator will be faced with an array of solitary prey. Chasing the evader also makes sense because only the slowest group members should leave first, signalling their weaknesses.

The plot thickens if the fastest individuals have a choice of escape route. Selecting a difficult route will certainly increase the vulnerability of the slowest group members, thus facilitating group breakup. However, this strategy may be counterproductive if individuals that leave the group can hide effectively from predators. Adopting a more challenging route, the fastest group members may thin their buffer of protection and increase their relative risk of capture. This scenario is eerily similar to the group membership games explored in the previous chapter. When the benefits of foraging alone are substantial, more dominant group members should increase the relative share of resources they make available to subordinates to ensure group unity. In the fleeing game, it is therefore conceivable that the fastest group members will actually select the riskiest escape route for themselves for the sake of retaining other group members. This model thus extends the selfish-herd argument by considering the interplay between predator and prey behaviour when an attack has been launched, leading to novel insights into escape behaviour by prey in groups.

9.2.3. Vigilance in Time

In this third example, I now turn to the situation where a predator encounters a prey but the prey is not immediately available. This sets up a type of arms race in which the outcome of an encounter between predator and prey may be manipulated by the prey to reduce the likelihood of capture. Burrowing animals, for instance, can escape from an approaching predator by hiding in their burrows. However, the longer individuals spend in their protected burrow the lower their food intake rate is. Emerging rapidly may solve the problem of finding food more quickly, but may expose the prey to the still waiting predator. The predator, on the other hand, must choose between staying at the burrow for the prey to emerge and resuming search elsewhere. Should the predator always remain at the burrow for a fixed period of time, the prey will likely adapt by emerging just slightly after. Similarly, prey individuals that always emerge at the same time will find the predator waiting just slightly longer. The evolutionarily stable solution in this war of attrition is selection for unpredictable waiting times for the predator and unpredictable emerging times for the prey (Hugie, 2003). Recent work indicates that in one avian predator species, individuals tend to return to patches that they visited earlier when the prey fish that escaped are more likely to emerge from hiding (Katz et al., 2013), suggesting that individuals in this species have the spatial and memory skills needed to engage in these predator–prey games.

Ruxton and I examined a similar waiting game between predators and prey in the context of antipredator vigilance (Beauchamp and Ruxton, 2012a). Many prey species travel from patch to patch in search of resources, but often have little information about current predation risk in each patch. In particular, prey species that arrive at a new patch must assess whether a predator lurks nearby. A familiar example is mammalian herbivores feeding or drinking on the open savannah where a lion may be hiding in tall vegetation (Périquet et al., 2010; Scheel, 1993).

If the hiding predator always breaks cover at the same time, prey animals will be selected to leave immediately before that time or to be extra-vigilant early on. Predators should thus keep the prey guessing about attack time. However, waiting too long may not be advantageous, because the prey animals may detect the predator or simply leave prematurely for unrelated reasons. Prey animals, in response to experiencing early attack time, may increase their antipredator vigilance. Nevertheless, this high vigilance comes at the cost of reduced food intake rate. Prey animals should also keep the predator guessing about when vigilance will decrease as the predator could simply wait for vigilance to come down before launching an attack. Considering the various costs and benefits of waiting (for the predator) and of high vigilance (for the prey), we found an evolutionarily stable solution in this game that holds for a broad range of ecological conditions: predators should attack at unpredictable times but typically early in the foraging bout, and the prey should adopt a high vigilance early and then switch unpredictably later to lower vigilance. The main message from this model is that antipredator vigilance may change as a function of time since arrival at a patch. Influential models of antipredator vigilance have never considered dynamic changes in vigilance through time (Chapter 4). Nevertheless, this arms race in which hiding predators and their prey plant uncertainty in the minds of each other has important consequences for the evolution of predator and prey tactics.

We found some evidence in support of dynamic changes in vigilance through time from field data. Gulls aggregate at loafing sites at high tide to preen and rest while waiting for the tide to recede. Sleeping is considered a low vigilant state, as animals are less responsive to external disturbances. At these loafing sites, the proportion of sleeping gulls tended to increase with time since arrival at the loafing site (Beauchamp and Ruxton, 2012b) (Fig. 9.2). Gulls probably do not face hiding predators as assumed in the above model, but they may still benefit from adjusting vigilance through time. Indeed, as time goes by without disruption, gulls may feel safer and progressively reduce their vigilance (Sirot and Pays, 2011). The prediction that vigilance ought to decrease with time remains to be tested in a system where predators hide before an attack.

Probable cases of reciprocal evolution in predator–prey relationships are quite numerous, but typically involve morphological or defensive traits like prey

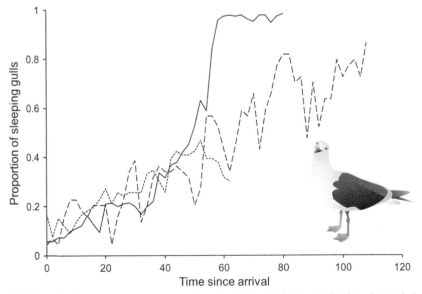

FIGURE 9.2   Vigilance in time: the proportion of sleeping gulls increases with time since arrival at the resting site, illustrating the dynamic nature of vigilance in animals. Sleeping represents a low vigilance state. Changes with time are illustrated for three different flocks at one field site. *Adapted from Beauchamp and Ruxton (2012).* (For colour version of this figure, the reader is referred to the online version of this book.)

toxicity (Brodie and Brodie, 1999). The above examples illustrate how reciprocal adaptations may also involve behavioural patterns. Many promising models have been presented; now we need empirical tests of their predictions. It is
important to realize, however, that interactions between predators and prey may
not always lead to an arms race. Predators may have access to other prey types,
allowing them to escape prey defences (Brodie and Brodie, 1999). Indeed, prey
species that become harder to capture may simply be avoided by a predator with
access to alternative prey, breaking the cycle of reciprocal adaptations. Similarly, prey may be subject to predation from more than one predator, reducing
the scope of specific adaptations to evade one type of predator. In general, co-
evolution is thought to be less likely when the relationship between a predator
and a prey species is not exclusive (Endler, 1991).

9.3. EVOLUTION OF SOCIAL PREDATION

The previous section dealt with a dynamic force that can drive the evolution of
predator and prey behaviour. In this section, I focus on how more static selection pressures from the environment may shape traits related to social predation.
Such analyses are based on the comparative method, which I will introduce
briefly before exploring case studies from a range of species.

9.3.1. Principles of Comparative Analyses

The question of why species hunt or forage in groups can be approached from
two different angles. The first approach consists in identifying the current costs
and benefits of sociality; this pinpoints the selection forces that may have acted
in the past to bring about the evolution of the trait. This functional approach is
best suited to determining the current adaptive value of a trait or the balance of
selection forces acting now to maintain the trait in a population.

The current function of a trait can be studied using an experimental approach.
If white plumage in birds, for instance, has evolved as a signal to facilitate the
recruitment of distant companions to food patches (Tickell, 2003), researchers
can experimentally manipulate plumage colouration to determine whether white
plumage plays its purported role.

A trait may have originated in the evolutionary past for other reasons than
those acting currently. For example, feathers originally evolved for thermoregu-
lation purposes but now also serve for flight. For this reason, a more historical
approach may be better suited to uncovering how a trait actually evolved over
time. The second approach to investigating adaptations seeks to uncover ecological correlates of the evolution of morphological or behavioural traits using historical information. As an example, variation in diet may be linked to variation
in group size among species as a means to infer past evolutionary processes,
and thus to identify selection pressures associated with the evolution of group
size. This retrospective approach, known as the comparative method (Harvey

and Pagel, 1991), takes advantage of past events to reconstruct the evolution of traits across time. In contrast to paleontology, where fossils leave physical proofs of evolution, behavioural ecology can only reconstruct past events using the distribution of traits among extant species.

Retrospective studies, which are essentially observational in nature, form the basis of research in a number of fields, including epidemiology, history, and astronomy, where experiments are either impossible or ethically suspect. To give one example, the best way to establish a link between smoking and lung cancer would be to simply run an experiment where human subjects are allocated randomly to a group where smoking is allowed or not. Obviously, such an experiment would be unethical, and so researchers rely on indirect approaches, such as retrospective studies, to establish prior exposure to potential risk factors associated with lung cancer (e.g. smoking). Instead of screening for risk factors in a large sample of the population, especially for uncommon diseases, researchers target sick individuals and match them with non-sick individuals that share the same characteristics. Ideally, sick and non-sick individuals will differ in only a few risk factors, which can then be associated with the expression of the disease.

Comparative analyses adapt this logic to species traits. Returning to the association between diet and group size, researchers may wish to contrast variation in group size among species to variation in their diet to determine whether evolutionary transitions in diet type are consistently associated with changes in group size across a broad range of species. Contrasts between species traits can be calculated using pairs of species known to be closely related to each other. This is known as the pairwise comparative method. In other cases, a phylogenetic tree may be available that maps the evolutionary relationships among many species through the ages based on their similarities. Nowadays, phylogenetic trees are typically built using genetic data extracted from specific genes. Once a phylogenetic tree becomes available, contrasts are calculated using estimated trait values from adjacent branches proceeding from the tip of the tree down to the root.

Comparative studies of animal behavioural patterns have a long history. An early example, which combines a large number of species and a phylogenetic framework, documented evolutionary shifts in habitat use and co-adapted morphological traits in an ant tribe (Brown and Wilson, 1959). At about the same time, a large-scale study of weaverbirds revealed a close relationship between diet, habitat, and social foraging (Crook, 1964). This study found that species that live in closed habitats, such as forests, generally eat insect prey and tend to be solitary. By contrast, species living in open habitats tend to be more gregarious and feed on seeds, which are more clumpily distributed than insects. The results imply that resource distribution is associated with foraging group size. Clumped resources like seeds and fruit tend to be unpredictably distributed in both space and in time, making group foraging a suitable strategy to exploit such resources (Chapter 1). They are also characterized by high within-clump food abundance, which reduces the costs of competing for resources. Dispersed food types, such as insects, are characterized by low within-clump abundance and a more even distribution, both of which are thought to deter group foraging.

Although uncovering associations between group foraging and environmental factors like habitat openness and diet support adaptive hypotheses for the evolution of social predation, more evidence is needed to assess the direction of the effect. In the weaverbirds example, selection pressures other than diet, such as predation avoidance, may have favoured the evolution of group foraging in the first place. Perhaps group foraging then forced bird species to specialize on a diet that can support a large number of competitors. In this case, diet represents the consequence rather than the cause of group foraging. Comparative analyses typically document the pattern of covariation between behavioural (or morphological) traits and ecological variables and less often the direction of the effects.

Modern comparative analyses have benefitted from advances in molecular phylogenetics, which allow a more precise mapping of the evolutionary relationships among extant species. In addition, the hypothesized ordering of events in the evolution of a trait can often be detected in these phylogenies, thereby addressing causality issues more directly. For instance, molecular phylogenetics showed that aposematism tended to appear prior to the evolution of sociality in lepidopteran larvae. This does not support the hypothesis that living in groups should facilitate the evolution of palatability signals (see Chapter 3).

9.3.2. Case Studies

In the following, I present four case studies to illustrate the insights that can emerge from comparative analyses of social predation. The first cases focus on types of species exposed to predators while the last case focuses on carnivorous species where sociality is expected to vary in response to the distribution of resources rather than predation pressure.

9.3.2.1. Group Size in African Antelopes

African antelopes vary tremendously in size, diet, habitat use, and group size (Leuthold, 1977), making this group ideally suited for a comparative analysis. Jarman (1974) used data from 75 species to uncover relationships between ecology and social tendencies. He hypothesized a number of different effects. In general, roughage-feeding species face little competition over their abundant food supply, which may favour the evolution of larger groups. Species that live in open habitats may be more exposed to predators, which again should select for larger group sizes. The need to cover a wide feeding area to track annual fluctuation in roughage quality and abundance may also favour a larger body size. Species living in small groups may have little defence against larger predators, and may thus seek to hide instead of running for cover or counter-attacking predators.

Jarman's analysis at the species level indicated that group size was positively associated with diet breadth and body mass (Fig. 9.3) and that species in larger groups tended to flee or counter-attack, rather than hide, when threatened by predators (Jarman, 1974). This study had a large impact on subsequent research despite lacking proper statistical testing. A detailed phylogeny of antelope species was lacking at the time, and no attempt was made to control for the fact that closely related species may show similar behaviour partly in response to shared ancestry.

A re-analysis of the data using statistical techniques to control for shared ancestry enables us to re-evaluate these previous qualitative conclusions (Brashares et al., 2000). In the re-analysis, both body size and group size still covary with diet. In addition, the relationship between group size and the tendency to rely on fleeing and counter-attacks when threatened by predators still stands. However, the relationship between group size and body size disappeared in the phylogenetically corrected analysis. Similar changes in the strength of a relationship before and after phylogenetic control have been noted in other comparative studies (Pagel, 1998), pointing to potential statistical artefacts in analyses based on species values. Inspection of Fig. 9.3 certainly suggests that the relationship between group size and body size is weaker within than between

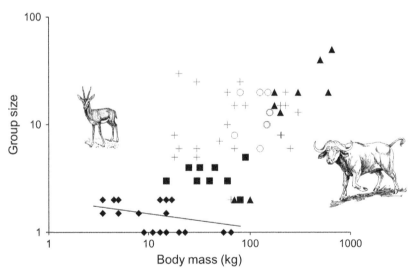

FIGURE 9.3   Covariation in group size, body mass, and diet in African antelopes: larger species of antelopes live in larger groups, and species that feed on rough vegetation, rather than specialized plant parts, also tend to live in larger groups. Selective browsers are represented by three groups (filled diamonds, filled squares, crosses) and less selective grazers are represented by two groups (open circles and filled triangles). A simple linear regression line was fitted for one selective browser group. The overall positive relationship between group size and body size did not persist in a phylogenetically controlled analysis. *Adapted from Brashares et al. (2000).*

subfamilies of antelopes, suggesting a relative lack of flexibility in group size among species with shared ancestry.

The pattern of covariation between body size, group size, diet, and habitat fits with the expectation that ecology can influence the evolution of behavioural traits. As noted earlier, however, the arrow of causation remains difficult to establish with simple correlations. For instance, it is not clear whether antelope species evolved larger group sizes after or before moving to more open habitats.

Further comparative analyses in African antelopes have sought to refine and extend these earlier findings. One comparative analysis in particular sought to uncover a relationship between group size and longevity. Jarman hypothesized that species living in larger groups can rely on one another to deter predation. Indeed, as we saw earlier, larger groups can detect predators sooner and counter-attack if needed. This makes larger groups more effective in avoiding predation, thereby reducing extrinsic mortality caused by predation. Life-history theory predicts that such a reduction in extrinsic causes of mortality would lead to an increase in lifespan (Abrams, 1993). In African antelopes, the comparative analysis revealed that lifespan does increase with group size (Bro-Jørgensen, 2012) when the known correlates of longevity are controlled. However, such a relationship between longevity and sociality is not apparent in other groups of species (Beauchamp, 2010c; Blumstein and Møller, 2008; Kamilar et al., 2010), suggesting that the pattern uncovered in antelopes may not be universal. Why longevity fails to covary with sociality in different groups of species remains to be investigated.

Other antipredator traits in antelopes and other ungulate species have since been related to group living (Caro et al., 2004). Many of these traits may function as signals to alert other group members of a predation threat. However, such signals may also be directed at the predator and serve to deter attack, in which case such traits may not be closely associated with group living. Tail-flagging in fleeing white-tailed deer represents a classic example. Adaptive interpretations of the behaviour consider that deer use tail-flagging to entice the predator into a futile chase (Smythe, 1970) or to alert companions about danger (Hirth and McCullogh, 1977). If it turns out that tail-flagging is consistently associated with group living in a large number of species, we can be more certain that the behaviour serves to transfer information about predation threats to other group members.

The results from this comparative analysis revealed several interesting associations between antipredator traits and group size in ungulates. For instance, ungulates that live in large groups or those that live in more open habitats tend to use tail-flagging more often. This result may indicate a need to alert companions or to communicate with predators in habitats where threats are more common. Predator inspection, bunching, and group attacks are more predominant in larger species and those living in larger groups.

9.3.2.2. Group Size in Primates

Primates have been a key group for understanding covariation between ecological variables and social organization. Crook and Gartlan extended their analysis of covariation between body size, group size, and habitat type in weaverbirds to primate species. This work stimulated much research and laid the foundation for what became known as primate socio-ecology (Clutton-Brock and Janson, 2012). In their analysis, the researchers separated species by the timing of their activity (nocturnal vs diurnal) and their habitat preferences (open vs closed) and observed that more solitary species tend to be nocturnal or forest dwellers, and that terrestrial species living in more open habitats tend to forage in larger groups probably because they are exposed to more predation threats and food tends to be more scattered in such habitats (Crook and Gartlan, 1966). Many of these qualitative findings have been corroborated in subsequent quantitative analyses (Clutton-Brock and Harvey, 1977).

The relative influence of predation pressure and resource distribution on primate sociality has been a hotly debated topic in primatology. An influential paper by Wrangham in the early 1980s argued that the ability of larger groups to displace smaller groups from defendable resources played a key role in the evolution of sociality in primates (Wrangham, 1980). According to this hypothesis, variation in group size in primates is not directly linked to predation pressure. However, this makes it difficult to explain why some species live in groups despite there being little resource competition. A subsequent study suggested that both predation pressure and resource distribution have contributed to the evolution of group size in primates (van Schaik, 1983). In the current consensus, larger group sizes are thought to reduce individual predation risk and allow greater access to contested resources (Clutton-Brock and Janson, 2012). The spatio-temporal distribution of resources influences the level of competition within a group and sets a limit to the number of individuals that can coexist within a group.

Comparative analyses of primate species have been useful in testing these hypotheses. In particular, an analysis at the species level revealed that more terrestrial species tend to live in larger groups, supporting the prediction that habitats with presumably higher predation risk are associated with larger group sizes (Janson and Goldsmith, 1995). To assess the relevance of within-group competition, Janson and Goldsmith (1995) studied daily travel costs in a range of primate species. They hypothesized that an increase in group size would increase competition and force groups to travel more each day to fulfil their food requirements. Group size was indeed associated with an increase in daily travel costs, as would be expected if group size influences the intensity of feeding competition within groups (Fig. 9.4). However, neither of these tests used a phylogenetic control or a broad range of species. Therefore, it is not really clear whether the above conclusions are robust, especially given that many features of social organization in primates show little flexibility within different clades (di Fiore and Rendall, 1994).

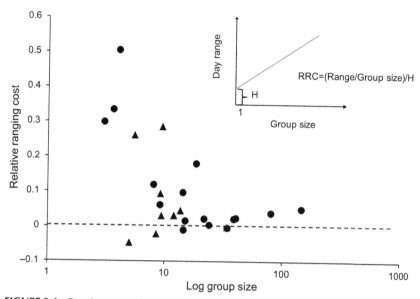

FIGURE 9.4 Ranging costs and group size in primates: relative ranging costs decrease in primate species living in larger groups (leaf-eating species: filled triangles; fruit-eating species: filled circles). Relative ranging cost is thought to reflect food competition within groups and represents the ratio between the slope of the relationship between day range and group size and the intercept when group size equals 1 (inset). When additional group members cause a large increase in day range in response to increased competition, the relative ranging cost will be high and group size will remain small. *Adapted from Janson and Goldsmith (1995). (For colour version of this figure, the reader is referred to the online version of this book.)*

It is important to bear in mind that several factors other than feeding competition may set a limit to group size. For instance, larger groups may be more easily detected by predators or face a greater risk of disease transmission, both of which would favour smaller groups. The importance of feeding competition as an evolutionary force shaping group size is also called into question by a recent comparative analysis of the reproductive consequences of daily travel costs (Pontzer and Kamilar, 2009). It was hypothesized that an increase in travel costs in response to competition would lead to a reduction in investment in maintenance and reproduction. However, this study found that mammalian species (including many primate species) which travel the most on a daily basis actually produce more offspring, suggesting that an increase in travel distances each day may represent a strategy to acquire more resources rather than a response to lower food availability caused by increased competition. Teasing apart the role of predation pressure, feeding competition and other factors in shaping variation in sociality remains a challenge for primate socio-ecology.

If terrestrial habitats are associated with higher predation risk, one might expect terrestrial species to have shorter lifespans in response to greater extrinsic mortality. Although arboreal species of mammals tend to live longer than terrestrial species of similar sizes, this is not true for primates (Shattuck and Williams, 2010). Terrestrial primates have evolved from arboreal ancestors, and arboreality represents a defining feature of most primate societies. The authors thus suggest that increased longevity has been inherited by all primate species, revealing phylogenetic inertia. However, the fact that longevity varies with group size in ungulates (see above), and that terrestrial species of primates tend to live in larger groups, suggests an alternative explanation: large group size in terrestrial primate species may be sufficient to decrease individual predation risk and maintain long lifespans, even in a riskier habitat. Therefore, I recommend extending this analysis to determine whether longevity does actually decrease in terrestrial species when controlling for both body size and group size.

Comparative analyses in primates have also shed light on the intriguing relationship between brain size, and indirectly cognitive abilities, and group size. The brain in many species of birds and mammals has grown over evolutionary time to become much larger than is needed to simply sustain physical needs. Nowhere is this more evident than in primate brains (Dunbar and Shultz, 2007a). Earlier hypotheses linked larger brain sizes in primates to body size and the need to form the detailed mental maps required to navigate large territories and locate unpredictable food supplies (Clutton-Brock and Harvey, 1980). Later, an intriguing hypothesis emerged: larger brains are linked to more complex social lives (Byrne and Whiten, 1988). Indeed, comparative analyses have shown that the relative size of the neocortex, the part of the brain most likely related to social skills, is positively correlated with social group size in many primate species (Dunbar, 1992) (Fig. 9.5).

The explanation for this effect is that individuals in groups face unique challenges, such as maintaining group cohesion in a social environment that poses many threats. This reasoning, dubbed the social brain hypothesis, recognizes that social life solves a host of ecological and behavioural challenges, such as predator avoidance or food finding, and that a larger brain solves the riddles of social life. Many ecological factors, such as diet and habitat type, are known to influence brain size directly. As is always the case with correlational approaches, teasing apart the factor(s) that directly led to enlarged brains from factors that merely acted as constraints is a significant challenge. One promising method is path analysis, a statistical technique that assigns a direction to purported associations. It turns out that several factors influence brain size, but it is relative neocortex size that is most closely associated with social factors, as would be predicted based on the social brain hypothesis (Dunbar and Shultz, 2007b).

The social brain hypothesis should apply to any group of species that faces social challenges, and indeed a positive association between relative neocortex size and group size has been documented in several mammalian taxa other than primates, including ungulates, bats, and carnivores (Dunbar and Bever, 1998; Pérez-Barberia et al., 2007; Shultz and Dunbar, 2005). However, an analysis of telencephalon size in birds, the avian counterpart of the neocortex, failed to provide support for the social brain hypothesis (Beauchamp and Fernández-Juricic,

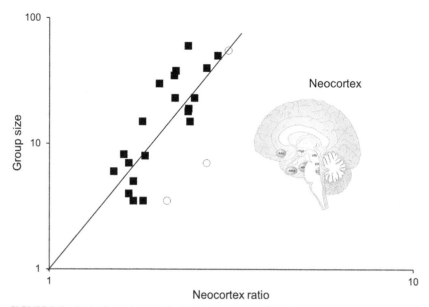

FIGURE 9.5   Brain size and group size in primates: group size and neocortex ratio are positively correlated in primates. Neocortex ratio equals neocortex volume divided by the volume of the rest of the brain. Similar trends in the evolution of brain size have been found for monkeys (filled squares) and apes (open circles). *Adapted from Dunbar and Schultz (2007)*. (For colour version of this figure, the reader is referred to the online version of this book.)

2004). Although this finding raises the possibility that the evolution of larger brains may have followed qualitatively different paths in birds and mammals, it may also be the case that adequate indices of social complexity are lacking in birds. Foraging group size, for instance, may be a poor proxy of social complexity in birds given the rather fluid nature of most avian groups. Testing the social brain hypothesis in birds may hold the most promise in cooperative-breeding species, which like most primates spend most of the year in stable social groups composed of closely related individuals (Ligon and Burt, 2004). An alternative hypothesis may be that selection pressures related to reproduction led to the initial evolution of larger brains in birds. Indeed, in birds and mammalian species other than primates, it appears that relatively larger brains have evolved in pair-bonded mating systems (Shultz and Dunbar, 2007), suggesting that the demands associated with maintaining stable pair bonds may also drive the evolution of larger brains.

Some recent studies are particularly exciting because they focus on identifying what specific traits associated with larger brains are enhancing fitness. If a larger brain allows for more cognitive complexity, it should be possible to determine the fitness consequences of variation in the expression of social cognition traits (Silk, 2007). Individuals that are best able to navigate their complex social environment should have a reproductive edge. The comparative method may also be useful in mapping the evolution of social cognition traits (MacLean et al., 2012), allowing us to infer ancestral states and determine associations with life history and environmental factors.

9.3.2.3. Group Size in Birds

The early work of Crook on weaverbirds related environmental factors to variation in group size among species. In general, flocking in birds is common in seed-eating species living in open habitats, presumably to facilitate the exploitation of locally abundant but spatially patchy resources. Other mostly anecdotal studies of various different species in specific areas have emphasized the role of predation pressure on the evolution of flocking. For instance, small species and those living in more exposed habitats tend to occur in larger groups providing more protection (Buskirk, 1976; Thiollay and Jullien, 1998). However, to apply such findings to a broader range of species, a phylogenetically corrected analysis is needed. I attempted to carry out a study, but immediately faced many challenges. In contrast to antelopes and primates, which are rather few in numbers and whose biology is quite well known, there are over 10,000 species of birds, and for many of these species we know little about their social behaviour and ecology. Phylogenetic relationships among species in many families were poorly known at the time (2002) and there is still much debate even on relationships at the family level (Hackett et al., 2008). For my study, I used the best phylogenetic evidence available at the time, aggregating information at the species level to make estimates of flocking tendency and ecological factors at the family level (Beauchamp, 2002a).

As is the case with ungulates and primates, I found that many factors have probably influenced the evolution of flocking in birds. Flocking evolved more often in families that forage for clumped resources, such as seeds or fruits, as would be predicted if flocking facilitates the exploitation of scattered resources. Flocking, however, evolved more often in families with larger body sizes and was not related to habitat openness, as would be predicted if flocking reduces predation risk. Pending further analysis at the species level, it would appear that, at the family level at least, resource distribution rather than predation pressure represents the driving force behind the evolution of flocking in birds.

Other comparative analyses in birds have focused on the evolution of traits associated with the formation and cohesion of groups, the latter being essential for group members to reap any benefits from living in groups. It has long been known that plumage in birds can serve communication purposes (Senar, 2006). For instance, strikingly coloured patches can signal social status to other group members (Rohwer, 1975). Similarly, flash marks, a patch of plumage that become conspicuous during flight, are thought to communicate alarm. As with tail-flagging in antelopes, flash marks may be aimed at predators, perhaps to increase confusion or deter pursuit. Flash marks may also alert companions and thereby facilitate escape by other group members.

A study of wading bird species that do and do not possess flash marks showed that flash marks were indeed more prevalent in flocking species (Brooke, 1998). This analysis has been extended to rails, marshland birds known for their secretive habits. Using a phylogenetic framework this time, it turns out that a white tail, a flash mark, is more common in species that flock at least part of the year and in species that live in more open habitats (Stang and McRae, 2009). Since adaptation to open habitats typically preceded the evolution of white tails in rails while gregariousness tended to evolve later, the authors concluded that white tails probably function as signals to deter predation rather than communicate alarm to group members.

Further evidence for the predator-deterrence function of plumage traits comes from the observation that pigeons with a white rump are more likely to evade capture by peregrine falcons than those without. During a chase, falcons may focus too intently on the white rump of a fleeing pigeon, missing abrupt changes in direction (Fig. 9.6) (Palleroni et al., 2005). Or it may be that white-rumped pigeons are simply better at escaping predators than other types of pigeons for some reason unrelated to their white rump. However, when the white rump of these pigeons was concealed experimentally, these pigeons, now dark-rumped, fell to attacks by falcons just as frequently as did the control birds with a naturally dark rump, suggesting that it was the white rump, rather than other physical attributes, that influenced predator success.

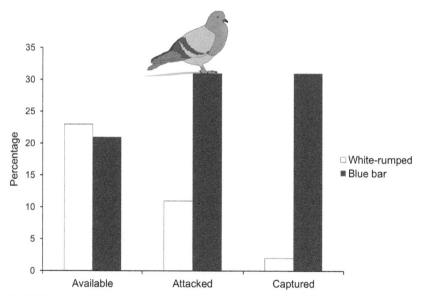

FIGURE 9.6   Prey plumage and predation: feral pigeons with a white rump are attacked less frequently by adult falcons than expected based on their availability in the local population, and are captured less frequently than blue bar pigeons, which lack a white rump. *Adapted from Palleroni et al. (2005).* (For interpretation of the references to colour in this figure legend, the reader is referred to the online version of this book.)

Although it is clear that some plumage traits in birds have evolved as communication signals aimed at predators, other traits appear to serve as signals directed at companions in the group. Two comparative studies in birds provide support for the conspecific-signalling hypothesis. In the first study, Heeb and I looked at the evolution of white plumage in birds. White colouration has been thought to increase the conspicuousness of individuals against a dark background. Hence, white plumage may serve as a passive signal to recruit distant foraging companions, thus facilitating the formation of flocks (Tickell, 2003). If whiter individuals obtain net benefits from attracting companions, directional selection should lead to whiter plumage over evolutionary time. We thus contrasted whiteness in the plumage of closely related pairs of species that markedly differ in flocking tendencies. Choosing closely related species increases the similarity between species and reduces the chances that other traits covaried with the change in flocking tendency. Our study revealed that species foraging in flocks tended to be whiter in general than their more solitary counterparts (Fig. 9.7) (Beauchamp and Heeb, 2001). Other traits failed to covary systematically with contrasting flocking tendencies in the species pairs, adding support to

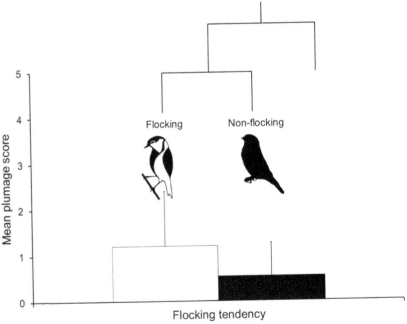

FIGURE 9.7　White plumage in birds and flocking: in pairs of closely related species of birds, flocking species tend to have whiter plumage than non-flocking counterparts ($n = 80$), supporting the hypothesis that white plumage has evolved as a signal to attract distant conspecifics for recruitment into the flock. Whiteness in the plumage of each species was ranked from one to five, with five being all white. The inset provides a hypothetical example of a pair of closely related species on a phylogenetic tree. Error bars show one standard deviation. *Adapted from Beauchamp and Heeb (2001).*

the conspecific-signalling hypothesis. These results immediately suggest future experiments, similar to those performed in pigeons, where the white plumage patches are darkened and the resulting effects on the recruitment of conspecifics observed.

The second piece of evidence for conspecific signalling comes from a study of delayed maturation in birds. It is common in birds for the onset of adult appearance to be delayed (Hawkins et al., 2012). Indeed, juvenile birds in many species moult into adult plumage often well after their first breeding season, resembling females from their first winter until their first breeding moult. Adaptive hypotheses for the evolution of delayed maturation focus on the benefits of advertising youth and inexperience. In the non-breeding season, non-adult appearance may be useful to signal social status (Rohwer and Butcher, 1988). Indeed, delayed maturation may increase survival by allowing individuals to avoid costly and mostly futile fights for resources with older, more experienced companions. Therefore, delayed maturation may be especially common in flocking species, where opportunities to interact with dominant companions are expected to be more frequent. Using the comparative pairwise approach, described in the previous paragraph, I identified pairs of species with contrasting modes of maturation and determined whether delayed maturation was more frequent in more social species (Beauchamp, 2003). For a broad range of species, I found that delayed plumage maturation was indeed more prevalent in flocking species. Although delayed plumage maturation may also play a role during the breeding season, when individuals compete for mating opportunities, my results suggest that advertising social status with non-adult traits can be beneficial when foraging in groups.

Comparative analyses in birds have also proven useful for teasing apart the relative contribution of predation pressure versus enhanced foraging efficiency to the evolution of group living. If living in groups represents primarily an adaptation to reduce predation, species facing little predation risk should not live in groups. This simple idea was formulated more than 40 years ago by Willis, who argued that species living on islands with reduced predation pressure should be solitary (Willis, 1972). The reasoning is that living in groups should disappear over evolutionary time in the absence of positive selection by predation pressure, especially if antipredator behaviour is costly (Wcislo and Danforth, 1997). A comparative analysis of island versus closely related mainland species may be able to determine whether this is indeed the case.

Previous studies have documented a wide range of morphological and behavioural changes in prey species living on predator-free islands, including the loss of antipredator behaviour (Blumstein and Daniel, 2005; Fullard, 1994). With respect to group size, one study showed that group size in long-tailed macaques is smaller on a predator-free island than on a nearby island with predators (van Schaik and van Noordwijk, 1986). In order to broaden the scope of the earlier analyses, I carried out a comparative analysis of flocking tendencies of birds living on many islands throughout the world, which should decrease the chances that the observed effects are specific to one particular population or island.

My comparative pairwise analysis showed that species that live on islands with reduced predation risk typically forage in smaller groups than do matched species that live on the mainland and face a fuller array of predators (Fig. 9.8) (Beauchamp, 2004). Nevertheless, it was rare that flocking disappeared totally on islands, suggesting that while predation pressure is probably a significant factor maintaining group living, it probably rarely acts alone.

There are a number of factors that may contribute to relaxed predation pressure over evolutionary time. In particular, a larger size is thought to provide a refuge from predation by reducing the number of potential predators (Cohen et al., 1993). Similarly, foraging in remote habitats or at times of day where predators are less active may also reduce predation pressure (Sih, 1987). Adopting the pairwise comparative method once more, I contrasted group size in closely related species that experience widely different levels of predation risk due to changes in body size or habitat use. As was the case for island species, species under relaxed predation pressure tend to live in smaller groups, but again flocking rarely disappeared completely (Beauchamp, 2010d). Similar studies are needed with other types of species to assess the generality of these conclusions.

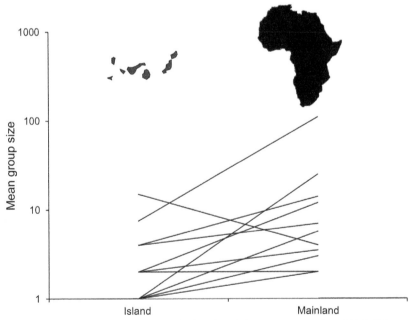

FIGURE 9.8 Group size on islands with relaxed predation pressure: species of birds living on islands with relaxed predation pressure tend to forage in smaller groups than their matched counterparts on the mainland. *Adapted from Beauchamp (2004).* (For colour version of this figure, the reader is referred to the online version of this book.)

9.3.2.4. Group Size in Dolphins

Some of the best known examples of social predation come from large carnivorous species, such as lions, wolves, and dolphins. Yet these species are the exception rather than the rule: most carnivorous species hunt alone. For this reason, it may be relatively easy to pinpoint the ecological factors underlying sociality in carnivorous species.

My discussion of group size in antelopes, primate, and birds relied considerably on the results of comparative analyses, where the focus is on variation in social behaviour among species. However, there is another approach and that is to analyse variation among populations within the same species. In fact, a number of species show substantial variation in group size from one population to another, and this flexibility in behavioural responses provides another opportunity to assess which factors drive sociality. The results of such an approach also have good generalizability because the relationship between group size and ecological factors is expected to be similar both within and between species. In contrast to comparative analyses at the species level, comparative studies within species need not be corrected for shared ancestry, thus facilitating the statistical analysis of the data.

Dolphins and porpoises are large aquatic carnivorous mammals that occupy a wide range of habitats and show large variation in group size between populations (Gygax, 2002). In a study of inter-population variation, group size was negatively correlated with body size and tended to be larger in more open habitats for many species of dolphins, but not for porpoises (Gygax, 2002). Environmental variables, such as temperature, which probably influence the availability of food, also influenced the pattern of variation in group size in some species. These results support the general finding, documented earlier in antelopes and birds, that both predation pressure and food availability influence grouping patterns in animals. However, the pattern of covariation between group size and specific ecological factors varied from species to species, thus precluding general conclusions. Different responses by different species to the same set of ecological factors are rather troublesome for comparative analyses at the species level and sometimes the results are unexpected. For instance, in killer whales, group size increased with habitat openness, an antipredator strategy that would not be expected in such a top predator. Habitat openness, which is usually considered a proxy for predation risk, may thus reflect variation in other selection pressures, such as food availability.

Other studies involving large carnivorous species, which tend to face few predation threats of their own, have sought to relate variation in group size to foraging efficiency and other traits specific to the species studied. In lions, the only species of cats that consistently hunts in groups, foraging success peaks when hunting alone or in groups of five to six. Yet, female lions often live in groups that would not maximize the daily *per capita* food intake (Packer et al., 1990). Evidence suggests that other factors have driven the evolution of group living in lions, including the ability to repel other lion groups (Mosser and Packer, 2009) and to protect cubs from infanticidal males (Packer et al., 1990). Similarly, both intra- and interspecific competition at kills have been invoked as factors contributing to larger group sizes in other large carnivorous species (Creel and Macdonald, 1995).

9.4. CONCLUDING REMARKS

Why species live in groups from an evolutionary standpoint has been the subject of much research. The social behaviour that we see today in animals has evolved over time. In this chapter, I attempted to identify some selection pressures that have shaped this behaviour in different species. Co-evolution paints a dynamic view of predator–prey relationships with both predator and prey adjusting their tactics in response to each other. While there is solid evidence for co-evolution in the case of morphological traits, this is less true for reciprocal adjustments in behavioural tactics by predators and their prey. Recent game-theory models of arms races between predators and prey have led to testable predictions, but empirical scrutiny is now needed.

Comparative analyses have proven very useful in identifying the ecological drivers of social tendencies and behaviour in animals. This approach, in contrast to the co-evolution paradigm, assumes that selection pressures are fixed over evolutionary time and that animals adapt to static ecological circumstances. For instance, some traits, like small body size, are thought to increase predation risk, thereby favouring the evolution of group living as an antipredator defence. The risk associated with group living is not thought to change over evolutionary time, generating directional selection pressure to evolve antipredator traits. Of course, this may not be the case if predators can adapt their behaviour to the antipredator responses of their prey. For example, predators may evolve or learn to overcome the confusion effect associated with the escape of their small prey in groups, which would make directional selection less likely.

Nevertheless, comparative analyses have been extremely useful in identifying the driving factors behind the evolution of group living in predators and their prey. Overall, the evidence that I have reviewed here shows that resource dispersion and predation pressure often combine to favour social foraging. In addition, many behavioural and morphological traits covary with group size, including brain size and longevity, which may reflect novel selection pressures that arise once animals live in groups. Much of the work on the evolution of group living in predators and their prey has focused on birds and mammals. We now need to extend comparative analyses to other animal groups, which is now possible thanks to recent developments in molecular phylogenetic studies. There is, in fact, considerable variation in gregariousness associated with foraging in fish (Hoare et al., 2000), and in many invertebrates as well, such as cockroaches (Grandcolas, 1998), spiders (Uetz, 1992), marine zooplankton (Ritz et al., 2011), and phytophagous insects (Costa and Pierce, 1997). Comparative analyses of these species groups can tell us whether there are general themes in the evolution of sociality across the animal kingdom.

Comparative analyses need not be restricted to inter specific variation. As I have shown, variation in gregariousness within species can also provide useful information on the ecological drivers of sociality (Lott, 1991). However, this approach faces several challenges. First of all, intra specific analyses are often restricted to a small number of populations where sufficient variation occurs in key ecological factors. This limits their statistical power. In addition, the assumption that no phylogenetic correction is required when analysing populations of the same species may not always be valid. For example, different populations may represent different subspecies that have evolved from a common ancestral population, and so they are not statistically independent. Another nagging issue is that differences among populations may not indicate adaptive variation, but rather simply be a by-product of differences in population demography. This would seem to be the case in the following two cases. First, intra specific variation in dominance structure in group-living primates probably reflects changes in the number of individuals available to form coalitions rather than adaptive variation in dominance style (Datta and Beauchamp, 1991). Second, intra specific variation in group size probably reflects to some extent variation in population size rather than adaptation to local ecological circumstances (Beauchamp, 2011c; Hensor et al., 2005). Given these caveats, it may be premature to focus too intensely on within-species variation, as advocated by some researchers (Chapman and Rothman, 2009).

A persistent problem in comparative analyses, both within and between species, revolves around the identification of habitat or morphological characteristics that influence predation risk. Although it is clear that an increase in predation risk may lead to the evolution of group living, it has been much harder to identify when individuals are expected to experience higher predation risk. For example, it has been predicted that predation risk in rodents should be higher in larger species (Ebensperger and Blumstein, 2006) while the reverse is typically predicted for birds (Buskirk, 1976; Thiollay and Jullien, 1998), primates (Isbell, 1994), and dolphins (Gygax, 2002). Interestingly, researchers have found no clear relationship between group size and body size in African antelopes, while larger species are, in fact, more likely to live in groups in birds (Beauchamp, 2002a) and primates (Janson and Goldsmith, 1995). Obviously, differences between groups of animals in the association between group size and body size may simply reflect the action of different predation pressures, but it may also be the case that our understanding of what constitutes predation risk is shaky.

A real measure of predation risk must be based on the likelihood that an individual dies over a given period of time when no time and energy is invested in antipredator defences (Hill and Dunbar, 1998). Predation risk should thus reflect factors that are not under the control of the prey, such as the number of predators present, the availability of alternative prey for the predators, and the presence of refuges for the prey. However, to get such estimates, one has to rely on current predator behaviour, which may have been shaped in the past by antipredator traits in their prey. To take an example, prey species may have evolved complex antipredator traits in very risky habitats forcing predators to hunt now in what were in the past low risk habitats, but which would be classified as high risk based on the current spatial distribution of predators. Refining estimates of predation risk should be a priority when undertaking comparative analyses involving selection pressures based on predation.

Conclusion

WHAT HAVE WE LEARNED?

In this book, I wanted to illustrate the various ways group living can be advantageous to predators and their prey. For predators, I showed that group members may detect prey faster, acquire larger prey, spend more time foraging, defend their prey or displace other groups more easily, harvest resources more efficiently, and make more accurate choices during foraging. However, foraging in groups may also be costly because individuals compete more intensely for resources, either directly or indirectly, produce resources at a lower rate than expected from the size of their groups or attract more predators. It is also no surprise that selfish, exploitative strategies have evolved when exploiting resources in groups. Indeed, scrounging resources obtained by companions in the group is expected to increase in frequency in larger groups, reducing the ability to find resources at the group level and decreasing the amount of resources obtained by each group member.

Various lines of defences against predators are available to a prey species that lives in groups, with some defences deployed early in the predation sequence while others come into play later as the predator closes in on the group. Animals in groups invest time in antipredator vigilance, and even though models predict that individuals should invest less time in vigilance in larger groups, typically more eyes and ears are available in a group to scan the surroundings for predation threats. Individuals in groups may thus be able to allocate more time to other fitness-enhancing activities, such as foraging and sleeping, and detect predators sooner by pooling information gathered from the senses of many individuals. Groups of prey may be encountered at a lower frequency by predators, and if predators can only capture one individual during an attack, prey in groups will also benefit from a dilution of predation risk. After an encounter with a predator, prey in groups may also reduce their relative predation risk by seeking more sheltered positions in the group. Fleeing in a group may create confusion in the predator and reduce capture rate. The confusion effect may favour homogeneity in the appearance and behaviour of prey group members. Animals in groups may also gain protection against predators by adopting defensive formations or by actually chasing predators away. Signals aimed at dissuading predators from attacking in the first place also appear quite common in prey that live in groups. Despite these various lines of defences, not all prey animals appear to survive better in groups, suggesting that other forces keep animals together.

In general, the tendency to form groups provides opportunities to increase foraging efficiency and decrease predation risk, but entails costs like increased competition. Striking a balance between the costs and benefits of foraging in groups should lead to the formation of groups whose size maximizes individual fitness. However, the expected group size will vary depending on whether entry in the group is free or restricted. In a free-entry system, groups are expected to be larger than the optimal size, while groups that control entry may be able to keep group size closer to the optimal value. The paradox that groups often tend to be larger than the optimal size in a free-entry system can be resolved through many means. In particular, relatedness, within-group competition and learning about habitat quality may all act to keep the size of a group closer to the optimal value.

Group composition has emerged as an important consideration when assessing the fitness consequences of group living. Segregation of groups along phenotypic traits, such as body size, has been documented in many species, and certainly suggests that individuals pay attention not only to the size but also to the composition of their groups. The formation of mixed-species groups also stresses the importance of group composition. In particular, species in these groups may pool their specific attributes to increase the detection of resources and predators. However, some species may be at a competitive disadvantage in mixed-species groups or may suffer proportionately more predation. If mixed-species groups have evolved to become functional units with clear advantages to participating species, the formation of such groups may represent an example of niche construction, with the intriguing possibility that specific traits may evolve in response to novel selection pressures that arise when joining other species. Leadership and mimicry in mixed-species groups represent potential cases of niche construction.

Why predators and prey live in groups from an evolutionary standpoint has been the object of much research. Co-evolution paints a dynamic view of predator–prey relationships, suggesting a struggle to adjust to tactics adopted by the predator or the prey within one generation or through evolutionary time. Co-evolution is expected to influence the expression of various traits in predators and their prey, suggesting that analyses that treat such traits as fixed may be oversimplified. More broadly from an evolutionary perspective, comparative analyses in various types of animals have identified ecological drivers of social tendencies and behavioural patterns, including the spatial and temporal distribution of resources and level of predation risk. Comparative analyses have also uncovered several correlates of group size in animals, such as brain size, antipredator defences, and visual signals.

WHERE DO WE GO FROM HERE?

I want to highlight challenges for future research on social predation. I first start with specific challenges for predator and prey species, respectively, and then proceed to more general issues applicable to all types of species.

PREDATORS

For species of predators subject to predation, the net value of living in groups will depend not only on their ability to find and exploit resources, but also on their ability to escape predation. How vulnerability to predation varies with group size clearly represents an area that deserves more attention. In particular, are species living in groups able to deflect predation by encountering predators less often, or do they attract more predators or pass along the risk to more vulnerable group members?

Overlap in search areas for predators in groups appears inevitable and yet little is known about this potential cost of group foraging. Adjustments in search paths may be able to decrease the extent to which foragers interfere with one another, but documenting search paths with and without interference will require exacting studies. At the moment, it is also difficult to determine whether the slower rate of prey discovery in larger groups results from overlapping search areas or more simply from a decrease in search effort caused either by a greater investment in scrounging or by the avoidance of threatening companions.

At the proximate level, predators in groups must move together and maintain cohesion while searching for prey. Different individuals in the group may vary in the quantity and quality of the information they possess and may also face unique challenges in relation to their phenotypic attributes. For instance, some individuals may be more vulnerable to predation or may be better at finding resources. With the recent surge of interest in collective animal behaviour and decision-making, it may be possible in the near future to determine how groups can reach a consensus about such collective decisions and maintain cohesion despite the various needs and challenges of different group members.

Exploitative strategies in predators searching for resources together have emerged as a major constraint on achieving increased foraging efficiency in groups. The study of producing and scrounging tells us that some individuals may invest relatively little in searching for resources and expect to benefit from the discoveries made by companions. Scrounging emerges as a cost of foraging in groups because time spent scrounging detracts from the ability to discover resources. However, the lack of evidence for a trade-off between producing and scrounging hampers our understanding of the tactics used in a group to obtain resources. Other research paradigms assume no incompatibility between search modes, and a challenge for future work will be to provide relevant empirical evidence in support of each paradigm.

PREY

One of the greatest challenges for prey species remains to tease apart the contribution of each mechanism available to increase individual safety in groups. More than one mechanism probably plays an important role in many species and several mechanisms, unfortunately, often make the same predictions.

Although it is quite clear that living in groups can reduce predation risk, this relationship is far from universal. A weak association between predation risk and group living may occur because limited resources are allocated to other fitness-enhancing traits like foraging. A difficulty at the moment is that while we have some evidence for fitness benefits associated with the encounter-dilution effect in largely immobile species, survival consequences in other species and in relation to other mechanisms are currently lacking.

In many species, more than one ploy probably plays a role in deterring predation. What is the best combination of ploys for deterring predation, and under which environmental conditions certain combinations work best, remain unanswered questions. The behavioural ecology approach may be particularly well suited to investigating the adaptive value of different ploy combinations.

Antipredator vigilance has received a lot of attention over the years and plays a crucial role in predator detection and in the trade-off between predation risk and foraging efficiency. Recent theoretical developments have made it possible to examine changes in vigilance in response to variation in a host of ecological factors, such as patch richness and energy levels. Testing predictions from these models clearly represents an area that deserves more attention.

Many empirical studies have questioned the validity of assumptions made in early models of antipredator vigilance, including randomness and independence of vigilance among group members. The occurrence of vigilance copying among group members, in particular, certainly complicates matters, as individual vigilance now becomes a function of the behaviour of many neighbours. The foundation for vigilance models including copying has been laid, but much theoretical work remains to be done. For instance, how copying should vary with factors such as group size and predation risk has yet to be addressed.

The key prediction that vigilance should decrease with group size has been documented in many species, and yet we still know little about why the magnitude of the effect of group size on vigilance varies from species to species. The group-size effect on vigilance also offers other opportunities for future research. For example, this effect relies on untested assumptions about the perception of group size, which should be examined more thoroughly.

PREDATORS AND PREY

Both predators and their prey live in groups whose size may be under selection pressure to maximize fitness. Our understanding of the factors that influence group size would greatly benefit by extending the range of species used for these analyses, which up to now have been mostly limited to large carnivorous species.

Segregation of groups along phenotypic traits suggests that individuals pay attention not only to the size but also to the composition of their groups. Perhaps the greatest impediment to our understanding of the factors that control joining preferences is the almost total lack of data from groups other than fish. Future work with other species should be able to establish the generality of the previous findings.

Optimal group size is based on the notion of interchangeable group members. However, the ability to form and maintain partnerships with particular companions may set a limit on the size of a group, which may differ considerably from the size predicted from simple considerations about food and predators. Social network analysis may prove useful to examine issues about relationships between group members and their consequences on optimal group size and composition in single- as well as mixed-species groups.

GENERAL ISSUES

Three general issues spring to mind to further the field of social predation. Succinctly, when investigating various aspects of social predation, past studies have used a narrow range of species, focused on a narrow range of explanations, and have made some rather simplifying assumptions about interactions between predators and prey. Because social predation studies tend to involve behavioural ecologists, it comes as no surprise that certain taxonomic groups, such as birds, and functional explanations have been favoured in the past (Owens, 2006).

NARROW TAXONOMIC FOCUS

To bring back glaring examples of narrow taxonomic focus, most studies on group composition have involved laboratory fish, studies on scrounging have relied mostly on captive bird species, and studies on vigilance and mixed-species grouping have concentrated on birds and mammals, while studies on optimal group size have been mostly concerned with large mammalian carnivores. Some of these biases reflect historical contingencies. For instance, work on scrounging originated in the 1980s and used captive bird species as study systems. Extension of this work thus naturally concentrated on birds in the laboratory. Similarly, influential models of antipredator vigilance were derived from the small-bird-in-winter paradigm, which explains why birds still attract much attention in vigilance research. Other biases reflect the types of approaches favoured by researchers. Experimental analyses, for one, can only be carried out with relatively small species in confined spaces, such as insects or fish. Work on collective behaviour and group size, which relies on sophisticated experimental approaches, thus tends to favour insects or fish species.

Narrow taxonomic focus also partly arises because scientists are busier than ever. It is indeed always a temptation to pay closer attention to research carried out with our own group of animals, which will tend to reinforce the association between particular questions and particular groups of animals. This is evident in my field of research on vigilance, where strikingly the vast majority of the papers cited in an article involve the taxonomic group under investigation.

Such biases, which may arise for a number of reasons, can have a negative impact on social predation studies. First and foremost, the quality of the inference drawn from studies limited to a narrow range of species under a narrow range of conditions may be similarly limited. For instance, studies of scrounging in the laboratory assume that patches discovered by companions can be reached instantaneously and at no cost to scroungers, which is not likely the case in the field where distance matters. As described earlier, predictions from scrounging models can differ quite remarkably in a spatially explicit world. As another example, factors that influence the size and composition of groups have been mostly explored in the laboratory with fish species. Although some of the conclusions reached in the laboratory have been supported in the field, others remain to be conclusively demonstrated in the field and with other species. Consider also that the fitness benefits associated with some antipredator ploys are only known for relatively sessile organisms, such as insect larvae or territorial fish. Whether such conclusions extend to other animal groups remains an open question.

Extending the focus of social predation studies to a broader range of species appears urgent to determine how closely our expectations fit with observations from the natural world. Although it will always be the case that some species or groups of species are more suited to a particular type of question, it is important to realize that many different study systems are amenable to investigate general issues.

NARROW EXPLANATIONS

Social predation studies typically ask questions related to the adaptive value of living in groups for predators and their prey. The history of the evolution of social predation has received less attention. Similarly, questions regarding mechanisms and ontogeny have been largely neglected. All of these questions are important to get a fuller understanding of any behavioural pattern. As an example, it has now become quite clear how animals can benefit from antipredator vigilance and how vigilance should be adjusted in response to variation in a host of ecological factors such as group size and predation risk. However, we know little about the driving factors in the evolution of vigilance. In addition, how animals perceive group size and how their sensory apparatus influences vigilance patterns is poorly known. How vigilance patterns change during the lifetime of a species has also been neglected. The same lopsided view applies to many antipredator ploys and predator strategies to exploit resources. A better balance between these various questions is certainly going to improve our understanding of social predation in animals. The recent realization that mechanisms and other types of questions matter in behavioural ecology will help to adopt a broader view when investigating social predation.

NARROW ASSUMPTIONS

Many developments in social predation research are based on the notion that predators and their prey are inflexible in their strategies. For instance, the ESS level of vigilance maintained by prey in a group is calculated assuming a fixed rate of attack by predators. Conversely, the rate of encounter with prey for a predator is assumed to be constant in a given habitat. The idea that predators and their prey may follow courses of action that depend on each other's behavioural patterns suggests a more dynamic view of their relationships. In this context, the following types of questions become highly relevant: for instance, is there anything to prevent a predator from learning or evolving the ability to overcome the confusion effect, or to prevent a prey species to learn or evolve the ability to alter encounter rates with predators through changes in vigilance or habitat use? Incorporating reciprocal selection pressures of this sort will certainly increase the complexity of theoretical models dealing with social predation, but may be able to shed more light on issues that are resolutely more complex than initially thought.

I now end on a more philosophical tone. It should be clear now that predators can increase their foraging efficiency by hunting in groups and that prey living in groups can experience a reduction in predation risk. In the end, who wins? Would we not expect a social predator to capture prey every time or a prey species to always evade predation if evolution produces ever better adapted individuals? Obviously, the perfect predator would be a victim of its own success by wiping out its resource base. So, perhaps, the predator species that we see today are the less than perfect ones on their way to better success. It may also be the case that faced with dwindling resources, predatory species must include other species in their diet, each with their own defences against which no single attack tactic may be effective, leaving us with imperfect predators. The perfect prey species may also become victim of its own success. The increase in population size, which results from a decrease in mortality rate through predation, may attract other predators, each with their own mode of attack against which no single defence may be effective. The chase to become more efficient may thus lead to the evolution of imperfect predators and prey.

Predators and their prey may not always be able to develop the traits or tactics that would increase their success. For example, reducing inter-individual distances to increase collective detection ability may well increase the ability to escape predation, but may increase foraging interference and reduce food intake rate so that in the end, less effective collective detection does maximize overall fitness. As another example, predators in groups may be economically unable to prevent scroungers from exploiting their food discoveries: resource defence may be too time consuming or the risk of injury too high. In the end, scrounging may prevent predators from achieving a higher foraging success.

In a co-evolutionary framework, it makes sense to find that predators and their prey are imperfect as each tries to catch up with the changes adopted by the other so that, at any one time, one must be lagging behind the other. Despite all the ways living in groups may benefit predators and their prey, there are several reasons to expect that no one should win. The fascinating question is to find out why.

References

Abrams, P.A., 1993. Does increased mortality favor the evolution of more rapid senescence? Evolution 47, 877–887.

Agrawal, A.A., 2001. Phenotypic plasticity in the interactions and evolution of species. Science 294, 321–326.

Ale, S.B., Brown, J.S., 2007. The contingencies of group size and vigilance. Evol. Ecol. Res. 9, 1263–1276.

Alexander, R.D., 1974. The evolution of social behavior. Ann. Rev. Ecol. Syst. 5, 325–383.

Allen, W.E., 1920a. Behavior of feeding mackerel. Ecology 1, 310.

Allen, W.E., 1920b. Behavior of loon and sardines. Ecology 1, 309–310.

Allen, J.A., Greenwood, J.J.D., 1988. Frequency-dependent selection by predators [and discussion]. Philos. Trans. R. Soc. London Ser. B. 319, 485–503.

Altmann, M., 1958. The flight distance in free ranging big mammals. J. Wildl. Manage. 22, 207–209.

Altmann, S.A., Altmann, J., 1970. Baboon Ecology: African Field Research. University of Chicago Press, Chicago.

Anderson, J.G.T., 1991. Foraging behavior of the American white pelican (*Pelecanus erythrorhynchos*) in western Nevada. Colon. Waterbirds 14, 166–172.

Andersson, M., 1976. Predation and kleptoparasitism by skuas in a Shetland seabird colony. Ibis 118, 208–217.

Andersson, M., Götmark, F., Wiklund, C.G., 1981. Food information in the black-headed gull, *Larus ridibundus*. Behav. Ecol. Sociobiol. 9, 199–202.

Ang, T.Z., Manica, A., 2010. Unavoidable limits on group size in a body size-based linear hierarchy. Behav. Ecol. 21, 819–825.

Anstey, M.L., Rogers, S.M., Ott, S.R., Burrows, M., Simpson, S.J., 2009. Serotonin mediates behavioral gregarization underlying swarm formation in desert locusts. Science 323, 627–630.

Arnegard, M.E., Carlson, B.A., 2005. Electric organ discharge patterns during group hunting by a mormyrid fish. Proc. R. Soc. London Ser. B. 272, 1305–1314.

Avilés, L., Tufiño, P., 1998. Colony size and individual fitness in the social spider *Anelosimus eximius*. Am. Nat. 152, 403–418.

Avilés, L., Fletcher, J.A., Cutter, A.D., 2004. The kin composition of social groups: trading group size for degree of altruism. Am. Nat. 164, 132–144.

Bahr, D.B., Bekoff, M., 1999. Predicting flock vigilance from simple passerine interactions: modelling with cellular automata. Anim. Behav. 58, 831–839.

Bailey, I., Myatt, J.P., Wilson, A.M., 2012. Group hunting within the carnivora: physiological, cognitive and environmental influences on strategy and cooperation. Behav. Ecol. Sociobiol. 1–17.

Baker, M.C., Belcher, C.S., Deutsch, L.C., Sherman, G.L., Thompson, D.B., 1981. Foraging success in junco flocks and the effects of social hierarchy. Anim. Behav. 29, 137–142.

Baker, D.J., Stillman, R.A., Smart, S.L., Bullock, J.M., Norris, K.J., 2011. Are the costs of routine vigilance avoided by granivorous foragers? Funct. Ecol. 25, 617–627.

Barber, I., Hoare, D., Krause, J., 2000. Effects of parasites on fish behaviour: a review and evolutionary perspective. Rev. Fish Biol. Fish 10, 131–165.

Barbosa, A., 1995. Foraging strategies and their influence on scanning and flocking behaviour of waders. J. Avian Biol. 26, 182–186.

Barbosa, A., 2002. Does vigilance always covary negatively with group size? Effects of foraging strategy. Acta. Ethol. 5, 51–55.

Barlow, A., Pook, C.E., Harrison, R.A., Wüster, W., 2009. Coevolution of diet and prey-specific venom activity supports the role of selection in snake venom evolution. Proc. R. Soc. London Ser. B. 276, 2443–2449.

Barnard, C.J., 1979. Predation and the evolution of social mimicry in birds. Am. Nat. 113, 613–618.

Barnard, C.J., 1980. Flock feeding and time budgets in the house sparrow (Passer domesticus L.). Anim. Behav. 28, 295–309.

Barnard, C.J., 1984. The evolution of food-scrounging strategies within and between species. In: Barnard, C.J. (Ed.), Strategies of Exploitation and Parasitism: Producers and Scroungers, Chapman & Hall, London, pp. 95–126.

Barnard, C.J., Sibly, R.M., 1981. Producers and scroungers: a general model and its application to captive flocks of house sparrows. Anim. Behav. 29, 543–550.

Barnard, C.J., Stephens, H., 1981. Prey selection by lapwings in lapwing-gull associations. Behaviour 77, 1–22.

Barnard, C.J., Stephens, H., 1983. Costs and benefits of single and mixed-species flocking in field-fares (Turdus pilaris) and redwings (T. iliacus). Behaviour 84, 91–123.

Barnard, C.J., Thompson, D.B.A., Stephens, H., 1982. Time budgets, feeding efficiency and flock dynamics in mixed-species flocks of lapwings, golden plovers and gulls. Behaviour 80, 44–69.

Barrette, M., Giraldeau, L.-A., 2006. Prey crypticity reduces the proportion of group members searching for food. Anim. Behav. 71, 1183–1189.

Barros, M., Alencar, C., Silva, M.A.d.S., Tomaz, C., 2008. Changes in experimental conditions alter anti-predator vigilance and sequence predictability in captive marmosets. Behav. Proc. 77, 351–356.

Barta, Z., Giraldeau, L.-A., 1998. The effect of dominance hierarchy on the use of alternative foraging tactics: a phenotype-limited producing-scrounging game. Behav. Ecol. Sociobiol. 42, 217–223.

Barta, Z., Giraldeau, L.-A., 2001. Breeding colonies as information centers: a re-appraisal of information-based hypotheses using the producer-scrounger game. Behav. Ecol. 12, 121–127.

Barta, Z., Flynn, R., Giraldeau, L.-A., 1997. Geometry for a selfish foraging group: A genetic algorithm approach. Proc. R. Soc. London Ser. B. 264, 1233–1238.

Barta, Z., Liker, A., Monus, F., 2004. The effects of predation risk on the use of social foraging tactics. Anim. Behav. 67, 301–308.

Bates, H.W., 1863. The Naturalist on the River Amazons. Murray Press, London.

Baylis, J.R., 1982. Avian vocal mimicry: its function and evolution. In: Kroodsma, D., Miller, E.H. (Eds.), Acoustic Communication in Birds, Academic Press, New York, pp. 51–83.

Bazazi, S., Pfennig, K.S., Handegard, N.O., Couzin, I.D., 2012. Vortex formation and foraging in polyphenic spadefoot toad tadpoles. Behav. Ecol. Sociobiol. 66, 879–889.

Beatty, C.D., Bain, R.S., Sherratt, T.N., 2005. The evolution of aggregation in profitable and unprofitable prey. Anim. Behav. 70, 199–208.

Beauchamp, G., 1998. The effect of group size on mean food intake rate in birds. Biol. Rev. 73, 449–472.

Beauchamp, G., 2000a. The effect of prior residence and pair bond on scrounging choices in flocks of zebra finches (Taeniopygia guttata). Behav. Proc. 52, 131–140.

Beauchamp, G., 2000b. Learning rules for social foragers: implications for the producer-scrounger game and ideal free distribution theory. J. Theor. Biol. 207, 21–35.

Beauchamp, G., 2001a. Consistency and flexibility in the scrounging behaviour of zebra finches. Can. J. Zool. 79, 540–544.

Beauchamp, G., 2001b. Should vigilance always decrease with group size? Behav. Ecol. Sociobiol. 51, 47–52.

Beauchamp, G., 2002a. Higher-level evolution of intraspecific flock-feeding in birds. Behav. Ecol. Sociobiol. 51, 480–487.

Beauchamp, G., 2002b. Little evidence for visual monitoring of vigilance in zebra finches. Can. J. Zool. 80, 1634–1637.

Beauchamp, G., 2003. Delayed maturation in birds in relation to social foraging and breeding competition. Evol. Ecol. Res. 5, 589–596.

Beauchamp, G., 2004. Reduced flocking by birds on islands with relaxed predation. Proc. R. Soc. London Ser. B. 271, 1039–1042.

Beauchamp, G., 2005. Does group foraging promote efficient exploitation of resources? Oikos 111, 403–407.

Beauchamp, G., 2006a. Nonrandom patterns of vigilance in flocks of the greater flamingo, *Phoenicopterus ruber*. Anim. Behav. 71, 593–598.

Beauchamp, G., 2006b. Phenotypic correlates of scrounging behavior in zebra finches: role of foraging efficiency and dominance. Ethology 112, 873–878.

Beauchamp, G., 2007a. Effect of group size on feeding rate when patches are exhaustible. Ethology 113, 57–61.

Beauchamp, G., 2007b. Vigilance in a selfish herd. Anim. Behav. 73, 445–451.

Beauchamp, G., 2008a. Risk factors for predation attempts by peregrine falcons (*Falco peregrinus*) on staging semipalmated sandpipers (*Calidris pusilla*). Waterbirds 31, 651–655.

Beauchamp, G., 2008b. A spatial model of producing and scrounging. Anim. Behav. 76, 1935–1942.

Beauchamp, G., 2008c. What is the magnitude of the group-size effect on vigilance? Behav. Ecol. 19, 1361–1368.

Beauchamp, G., 2009a. How does food density influence vigilance in birds and mammals? Anim. Behav. 78, 223–231.

Beauchamp, G., 2009b. Sleeping gulls monitor the vigilance behaviour of their neighbours. Biol. Lett. 5, 9–11.

Beauchamp, G., 2010a. A comparative analysis of vigilance in birds. Evol. Ecol. 24, 1267–1276.

Beauchamp, G., 2010b. Determinants of false alarms in staging flocks of semipalmated sandpipers. Behav. Ecol. 21, 584–587.

Beauchamp, G., 2010c. Group-foraging is not associated with longevity in North American birds. Biol. Lett. 6, 42–44.

Beauchamp, G., 2010d. Relaxed predation risk reduces but does not eliminate sociality in birds. Biol. Lett. 6, 472–474.

Beauchamp, G., 2011a. Collective waves of sleep in gulls (*Larus* spp.). Ethology 117, 326–331.

Beauchamp, G., 2011b. Fit of aggregation models to the distribution of group sizes in Northwest Atlantic seabirds. Mar. Ecol. Prog. Ser. 425, 261–268.

Beauchamp, G., 2011c. Functional relationship between group size and population density in Northwest Atlantic seabirds. Mar. Ecol. Prog. Ser. 435, 225–233.

Beauchamp, G., 2012a. Flock size and density influence speed of escape waves in semipalmated sandpipers. Anim. Behav. 83, 1125–1129.

Beauchamp, G., 2012b. Foraging speed in staging flocks of semipalmated sandpipers: evidence for scramble competition. Oecologia 169, 975–980.

Beauchamp, G., 2013. Is the magnitude of the group-size effect on vigilance underestimated? Anim. Behav. 85, 281–285.

Beauchamp, G., Fernández-Juricic, E., 2004. Is there a relationship between forebrain size and group size in birds? Evol. Ecol. Res. 6, 833–842.

Beauchamp, G., Fernández-Juricic, E., 2005. The group-size paradox: effects of learning and patch departure rules. Behav. Ecol. 16, 352–357.

Beauchamp, G., Giraldeau, L.-A., 1997. Patch exploitation in a producer-scrounger system: test of a hypothesis using flocks of spice finches (Lonchura punctulata). Behav. Ecol. 8, 54–59.

Beauchamp, G., Goodale, E., 2011. Plumage mimicry in avian mixed-species flocks: more or less than meets the eye? Auk 128, 487–496.

Beauchamp, G., Heeb, P., 2001. Social foraging and the evolution of white plumage. Evol. Ecol. Res. 3, 703–720.

Beauchamp, G., Livoreil, B., 1997. The effect of group size on vigilance and feeding rate in spice finches (Lonchura punctulata). Can. J. Zool. 75, 1526–1531.

Beauchamp, G., Ruxton, G.D., 2003. Changes in vigilance with group size under scramble competition. Am. Nat. 161, 672–675.

Beauchamp, G., Ruxton, G.D., 2005. Harvesting resources in groups or alone: the case of renewing patches. Behav. Ecol. 16, 989–993.

Beauchamp, G., Ruxton, G.D., 2007a. Dilution games: use of protective cover can cause a reduction in vigilance for prey in groups. Behav. Ecol. 18, 1040–1044.

Beauchamp, G., Ruxton, G.D., 2007b. False alarms and the evolution of antipredator vigilance. Anim. Behav. 74, 1199–1206.

Beauchamp, G., Ruxton, G.D., 2008. Disentangling risk dilution and collective detection in the antipredator vigilance of semipalmated sandpipers in flocks. Anim. Behav. 75, 1837–1842.

Beauchamp, G., Ruxton, G.D., 2012a. Changes in anti-predator vigilance over time caused by a war of attrition between predator and prey. Behav. Ecol. 23, 368–374.

Beauchamp, G., Ruxton, G.D., 2012b. Vigilance decreases with time at loafing sites in gulls (Larus spp.). Ethology 118, 733–739.

Beauchamp, G., Alexander, P., Jovani, R., 2012. Consistent waves of collective vigilance in groups using public information about predation risk. Behav. Ecol. 23, 368–374.

Bednarz, J.C., 1988. Cooperative hunting in Harris' hawks (Parabuteo unicinctus). Science 239, 1525–1527.

Bednekoff, P.A., Blumstein, D.T., 2009. Peripheral obstructions influence marmot vigilance: integrating observational and experimental results. Behav. Ecol. 20, 1111–1117.

Bednekoff, P.A., Lima, S.L., 1998a. Randomness, chaos and confusion in the study of antipredator vigilance. Trends Ecol. Evol. 13, 284–287.

Bednekoff, P.A., Lima, S.L., 1998b. Re-examining safety in numbers: interactions between risk dilution and collective detection depend upon predator targeting behaviour. Proc. R. Soc. London Ser. B. 265, 2021–2026.

Bednekoff, P.A., Lima, S.L., 2002. Why are scanning patterns so variable? An overlooked question in the study of anti-predator vigilance. J. Avian Biol. 33, 143–149.

Bednekoff, P.A., Lima, S.L., 2004. Risk allocation and competition in foraging groups: reversed effects of competition if group size varies under risk of predation. Proc. R. Soc. London Ser. B. 271, 1491–1496.

Bednekoff, P.A., Lima, S.L., 2005. Testing for peripheral vigilance: do birds value what they see when not overtly vigilant? Anim. Behav. 69, 1165–1171.

Bednekoff, P.A., Ritter, R., 1994. Vigilance in Nxai Pan springbok, Antidorcas marsupialis. Behaviour 129, 1–11.

Bekoff, M., 1996. Cognitive ethology, vigilance, information gathering, and representation: who might know what and why? Behav. Proc. 35, 225–237.

Belmaker, A., Motro, U., Feldman, M.W., Lotem, A., 2012. Learning to choose among social foraging strategies in adult house sparrows (Passer domesticus). Ethology 118, 1111–1121.

Belt, T.W., 1874. The Naturalist in Nicaragua. Murray Press, London.

Benoit-Bird, K.J., Au, W.W.L., 2009. Cooperative prey herding by the pelagic dolphin, Stenella longirostris. J. Acoust. Soc. Am. 125, 125–137.

Berdahl, A., Torney, C.J., Ioannou, C.C., Faria, J.J., Couzin, I.D., 2013. Emergent sensing of complex environments by mobile animal groups. Science 339, 574–576.

Berger, J., 1978. Group size, foraging, and antipredator ploys: an analysis of bighorn sheep decisions. Behav. Ecol. Sociobiol. 4, 91–99.

Berger, J., Cunningham, C., 1988. Size-related effects on search times in North American grassland female ungulates. Ecology 69, 177–183.

Bernstein, C., Kacelnik, A., Krebs, J.R., 1988. Individual decisions and the distribution of predators in a patchy environment. J. Anim. Ecol. 57, 1007–1026.

Berryman, A.A., Dennis, B., Raffa, K.F., Stenseth, N.C., 1985. Evolution of optimal group attack, with particular reference to bark beetles (Coleoptera: Scolytidae). Ecology 66, 898–903.

Bertram, B.C.R., 1978. Living in groups: predator and prey. In: Krebs, J.R., Davies, N.B. (Eds.), Behavioural Ecology, Blackwell, Oxford, pp. 64–96.

Bertram, B.C.R., 1980. Vigilance and group size in ostriches. Anim. Behav. 28, 278–286.

Bijleveld, A.I., Egas, M., van Gils, J.A., Piersma, T., 2010. Beyond the information centre hypothesis: communal roosting for information on food, predators, travel companions and mates? Oikos 119, 277–285.

Bijleveld, A.I., Folmer, E.O., Piersma, T., 2012. Experimental evidence for cryptic interference among socially foraging shorebirds. Behav. Ecol. 23, 806–814.

Biondolillo, K., Stamp, C., Woods, J., Smith, R., 1997. Working and scrounging by zebra finches in an operant task. Behav. Proc. 39, 263–269.

Blanchard, P., Fritz, H., 2007. Induced or routine vigilance while foraging. Oikos 116, 1603–1608.

Blumstein, D.T., 1996. How much does social group size influence golden marmot vigilance? Behaviour 133, 1133–1151.

Blumstein, D.T., Daniel, J.C., 2002. Isolation from mammalian predators differentially affects two congeners. Behav. Ecol. 13, 657–663.

Blumstein, D.T., Daniel, J.C., 2003. Foraging behavior of three Tasmanian macropodid marsupials in response to present and historical predation threat. Ecography 26, 585–594.

Blumstein, D.T., Daniel, J.C., 2005. The loss of anti-predator behaviour following isolation on islands. Proc. R. Soc. London Ser. B. 272, 1663–1668.

Blumstein, D.T., Møller, A.P., 2008. Is sociality associated with high longevity in North American birds? Biol. Lett. 4, 146–148.

Blumstein, D.T., Evans, C.S., Daniel, J.C., 1999. An experimental study of behavioural group size effects in tammar wallabies, Macropus eugenii. Anim. Behav. 58, 351–360.

Blumstein, D.T., Daniel, J.C., Griffin, A.S., Evans, C.S., 2000. Insular tammar wallabies (Macropus eugenii) respond to visual but not acoustic cues from predators. Behav. Ecol. 11, 528–535.

Blumstein, D.T., Daniel, J.C., Ardron, J.G., Evans, C.S., 2002. Does feeding competition influence tammar wallaby time allocation? Ethology 108, 937–945.

Blumstein, D.T., Daniel, J.C., Springett, B.P., 2004a. A test of the multi-predator hypothesis: rapid loss of antipredator behavior after 130 years of isolation. Ethology 110, 919–934.

Blumstein, D.T., Verneyre, L., Daniel, J.C., 2004b. Reliability and the adaptive utility of discrimination among alarm callers. Proc. R. Soc. London Ser. B. 271, 1851–1857.

Blumstein, D.T., 2013. Yellow-bellied marmots: insights from an emergent view of sociality. Philos. Trans. R. Soc. London Ser. B. 368, 2012349.

Blurton-Jones, N.G., 1984. A selfish origin for human food sharing: tolerated theft. Ethol. Sociobiol. 5, 1–3.

Bohlin, T., Johnsson, J.I., 2004. A model on foraging activity and group size: can the relative contribution of predation risk dilution and competition be evaluated experimentally? Anim. Behav. 68, F1–F5.

Boland, C.R.J., 2003. An experimental test of predator detection rates using groups of free-living emus. Ethology 109, 209–222.

Bonabeau, E., Dagorn, L., Fréon, P., 1999. Scaling in animal group-size distributions. Proc. Natl. Acad. Sci. 96, 4472–4477.

Booth, D.J., 1995. Juvenile groups of coral reef damselfish: density-dependent effects of individual fitness and population demography. Ecology 76, 91–106.

Botham, M., Kerfoot, C., Louca, V., Krause, J., 2005. Predator choice in the field; grouping guppies, *Poecilia reticulata*, receive more attacks. Behav. Ecol. Sociobiol. 59, 181–184.

Bourke, A.F.G., Franks, N.R., 1995. Social Evolution in Ants. Princeton University Press, Princeton.

Bouskila, A., Blumstein, D.T., 1992. Rules of thumb for hazard assessment: predictions from a dynamic model. Am. Nat. 139, 161–176.

Brashares, J.S., Garland, T., Arcese, P., 2000. Phylogenetic analysis of coadaptation in behavior, diet, and body size in the African antelope. Behav. Ecol. 11, 452–463.

Bremset, G., Berg, O.K., 1999. Three-dimensional microhabitat use by young pool-dwelling Atlantic salmon and brown trout. Anim. Behav. 58, 1047–1059.

Bro-Jørgensen, J., 2012. Longevity in bovids is promoted by sociality, but reduced by sexual selection. PLoS One 7, e45769.

Brockmann, H.J., Barnard, C.J., 1979. Kleptoparasitism in birds. Anim. Behav. 27, 487–514.

Brodie, E.D., Brodie, E.D., 1999. Predator–prey arms races. Bioscience 49, 557–568.

Brooke, M.d.L., 1998. Ecological factors influencing the occurrence of 'flash marks' in wading birds. Funct. Ecol. 12, 339–346.

Brouwer, L., Richardson, D.S., Eikenaar, C.A.S., Komdeur, J.A.N., 2006. The role of group size and environmental factors on survival in a cooperatively breeding tropical passerine. J. Anim. Ecol. 75, 1321–1329.

Brown, C.R., 1988. Social foraging in cliff swallows: local enhancement, risk sensitivity, competition and the avoidance of predators. Anim. Behav. 36, 780–792.

Brown, C.R., Hoogland, J.L., 1988. Risk of mobbing for solitary and colonial swallows. Anim. Behav. 34, 1319–1323.

Brown, G.E., Brown, J.A., 1996. Kin discrimination in salmonids. Rev. Fish Biol. Fish 6, 210–219.

Brown, G.E., Godin, J.G.J., Pedersen, J., 1999. Fin-flicking behaviour: a visual antipredator alarm signal in a characin fish. Anim. Behav. 58, 469–475.

Brown, W.L., Wilson, E.O., 1959. The evolution of the dacetine ants. Q. Rev. Biol. 34, 278–294.

Brunton, B.J., Booth, D.J., 2003. Density- and size-dependent mortality of a settling coral-reef damselfish (*Pomacentrus moluccensis* Bleeker). Oecologia 137, 377–384.

Bshary, R., Noë, R., 1997. Anti-predation behaviour of red colobus monkeys in the presence of chimpanzees. Behav. Ecol. Sociobiol. 41, 321–333.

Bugnyar, T., Kotrschal, K., 2002. Scrounging tactics in free-ranging ravens, Corvus corax. Ethology 108, 993–1009.

Burger, J., Gochfeld, M., 1981. Age-related differences in piracy behaviour of four species of gulls. Larus. Behav. 77, 242–267.

Burger, J., Gochfeld, M., 1994. Vigilance in African mammals: differences among mothers, other females, and males. Behaviour 131, 153–169.

Burtt, E.H., Gatz, A.J., 1982. Color convergence: is it only mimetic? Am. Nat. 119, 738–740.

Buskirk, W.H., 1976. Social systems in a tropical forest avifauna. Am. Nat. 110, 293–310.

Butler, M., MacDiarmid, A.B., Booth, J.D., 1999. The cause and consequence of ontogenetic changes in social aggregation in New Zealand spiny lobsters. Mar. Ecol. Prog. Ser. 188, 179–191.

Byrkjedal, I., Kålås, J.A., 1983. Plover's page turn into plover's parasite: a look at the dunlin/plover association. Ornis Fenn. 60, 10–15.

Byrne, R.W., Whiten, A., 1988. Machiavellian Intelligence: Social Expertise and the Evolution of Intellect in Monkeys, Apes and Humans. Clarendon Press, Oxford.

Caldwell, G.S., 1986. Predation as a selective force on foraging herons: effects of plumage color and flocking. Auk 103, 494–505.

Caley, M.J., Schluter, D., 2003. Predators favour mimicry in a tropical reef fish. Proc. R. Soc. London Ser. B. 270, 667–672.

Camazine, S., Deneubourg, J., Franks, N.R., Sneyd, J., Theraulaz, G., Bonabeau, E., 2001. Self-Organization in Biological Systems. Princeton University Press, Princeton.

Cameron, E.Z., Du Toit, J.T., 2005. Social influences on vigilance behaviour in giraffes, *Giraffa camelopardalis*. Anim. Behav. 69, 1337–1344.

Cangialosi, K.R., 1990. Social spider defense against kleptoparasitism. Behav. Ecol. Sociobiol. 27, 49–54.

Caraco, T., 1979a. Time budgeting and group size: a test of theory. Ecology 60, 618–627.

Caraco, T., 1979b. Time budgeting and group size: a theory. Ecology 60, 611–617.

Caraco, T., 1980. Stochastic dynamics of avian foraging flocks. Am. Nat. 115, 262–275.

Caraco, T., 1981. Risk-sensitivity and foraging groups. Ecology 62, 527–531.

Caraco, T., 1982. Flock size and the organisation of behavioral sequences in juncos. Condor 84, 101–105.

Caraco, T., Giraldeau, L.-A., 1991. Social foraging: producing and scrounging in a stochastic environment. J. Theor. Biol. 153, 559–583.

Caraco, T., Wolf, L.L., 1975. Ecological determinant of group size of foraging lions. Am. Nat. 109, 343–352.

Caraco, T., Barkan, C., Beacham, J.L., Brisbin, L., Lima, S., Mohan, A., Newman, J.A., Webb, W., Withiam, M.L., 1989. Dominance and social foraging: a laboratory study. Anim. Behav. 38, 41–58.

Carbone, C., DuToit, J.T., Gordon, I.J., 1997. Feeding success in African wild dogs: does kleptoparasitism by spotted hyenas influence hunting group size? J. Anim. Ecol. 66, 318–326.

Carere, C., Montanino, S., Moreschini, F., Zoratto, F., Chiarotti, F., Santucci, D., Alleva, E., 2009. Aerial flocking patterns of wintering starlings, *Sturnus vulgaris*, under different predation risk. Anim. Behav. 77, 101–107.

Caro, T.M., 1986. The functions of stotting in Thomson's gazelles: some tests of the predictions. Anim. Behav. 34, 663–684.

Caro, T.M., 1994. Cheetahs of the Serengeti Plains: Group Living in an Asocial Species. Chicago University Press, Chicago.

Caro, T.M., 2005. Antipredator Defenses in Birds and Mammals. University of Chicago Press, Chicago.

Caro, T.M., Graham, C.M., Stoner, C.J., Vargas, J.K., 2004. Adaptive significance of antipredator behaviour in artiodactyls. Anim. Behav. 67, 205–228.

Carro, M.E., Fernandez, G.J., 2009. Scanning pattern of greater rheas, *Rhea americana*: collective vigilance would increase the probability of detecting a predator. J. Ethol. 27, 429–436.

Carter, A.J., Pays, O., Goldizen, A.W., 2009. Individual variation in the relationship between vigilance and group size in eastern grey kangaroos. Behav. Ecol. Sociobiol. 64, 237–245.

Carter, K.D., Seddon, J.M., Frere, C.H., Carter, J.K., Goldizen, A.W., 2013. Fission-fusion dynamics in wild giraffes may be driven by kinship, spatial overlap and individual social preferences.
Anim. Behav. 85, 385–394.

Cash, K.J., McKee, M.H., Wrona, F.J., 1993. Short- and long-term consequences of grouping and
group foraging in the free-living flatworm *Dugesia tigrina*. J. Anim. Ecol. 62, 529–535.

Catterall, C.P., Elgar, M.A., Kikkawa, J., 1992. Vigilance does not covary with group size in an
island population of silvereyes (*Zosterops lateralis*). Behav. Ecol. 3, 207–210.

Chamberlain, S.A., Hovick, S.M., Dibble, C.J., Rasmussen, N.L., Van Allen, B.G., Maitner, B.S.,
Ahern, J.R., Bell-Dereske, L.P., Roy, C.L., Meza-Lopez, M., Carrillo, J., Siemann, E.,
Lajeunesse, M.J., Whitney, K.D., 2012. Does phylogeny matter? Assessing the impact of
phylogenetic information in ecological meta-analysis. Ecol. Lett. 15, 627–636.

Chance, M.R.A., Russell, W.M.S., 1959. Protean displays: a form of allaesthetic behaviour. Proc.
Zool. London 132, 65–70.

Chapman, C.A., Chapman, L.J., 1996. Mixed-species primate groups in the Kibale forest: ecological
constraints on association. Int. J. Primatol. 17, 31–50.

Chapman, C.A., Chapman, L.J., 2000. Determinants of group size in primates: the importance of
travel costs. In: Boinski, S., Gerber, P.A. (Eds.), On the Move: How and Why Animals Travel
in Groups, University of Chicago Press, Chicago, pp. 24–42.

Chapman, C.A., Rothman, J.M., 2009. Within-species differences in primate social structure: evolution of plasticity and phylogenetic constraints. Primates 50, 12–22.

Cheney, K.L., Marshall, N.J., 2009. Mimicry in coral reef fish: how accurate is this deception in
terms of color and luminance? Behav. Ecol. 20, 459–468.

Cheney, D.L., Seyfarth, R.M., 1990. How Monkeys See the World. University of Chicago Press,
Chicago.

Childress, M.J., Herrnkind, W.F., 2001. The guide effect influence on the gregariousness of juvenile
Caribbean spiny lobsters. Anim. Behav. 62, 465–472.

Childress, M.J., Lung, M.A., 2003. Predation risk, gender and the group size effect: does elk vigilance depend upon the behaviour of conspecifics? Anim. Behav. 66, 389–398.

Chippaux, J.P., 1998. Snake-bites: appraisal of the global situation. Bull. WHO 76, 515–524.

Chivers, D.P., Brown, G.E., Smith, R.J.F., 1995. Familiarity and shoal composition in fathead
minnows (*Pimephales promelas*)—implications for antipredator behaviour. Can. J. Zool. 73,
955–960.

Clark, C.W., 1987. The lazy, adaptable lion: a Markovian model of group foraging. Anim. Behav.
35, 361–368.

Clark, C.W., Mangel, M., 1984. Foraging and flocking strategies: information in an uncertain environment. Am. Nat. 123, 626–641.

Clark, C.W., Mangel, M., 1986. The evolutionary advantages of group foraging. Theor. Popul. Biol.
30, 45–75.

Clifton, K.E., 1991. Subordinate group members act as food-finders within striped parrotfish
territories. J. Exp. Mar. Biol. Ecol. 145, 141–148.

Clutton-Brock, T.H., Harvey, P.H., 1977. Primate ecology and social organization. J. Zool. 183,
1–39.

Clutton-Brock, T.H., Harvey, P.H., 1980. Primates, brains and ecology. J. Zool. 190, 309–323.

Clutton-Brock, T.H., Janson, C., 2012. Primate socio-ecology at the crossroads: past, present, and
future. Evol. Anthropol. 21, 136–150.

Clutton-Brock, T.H., Gaynor, D., McIlrath, G.M., Maccoll, A.D.C., Kansky, R., Chadwick, P.,
Manser, M., Skinner, J.D., Brotherton, P.N.M., 1999a. Predation, group size and mortality in a
cooperative mongoose, *Suricata suricatta*. J. Anim. Ecol. 68, 672–683.

Clutton-Brock, T.H., O'Riain, M.J., Brotherton, P.N.M., Gaynor, D., Kansky, R., Griffin, A.S., Manser, M., 1999b. Selfish sentinels in cooperative mammals. Science 284, 1640–1644.

Cody, M.L., 1971. Finch flocks in the Mojave desert. Theor. Popul. Biol. 2, 142–158.

Cody, M.L., 1973. Character convergence. Ann. Rev. Ecol. Syst. 4, 189–211.

Cohen, J.E., 1971. Casual Groups of Monkeys and Men: Stochastic Models of Elemental Social Systems. Harvard University Press, Cambridge.

Cohen, J.E., Pimm, S.L., Yodzis, P., Sadana, J., 1993. Body sizes of animal predators and animal prey in food webs. J. Anim. Ecol. 62, 67–78.

Coleman, S.W., 2008. Mourning dove (Zenaida macroura) wing-whistles may contain threat-related information for con- and hetero-specifics. Naturwissenschaften 95, 981–986.

Connell, S.D., 2000. Is there safety-in-numbers for prey? Oikos 88, 527–532.

Conradt, L., 1998. Could asynchrony in activity between the sexes cause intersexual social segregation in ruminants? Proc. R. Soc. London Ser. B. 265, 1–5.

Conradt, L., Roper, T.J., 2005. Consensus decision making in animals. Trends Ecol. Evol. 20, 449–456.

Consla, D.J., Mumme, R.L., 2012. Response of captive raptors to avian mobbing calls: the roles of mobber size and raptor experience. Ethology 118, 1063–1071.

Coolen, I., 2002. Increasing foraging group size increases scrounger use and reduces searching efficiency in nutmeg mannikins (Lonchura punctulata). Behav. Ecol. Sociobiol. 52, 232–238.

Coolen, I., Giraldeau, L.-A., 2003. Incompatibility between antipredatory vigilance and scrounger tactic in nutmeg mannikins, Lonchura punctulata. Anim. Behav. 66, 657–664.

Coolen, I., Giraldeau, L.-A., Lavoie, M., 2001. Head position as an indicator of producer and scrounger tactics in a ground-feeding bird. Anim. Behav. 61, 895–903.

Cooper, S.M., 1991. Optimal hunting group size: the need for lions to defend their kills against loss to spotted hyaenas. Afr. J. Ecol. 29, 130–136.

Cooper, W.E., Pérez-Mellado, V., Hawlena, D., 2007. Number, speeds, and approach paths of preda-tors affect escape behavior by the Balearic lizard, Podarcis lilfordi. J. Herpetol. 41, 197–204.

Cords, M., 1990a. Mixed-species association of East African guenons: general patterns or specific examples? Am. J. Primatol. 21, 101–114.

Cords, M., 1990b. Vigilance and mixed-species association of some East African forest monkeys. Behav. Ecol. Sociobiol. 26, 297–300.

Cords, M., 2000. Mixed-species association and group movement. In: Boinski, S., Gerber, P.A. (Eds.), On the Move: How and Why Animals Travel in Groups, University of Chicago Press, Chicago, pp. 73–99.

Cornell, J.C., Stamp, N.E., Bowers, M.D., 1987. Developmental change in aggregation, defense, and escape behavior of buckmoth caterpillars, Hemileuca lucina (Saturniidae). Behav. Ecol. Sociobiol. 20, 383–388.

Cornell, H.N., Marzluff, J.M., Pecoraro, S., 2012. Social learning spreads knowledge about danger-ous humans among American crows. Proc. R. Soc. London Ser. B. 279, 499–508.

Coss, R.G., 1999. Effects of relaxed natural selection on the evolution of behavior. In: Foster, S.A., Endler, J.A. (Eds.), Geographic Variation in Behavior: Perspectives on Evolutionary Mecha-nisms, Oxford University Press, Oxford, pp. 180–208.

Costa, J.T., 1997. Caterpillars as social insects. Am. Sci. 85, 150–159.

Costa, J.T., Pierce, N., 1997. Social evolution in the Lepidoptera: ecological context and communi-cation in larval societies. In: Choe, J.C., Crespi, B.J. (Eds.), The Evolution of Social Behavior in Insects and Arachnids, Cambridge University Press, Cambridge.

Cote, J., Fogarty, S., Sih, A., 2012. Individual sociability and choosiness between shoal types. Anim. Behav. 83, 1469–1476.

Cott, H.B., 1940. Adaptive Coloration in Animals. Methuen, London.

Couchoux, C., Cresswell, W., 2012. Personality constraints versus flexible antipredation behaviors: how important is boldness in risk management of redshanks (*Tringa totanus*) foraging in a natural system? Behav. Ecol. 23, 290–301.

Courchamp, F., Clutton-Brock, T.H., Grenfell, B., 1999. Inverse density dependence and the Allee effects. Trends Ecol. Evol. 14, 405–410.

Couzin, I.D., Krause, J., Franks, N.R., Levin, S.A., 2005. Effective leadership and decision-making in animal groups on the move. Nature 433, 513–516.

Cowlishaw, G., 1994. Vulnerability to predation in baboon populations. Behaviour 131, 293–304.

Cowlishaw, G., Lawes, M.J., Lightbody, M., Martin, A., Pettifor, R., Rowcliffe, J.M., 2004. A simple rule for the costs of vigilance: empirical evidence from a social forager. Proc. R. Soc. London Ser. B. 271, 27–33.

Crane, A., Mathis, A., McGrane, C., 2012. Socially facilitated antipredator behavior by ringed salamanders (*Ambystoma annulatum*). Behav. Ecol. Sociobiol. 66, 811–817.

Creel, S., Creel, N.M., 1995. Communal hunting and pack size in African wild dogs, *Lycaon pictus*. Anim. Behav. 50, 1325–1339.

Creel, S., Creel, N.M., 2002. The African Wild Dog: Behavior, Ecology and Conservation. Princeton University Press, Princeton.

Creel, S., Macdonald, D.W., 1995. Sociality, group size, and reproductive suppression among carnivores. Adv. Stud. Behav. 24, 203–257.

Cresswell, W., 1994. Flocking is an effective anti-predation strategy in redshanks, *Tringa totanus*. Anim. Behav. 47, 433–442.

Cresswell, W., 1996. Surprise as a winter hunting strategy in sparrowhawks *Accipiter nisus*, peregrines *Falco peregrinus* and merlins *F. columbarius*. Ibis 138, 684–692.

Cresswell, W., 1997. Interference competition at low competitor densities in blackbirds *Turdus merula*. J. Anim. Ecol. 66, 461–471.

Cresswell, W., Quinn, J.L., 2004. Faced with a choice, sparrowhawks more often attack the more vulnerable prey group. Oikos 104, 71–76.

Cresswell, W., Hilton, G.M., Ruxton, G.D., 2000. Evidence for a rule governing the avoidance of superfluous escape flights. Proc. R. Soc. London Ser. B. 267, 733–737.

Crofoot, M.C., 2012. Why mob? Reassessing the costs and benefits of primate predator harassment. Folia Primatol. 83, 252–273.

Crofoot, M.C., Gilby, I.C., Wikelski, M.C., Kays, R.W., 2008. Interaction location outweighs the competitive advantage of numerical superiority in *Cebus capucinus* intergroup contests. Proc. Natl. Acad. Sci. 105, 577–581.

Croft, D.P., Arrowsmith, B.J., Bielby, J., Skinner, K., White, E., Couzin, I.D., Magurran, A.E., Ramnarine, I., Krause, J., 2003. Mechanisms underlying shoal composition in the Trinidadian guppy, *Poecilia reticulata*. Oikos 100, 429–438.

Croft, D.P., Darden, S.K., Ruxton, G.D., 2009. Predation risk as a driving force for phenotypic assortment: a cross-population comparison. Philos. Trans. R. Soc. London Ser. B. 276, 1899–1904.

Croft, D.P., Hamilton, P.B., Darden, S.K., Jacoby, D.M.P., James, R., Bettaney, E.M., Tyler, C.R., 2012. The role of relatedness in structuring the social network of a wild guppy population. Oecologia 170, 955–963.

Crook, J.H., 1960. Studies on the social behaviour of *Quelea q. quelea* in French West Africa. Behaviour 16, 1–55.

Crook, J.H., 1964. The evolution of social organisation and visual communication in the weaver birds (Ploceinae). Behaviour (Suppl. 10), 1–178.

Crook, J.H., 1965. The adaptive significance of avian social organisations. Symp. Zool. Soc. London 14, 181–218.

Crook, J.H., Gartlan, J.S., 1966. Evolution of primate societies. Nature 210, 1200–1203.

Cullen, J.M., 1960. Some adaptations in the nesting behaviour of terns. Proc. Int. Ornithol. Congr. 12, 153–157.

Curio, E., 1976. The Ethology of Predation. Springer-Verlag, New York.

Curio, E., 1978. The adaptive significance of avian mobbing: I. Teleonomic hypotheses and predictions. Z. Tierpsychol. 48, 175–183.

Cuthill, I.C., Bennett, A.T.D., 1993. Mimicry and the eye of the beholder. Proc. R. Soc. London Ser. B. 253, 203–204.

Daly, D., Higginson, A.D., Chen, D., Ruxton, G.D., Speed, M.P., 2012. Density-dependent investment in costly anti-predator defences: an explanation for the weak survival benefit of group living. Ecol. Lett. 15, 576–583.

Danchin, E., Richner, H., 2001. Viable and unviable hypotheses for the evolution of raven roosts. Anim. Behav. 61, 7–11.

Darling, F.F., 1938. Bird Flocks and the Breeding Cycle: A Contribution to the Study of Avian Sociality. Cambridge University Press, Cambridge.

Dart, R.A., 1959. Adventures with the Missing Link. Harper and Brothers, New York.

Datta, S.B., Beauchamp, G., 1991. Demography and female dominance patterns in primates. 1. Mother-daughter and sister-sister relationships. Am. Nat. 138, 201–226.

David, M., Giraldeau, L.-A., 2012. Zebra finches in poor condition produce more and consume more food in a producer-scrounger game. Behav. Ecol. 23, 174–180.

Davis, J.M., 1975. Socially induced flight reactions in pigeons. Anim. Behav. 23, 597–601.

Dawkins, M.S., 2002. What are birds looking at? Head movements and eye use in chickens. Anim. Behav. 63, 991–998.

Dawkins, M.S., Guilford, T., 1995. An exaggerated preference for simple neural network models of signal evolution? Proc. R. Soc. London Ser. B. 261, 357–360.

De Vos, A., O'Riain, M.J., 2010. Sharks shape the geometry of a selfish seal herd: experimental evidence from seal decoys. Biol. Lett. 6, 48–50.

De Vos, A., O'Riain, M.J., 2012. Movement in a selfish seal herd: do seals follow simple or complex movement rules? Behav. Ecol. 24, 190–197.

Dechmann, D.K.N., Heucke, S.L., Giuggioli, L., Safi, K., Voigt, C.C., Wikelski, M., 2009. Experimental evidence for group hunting via eavesdropping in echolocating bats. Proc. R. Soc. London Ser. B. 276, 2721–2728.

Dehn, M.M., 1990. Vigilance for predators: detection and dilution effects. Behav. Ecol. Sociobiol. 26, 337–342.

Dekker, D., Dekker, I., Christie, D., Ydenberg, R., 2011. Do staging semipalmated sandpipers spend the high-tide period in flight over the ocean to avoid falcon attacks along shore? Waterbirds 34, 195–201.

Denno, R.F., Benrey, B., 1997. Aggregation facilitates larval growth in the neotropical nymphalid butterfly Chlosyne janais. Ecol. Entomol. 22, 133–141.

Dermody, B.J., Tanner, C.J., Jackson, A.L., 2011. The evolutionary pathway to obligate scavenging in gyps vultures. PLoS One 6, e24635.

Dewar, D., 1905. King crows and mynahs as mess-mates. J. Bombay Nat. Hist. Soc. 16, 364–366.

di Bitetti, M.S., Janson, C.H., 2001. Social foraging and the finder's share in capuchin monkeys, Cebus apella. Anim. Behav. 62, 47–56.

di Fiore, A., Rendall, D., 1994. Evolution of social organization: a reappraisal for primates by using phylogenetic methods. Proc. Natl. Acad. Sci. 91, 9941–9945.

Diamond, J.M., 1981. Mixed-species foraging groups. Nature 292, 408–409.

Diamond, J.M., 1982. Mimicry of friarbirds by orioles. Auk 99, 187–196.

Diamond, J., 1987. Flocks of brown and black New Guinean birds: a bicolored mixed-species foraging association. Emu 87, 201–211.

Dimond, S., Lazarus, J., 1974. The problem of vigilance in animal life. Brain Behav. Evol. 9, 60–79.

Dominguez, J., Vidal, M., 2007. Vigilance behaviour of preening black-tailed godwit Limosa limosa in roosting flocks. Ardeola 54, 227–235.

Du Toit, J.T., Yetman, C.A., 2005. Effects of body size on the diurnal activity budgets of African browsing ruminants. Oecologia 143, 317–325.

Duarte, A., Weissing, F.J., Pen, I., Keller, L., 2011. An evolutionary perspective on self-organized division of labor in social insects. Ann. Rev. Ecol. Syst. 42, 91–110.

Dubois, F., Giraldeau, L.-A., Grant, J.W.A., 2003. Resource defense in a group-foraging context. Behav. Ecol. 14, 2–9.

Dugatkin, L.A., Reeve, H.K., 1998. Game Theory and Animal Behavior. Oxford University Press.

Dugatkin, L.A., Wilson, D.S., 2000. Assortative interactions and the evolution of cooperation during predator inspection in guppies (Poecilia reticulata). Evol. Ecol. Res. 2, 761–767.

Dumbacher, J.P., Deiner, K., Thompson, L., Fleischer, R.C., 2008. Phylogeny of the avian genus Pitohui and the evolution of toxicity in birds. Mol. Phylogenet. Evol. 49, 774–781.

Dunbar, R.I.M., 1992. Neocortex size as a constraint on group size in primates. J. Hum. Evol. 20, 469–493.

Dunbar, R.I.M., Bever, J., 1998. Neocortex size determines group size in carnivores and some insectivores. Ethology 104, 695–708.

Dunbar, R.I.M., Shultz, S., 2007a. Evolution in the social brain. Science 317, 1344–1347.

Dunbar, R.I.M., Shultz, S., 2007b. Understanding primate brain evolution. Philos. Trans. R. Soc. London Ser. B. 362, 649–658.

Ebensperger, L.A., Blumstein, D.T., 2006. Sociality in new world hystricognath rodents is linked to predators and burrow digging. Behav. Ecol. 17, 410–418.

Eilam, D., Izhar, R., Mort, J., 2011. Threat detection: behavioral practices in animals and humans. Neurosic. Biobehav. Rev. 35, 999–1006.

Eiserer, L.A., 1984. Communal roosting in birds. Bird Behav. 5, 61–80.

Elcavage, P., Caraco, T., 1983. Vigilance behaviour in house sparrow flocks. Anim. Behav. 31, 303–304.

Elgar, M.A., 1986. Scanning, pecking and alarm flights in house sparrows. Anim. Behav. 34, 1892–1894.

Elgar, M.A., 1989. Predator vigilance and group size in mammals and birds. Biol. Rev. 64, 13–33.

Elgar, M.A., Catterall, C.P., 1982. Flock size and feeding efficiency in house sparrows. Emu 82, 109–111.

Elgar, M.A., Burren, P.J., Posen, M., 1984. Vigilance and perception of flock size in foraging house sparrows Passer domesticus L. Behaviour 90, 215–223.

Ellis, D.H., Bednarz, J.C., Smith, D.G., Flemming, S.P., 1993. Social foraging classes in raptorial birds. BioScience 43, 14–20.

Emlen, J.T., 1952. Flocking behavior in birds. Auk 69, 160–170.

Endler, J.A., 1991. Interactions between predators and prey. In: Krebs, J.R., Davies, N.B. (Eds.), Behavioural Ecology, Blackwell Scientific, Oxford, pp. 169–196.

Endler, J.A., Mielke, P.W., 2005. Comparing entire colour patterns as birds see them. Biol. J. Linn. Soc. 86, 405–431.

Erlandsson, A., 1988. Food-sharing vs. monopolising prey: a form of kleptoparasitism in Velia caprai (Heteroptera). Oikos 53, 23–26.

Eshel, I., Sansone, E., Shaked, A., 2011. On the evolution of group-escape strategies of selfish prey. Theor. Popul. Biol. 80, 150–157.

Evans, S.R., Finnie, M., Manica, A., 2007. Shoaling preferences in decapod crustacea. Anim. Behav. 74, 1691–1696.

Fairbanks, B., Dobson, F.S., 2007. Mechanisms of the group-size effect on vigilance in Columbian ground squirrels: dilution versus detection. Anim. Behav. 73, 115–123.

Fanshawe, J., FitzGibbon, C.D., 1993. Factors influencing the hunting success of an African wild dog pack. Anim. Behav. 45, 479–490.

Farine, D.R., Garroway, C.J., Sheldon, B.C., 2012. Social network analysis of mixed-species flocks: exploring the structure and evolution of interspecific social behaviour. Anim. Behav. 84, 1271–1277.

Fauchald, P., Rødven, R., Bårdsen, B.J., Langeland, K., Tveraa, T., Yoccoz, N.G., Ims, R.A., 2007. Escaping parasitism in the selfish herd: age, size and density-dependent warble fly infestation in reindeer. Oikos 116, 491–499.

Favreau, F.R., Goldizen, A.W., Pays, O., 2010. Interactions among social monitoring, anti-predator vigilance and group size in eastern grey kangaroos. Proc. R. Soc. London Ser. B. 277, 2089–2095.

Fernández-Juricic, E., 2012. Sensory basis of vigilance behavior in birds: synthesis and future prospects. Behav. Proc. 89, 143–152.

Fernández-Juricic, E., Kacelnik, A., 2004. Information transfer and gain in flocks: the effects of quality and quantity of social information at different neighbour distances. Behav. Ecol. Sociobiol. 55, 502–511.

Fernández-Juricic, E., Kowalski, V., 2011. Where does a flock end from an information perspective? A comparative experiment with live and robotic birds. Behav. Ecol. 22, 1304–1311.

Fernández-Juricic, E., Erichsen, J.T., Kacelnik, A., 2004a. Visual perception and social foraging in birds. Trends Ecol. Evol. 19, 25–31.

Fernández-Juricic, E., Kerr, B., Bednekoff, P.A., Stephens, D.W., 2004b. When are two heads better than one? Visual perception and information transfer affect vigilance coordination in foraging groups. Behav. Ecol. 15, 898–906.

Fernández-Juricic, E., Siller, S., Kacelnik, A., 2004c. Flock density, social foraging, and scanning: an experiment with starlings. Behav. Ecol. 15, 371–379.

Fernández-Juricic, E., Gilak, N., McDonald, J.C., Pithia, P., Valcarcel, A., 2006. A dynamic method to study the transmission of social foraging information in flocks using robots. Anim. Behav. 71, 901–911.

Fernández-Juricic, E., Gall, M.D., Dolan, T., Tisdale, V., Martin, G.R., 2008. The visual fields of two ground-foraging birds, house finches and house sparrows, allow for simultaneous foraging and anti-predator vigilance. Ibis 150, 779–787.

Fernández-Juricic, E., Delgado, J.A., Remacha, C., Jiménez, M.D., Garcia, V., Hori, K., 2009. Can a solitary avian species use collective detection? An assay in semi-natural conditions. Behav. Proc. 82, 67–74.

Fernández-Juricic, E., Beauchamp, G., Treminio, R., Hoover, M., 2011. Making heads turn: association between head movements during vigilance and perceived predation risk in brown-headed cowbird flocks. Anim. Behav. 82, 573–577.

Ferrari, M.C.O., Chivers, D.P., 2008. Cultural learning of predator recognition in mixed-species assemblages of frogs: the effect of tutor-to-observer ratio. Anim. Behav. 75, 1921–1925.

Ferriere, R., Cazelles, B., Cézilly, F., Desportes, J.P., 1996. Predictability and chaos in bird vigilant behaviour. Anim. Behav. 52, 457–472.

Finkbeiner, S.D., Briscoe, A.D., Reed, R.D., 2012. The benefit of being a social butterfly: communal roosting deters predation. Proc. R. Soc. London Ser. B. 279, 2769–2776.

Fisher, R.A., 1930. The Genetical Theory of Natural Selection. Clarendon Press, Oxford.

FitzGibbon, C.D., 1989. A cost to individual with reduced vigilance in groups of Thomson's gazelles hunted by cheetahs. Anim. Behav. 37, 508–510.

FitzGibbon, C.D., 1990. Mixed species grouping in Thompson's and Grant's gazelles: the antipredator benefits. Anim. Behav. 39, 1116–1126.

FitzGibbon, C.D., 1994. The costs and benefits of predator inspection behaviour in Thomson's gazelles. Behav. Ecol. Sociobiol. 34, 139–148.

Flasskamp, A., 1994. The adaptive significance of avian mobbing V. An experimental test of the 'move on' hypothesis. Ethology 96, 322–333.

Flower, T., 2011. Fork-tailed drongos use deceptive mimicked alarm calls to steal food. Proc. R. Soc. London Ser. B. 278, 1548–1555.

Flynn, R.E., Giraldeau, L.-A., 2001. Producer-scrounger games in a spatially explicit world: tactic use influences flock geometry of spice finches. Ethology 107, 249–257.

Fordyce, J.A., Agrawal, A.A., 2001. The role of plant trichomes and caterpillar group size on growth and defence of the pipevine swallowtail *Battus philenor*. J. Anim. Ecol. 70, 997–1005.

Forsman, J.T., Mönkkönen, M., 2001. Responses by breeding birds to heterospecific song and mobbing call playbacks under varying predation risk. Anim. Behav. 62, 1067–1073.

Fortin, D., Boyce, M.S., Merrill, E.H., 2004a. Multi-tasking by mammalian herbivores: overlapping processes during foraging. Ecology 85, 2312–2322.

Fortin, D., Boyce, M.S., Merrill, E.H., Fryxell, J.M., 2004b. Foraging costs of vigilance in large mammalian herbivores. Oikos 107, 172–180.

Foster, W.A., Treherne, J.E., 1981. Evidence for the dilution effect in the selfish herd from fish predation on a marine insect. Nature 293, 466–467.

Frank, S.A., 1998. Foundations of Social Evolution. Princeton University Press, Princeton.

Franks, D.W., Ruxton, G.D., 2008. How robust are neural network models of stimulus generalization? Biosystems 92, 175–181.

Frantzis, A., Hertzing, D.L., 2002. Mixed-species associations of striped dolphins (*Stenella coeruleoalba*), short-beaked common dolphins (*Delphinus delphis*), and Risso's dolphins (*Grampus griseus*) in the Gulf of Corinth (Greece, Mediterranean Sea). Aquat. Mamm. 28, 188–197.

Fretwell, S.D., Lucas, H.L.J., 1970. On territorial behavior and other factors influencing habitat distribution in birds. I. Theoretical development. Acta Biotheor. 19, 16–36.

Frid, A., 1997. Vigilance by female Dall's sheep: interaction between predation risk factors. Anim. Behav. 53, 799–808.

Fritz, H., Guillemain, M., Durant, D., 2002. The cost of vigilance for intake rate in the mallard (*Anas platyrhynchos*): an approach through foraging experiments. Ethol. Ecol. Evol. 14, 91–97.

Fry, B.G., Vidal, N., Norman, J.A., Vonk, F.J., Scheib, H., Ramjan, S.F.R., Kuruppu, S., Fung, K., Hedges, S.B., Richardson, M.K., Hodgson, W.C., Ignjatovic, V., Summerhayes, R., Kochva, E., 2006. Early evolution of the venom system in lizards and snakes. Nature 439, 584–588.

Fudenberg, D., Levine, D.K., 1998. The Theory of Learning in Games. The MIT Press, Cambridge.

Fullard, J.H., 1994. Auditory changes in noctuid moths endemic to a bat-free habitat. J. Evol. Biol. 7, 435–445.

Gagliardo, A., Guilford, T., 1993. Why do warningly-coloured prey live gregariously? Proc. R. Soc. London Ser. B. 251, 69–74.

Galef, B.G., Giraldeau, L.-A., 2001. Social influences on foraging in vertebrates: causal mechanisms and adaptive functions. Anim. Behav. 61, 3–15.

Galton, F., 1871. Gregariousness in cattle and men. MacMillan's Mag. 23, 353–357.

Galton, F., 1883. Inquiries into Human Faculty and Its Development. MacMillan, London.

Garber, P.A., 1988. Diet, foraging patterns, and resource defense in a mixed species troop of *Saguinus mystax* and *Saguinus fuscicollis* in Amazonian Peru. Behaviour 105, 18–34.

Gauthier-Hion, A., Tutin, C.E.G., 1988. Simultaneous attack by adult males of a polyspecific troop of monkeys against a crowned hawk eagle. Folia Primatol. 51, 149–151.

Gauthier-Hion, A., Quris, R., Gauthier, J.P., 1983. Monospecific vs. polyspecific life: a comparative study of foraging and antipredatory tactics in a community of *Cercopithecus* monkeys. Behav. Ecol. Sociobiol. 12, 325–335.

Gaynor, K.M., Cords, M., 2012. Antipredator and social monitoring functions of vigilance behaviour in blue monkeys. Anim. Behav. 84, 531–537.

Gazda, S.K., Connor, R.C., Edgar, R.K., Cox, F., 2005. A division of labour with role specialization in group-hunting bottlenose dolphins (*Tursiops truncatus*) off Cedar Key, Florida. Proc. R. Soc. London Ser. B. 272, 135–140.

Ge, C., Beauchamp, G., Li, Z., 2011. Coordination and synchronisation of anti-predation vigilance in two crane species. PLoS One 6, e26447.

Geist, C., Liao, J., Libby, S., Blumstein, D.T., 2005. Does intruder group size and orientation affect flight initiation distance in birds? Anim. Biodiversity Conserv. 28, 69–73.

Gerkema, M.P., Verhulst, S., 1990. Warning against an unseen predator: a functional aspect of synchronous feeding in the common vole, *Microtus arvalis*. Anim. Behav. 40, 1169–1176.

Gerlotto, F., Bertrand, S., Bez, N., Gutierrez, M., 2006. Waves of agitation inside anchovies school observed with multibeam sonar: a way to transmit information in response to predation. ICES J. Mar. Sci. 63, 1405–1417.

Ghent, A.W., 1960. A study of the group-feeding behaviour of the jack pine sawfly, *Neodiprion pratti banksianae* Roh. Behaviour 16, 110–148.

Gibson, R., Baker, A., 2012. Multiple gene sequences resolve phylogenetic relationships in the shorebird suborder Scolopaci (Aves: Charadriiformes). Mol. Phylogenet. Evol. 64, 66–72.

Gillespie, T.R., Chapman, C.A., 2001. Determinants of group size in the red colobus monkey (*Procolobus badius*): an evaluation of the generality of the ecological-constraint model. Behav. Ecol. Sociobiol. 50, 329–338.

Giraldeau, L.-A., Caraco, T., 1993. Genetic relatedness and group size in an aggregation economy. Evol. Ecol. 7, 429–438.

Giraldeau, L.-A., Caraco, T., 2000. Social Foraging Theory. Princeton University Press, Princeton.

Giraldeau, L.-A., Dubois, F., 2008. Social foraging and the study of exploitative behavior. Adv. Stud. Behav. 38, 59–104.

Giraldeau, L.-A., Gillis, D., 1985. Optimal group size can be stable: a reply to sibly. Anim. Behav. 33, 666–667.

Giraldeau, L.-A., Gillis, D., 1988. Do lions hunt in group sizes that maximize hunters' daily food returns? Anim. Behav. 36, 611–613.

Giraldeau, L.-A., Lefebvre, L., 1986. Exchangeable producer and scrounger roles in a captive flock of feral pigeons: a case for the skill pool effect. Anim. Behav. 34, 777–783.

Giraldeau, L.-A., Livoreil, B., 1998. Game theory and social foraging. In: Dugatkin, L.A., Reeve, H.K. (Eds.), Game Theory and Animal Behavior, Oxford University Press, New York, pp. 16–37.

Giraldeau, L.-A., Hogan, J.A., Clinchy, M.J., 1990. The payoffs to producing and scrounging: what happens when patches are divisible? Ethology 85, 132–146.

Giraldeau, L.-A., Soos, C., Beauchamp, G., 1994. A test of the producer-scrounger foraging game in captive flocks of spice finches, *Lonchura punctulata*. Behav. Ecol. Sociobiol. 34, 251–256.

Giraldeau, L.-A., Valone, T.J., Templeton, J.J., 2002. Potential disadvantages of using socially acquired information. Phil. Trans. R. Soc. Lond. B. Biol. Sci. 357, 1559–1566.

Gittleman, J.L., 1989. Carnivore group living: comparative trends. In: Gittleman, J.L. (Ed.), Carnivore Behavior, Ecology, and Evolution, Cornell University Press, New York, pp. 183–207.

Godin, J.G.J., Davis, S.A., 1995. Who dares, benefits: predator approach behaviour in the guppy (*Poecilia reticulata*) deters predator pursuit. Proc. R. Soc. London Ser. B. 259, 193–200.

Godin, J.G.J., Morgan, M.J., 1985. Predator avoidance and school size in a cyprinodontid fish, the banded killifish (*Fundulus diaphanus* Lesueur). Behav. Ecol. Sociobiol. 16, 105–110.

Goodale, E., Beauchamp, G., 2010. The relationship between leadership and gregariousness in mixed-species bird flocks. J. Avian Biol. 41, 99–103.

Goodale, E., Kotagama, S.W., 2005a. Alarm calling in Sri Lankan mixed-species bird flocks. Auk 122, 108–120.

Goodale, E., Kotagama, S.W., 2005b. Testing the roles of species in mixed-species bird flocks of a Sri Lankan rainforest. J. Trop. Ecol. 21, 669–676.

Goodale, E., Kotagama, S.W., 2006. Vocal mimicry by a passerine bird attracts other species involved in mixed-species flocks. Anim. Behav. 72, 471–477.

Goodale, E., Kotagama, S.W., 2008. Response to conspecific and heterospecific alarm calls in mixed-species bird flocks of a Sri Lankan rainforest. Behav. Ecol. 19, 887–894.

Goodale, E., Beauchamp, G., Magrath, R.D., Nieh, J.C., Ruxton, G.D., 2010. Interspecific information transfer influences animal community structure. Trends Ecol. Evol. 25, 354–361.

Goodale, E., Goodale, U., Mana, R., 2012. The role of toxic pitohuis in mixed-species flocks of lowland forest in Papua New Guinea. Emu 112, 9–16.

Goodson, J.L., Kingsbury, M.A., 2011. Nonapeptides and the evolution of social group sizes in birds. Front. Neuroanat. 5, 13.

Goodson, J.L., Wang, Y., 2006. Valence-sensitive neurons exhibit divergent functional profiles in gregarious and asocial species. Proc. Natl. Acad. Sci. 103, 17013–17017.

Goodson, J.L., Schrock, S.E., Klatt, J.D., Kabelik, D., Kingsbury, M.A., 2009. Mesotocin and nonapeptide receptors promote estrildid flocking behavior. Science 325, 862–866.

Goodson, J.L., Wilson, L.C., Schrock, S.E., 2012. To flock or fight: neurochemical signatures of divergent life histories in sparrows. Proc. Natl. Acad. Sci. 109, 10685–10692.

Gorman, M.L., Mills, M.G.L., Raath, J.P., Speakman, J.R., 1998. High hunting costs make African wild dogs vulnerable to kleptoparasitism by hyaenas. Nature 391, 479–481.

Goss-Custard, J.D., Durell, S.E.A.l.V.d., Ens, B.J., 1982. Individual differences in aggressiveness and food stealing among wintering oystercatchers, *Haematopus ostralegus* L. Anim. Behav. 30, 917–928.

Goss-Custard, J.D., 1996. The Oystercatcher: From Individuals to Population. Oxford University Press, Oxford.

Gosselin-Ildari, A.D., Koenig, A., 2012. The effects of group size and reproductive status on vigilance in captive *Callithrix jacchus*. Am. J. Primatol. 74, 613–621.

Götmark, F., Winkler, D.W., Andersson, M., 1986. Flock-feeding on fish schools increases individual success in gulls. Nature 319, 589–591.

Grand, T.C., Dill, L.M., 1999. The effect of group size on the foraging behaviour of juvenile coho salmon: reduction of predation risk or increased competition? Anim. Behav. 58, 443–451.

Grandcolas, P., 1998. The evolutionary interplay of social behavior, resource use and anti-predator behavior in Zetorinae + Blaberinae + Gyninae + Diplopterinae cockroaches: a phylogenetic analysis. Cladistics 14, 117–127.

Grant, J.W.A., 1993. Whether or not to defend? The influence of resource distribution. Mar. Behav. Physiol. 23, 137–153.

Graves, G.R., Gotelli, N.J., 1993. Assembly of avian mixed species flocks in Amazonia. Proc. Natl. Acad. Sci. 90, 1388–1391.

Greenberg, R., 2000. Birds of many feathers: the formation and structure of mixed-species flocks of forest birds. In: Boinski, S., Gerber, P.A. (Eds.), On the Move: How and Why Animals Travel in Groups, University of Chicago Press, Chicago, pp. 521–558.

Greig-Smith, P.W., 1981. The role of alarm responses in the formation of mixed-species flocks of heathland birds. Behav. Ecol. Sociobiol. 8, 7–10.

Grether, G.F., Switzer, P.V., 2000. Mechanisms for the formation and maintenance of traditional night roost aggregations in a territorial damselfly. Anim. Behav. 60, 569–579.

Griesser, M., 2009. Mobbing calls signal predator category in a kin group-living bird species. Proc. R. Soc. London Ser. B. 276, 2887–2892.

Griffiths, R.A., Foster, J.P., 1998. The effect of social interactions on tadpole activity and growth in the British anuran amphibians (*Bufo bufo, B. calamita*, and *Rana temporaria*). J. Zool. 245, 431–437.

Griffiths, S.W., Armstrong, J.D., Metcalfe, N.B., 2003. The cost of aggregation: juvenile salmon avoid sharing winter refuges with siblings. Behav. Ecol. 14, 602–606.

Grinnell, J., 1903. Call notes of the bushtit. Condor 5, 85–87.

Gross, M.R., 1996. Alternative reproductive strategies and tactics: diversity within sexes. Trends Ecol. Evol. 11, 92–98.

Gueron, S., Levin, S.A., Rubenstein, D., 1996. The dynamics of herds: from individuals to aggregations. J. Theor. Biol. 182, 85–98.

Guillemain, M., Martin, G.R., Fritz, H., 2002. Feeding methods, visual fields and vigilance in dabbling ducks (Anatidae). Funct. Ecol. 16, 522–529.

Gursky, S., 2006. Function of snake mobbing in spectral tarsiers. Am. J. Phys. Anthropol. 129, 601–608.

Gurven, M., 2004. To give and to give not: the behavioral ecology of human food transfers. Behav. Brain Sci. 27, 543.

Gygax, L., 2002. Evolution of group size in the dolphins and porpoises: interspecific consistency of intraspecific patterns. Behav. Ecol. 13, 583–590.

Ha, R.R., Ha, J.C., 2003. Effects of ecology and prey characteristics on the use of alternative social foraging tactics in crows, *Corvus caurinus*. Anim. Behav. 66, 309–316.

Ha, R.R., Bentzen, P., Marsh, J., Ha, J.C., 2003. Kinship and association in social foraging Northwestern crows (*Corvus caurinus*). Bird Behav. 15, 65–75.

Hackett, S.J., Kimball, R.T., Reddy, S., Bowie, R.C.K., Braun, E.L., Braun, M.J., Chojnowski, J.L., Cox, W.A., Han, K.L., Harshman, J., Huddleston, C.J., Marks, B.D., Miglia, K.J., Moore, W.S., Sheldon, F.H., Steadman, D.W., Witt, C.C., Yuri, T., 2008. A phylogenomic study of birds reveals their evolutionary history. Science 320, 1763–1768.

Hake, M., Ekman, J., 1988. Finding and sharing depletable patches: when group foraging decreases intake rate. Ornis Scand. 19, 275–279.

Halloy, J., Sempo, G., Caprari, G., Rivault, C., Asadpour, M., Tache, F., Said, I., Durier, V., Canonge, S., Ame, J.M., Detrain, C., Correll, N., Martinoli, A., Mondada, F., Siegwart, R., Deneubourg, J.L., 2007. Social integration of robots into groups of cockroaches to control self-organized choices. Science 318, 1155–1158.

Hamblin, S., Giraldeau, L.-A., 2009. Finding the evolutionarily stable learning rule for frequency-dependent foraging. Anim. Behav. 78, 1343–1350.

Hamilton, W.D., 1964. The genetical evolution of social behavior. J. Theor. Biol. 7, 1–52.

Hamilton, W.D., 1971. Geometry for the selfish herd. J. Theor. Biol. 31, 295–311.

Hamilton, I.M., 2000. Recruiters and joiners: using optimal skew theory to predict group size and the division of resources within groups of social foragers. Am. Nat. 155, 684–695.

Hamilton, I.M., Dill, L.M., 2003. Group foraging by a kleptoparasitic fish: a strong inference test of social foraging models. Ecology 84, 3349–3359.

Handegard, N.O., Boswell, K.M., Ioannou, C.C., Leblanc, S.P., Tjøstheim, D.B., Couzin, I.D., 2012. The dynamics of coordinated group hunting and collective information transfer among schooling prey. Curr. Biol. 22, 1213–1217.

Haney, J.C., Fristrup, K.M., Lee, D.S., 1992. Geometry of visual recruitment by seabirds to ephemeral foraging flocks. Ornis Scand. 23, 49–62.

Hansen, A.J., 1986. Fighting behavior in bald eagles: a test of game theory. Ecology 67, 787–797.

Hardie, S.M., Buchanan-Smith, H.M., 1997. Vigilance in single- and mixed-species groups of tamarins (*Saguinus labiatus* and *Saguinus fuscicollis*). Int. J. Primatol. 18, 219–232.

Hardin, G., 1968. The tragedy of the commons. Science 162, 1243–1248.

Harley, C.B., 1981. Learning the evolutionarily stable strategy. J. Theor. Biol. 89, 611–633.

Harrison, N.M., Whitehouse, M.J., 2011. Mixed-species flocks: an example of niche construction? Anim. Behav. 81, 675–682.

Hart, A., Lendrem, D.W., 1984. Vigilance and scanning patterns in birds. Anim. Behav. 32, 1216–1224.

Harvey, P.H., Pagel, M.D., 1991. The Comparative Method in Evolutionary Biology. Oxford University Press, Oxford.

Hassell, M.P., 1978. The Dynamics of Arthropod Predator–Prey Systems. Princeton University Press, Princeton.

Haupt, R.L., Haupt, S.E., 2004. Practical Genetic Algorithms. John Wiley & Sons, Hoboken.

Hawkins, B.A., 1994. Patterns and Process in Host-Parasitoids Systems. Cambridge University Press, Cambridge.

Hawkins, G.L., Hill, G.E., Mercadante, A., 2012. Delayed plumage maturation and delayed reproductive investment in birds. Biol. Rev. 87, 257–274.

Haykin, S., 1994. Neural Networks: A Comprehensive Foundation. MacMillan, New York.

Hebblewhite, M., Pletscher, D., 2002. Effects of elk group size on predation by wolves. Can. J. Zool. 80, 800–809.

Heinsohn, R., Packer, C., 1995. Complex cooperative strategies in group-territorial African lions. Science 269, 1260–1262.

Held, S., Mendl, M., Devereux, C., Byrne, R.W., 2000. Social tactics of pigs in a competitive foraging task: the 'informed forager' paradigm. Anim. Behav. 59, 569–576.

Held, S., Mendl, M., Devereux, C., Byrne, R.W., 2002. Foraging pigs alter their behaviour in response to exploitation. Anim. Behav. 64, 157–166.

Held, S.D.E., Byrne, R.W., Jones, S., Murphy, E., Friel, M., Mendl, M.T., 2010. Domestic pigs, *Sus scrofa*, adjust their foraging behaviour to whom they are foraging with. Anim. Behav. 79, 857–862.

Hensor, E., Couzin, I.D., James, R., Krause, J., 2005. Modelling density-dependent fish shoal distributions in the laboratory and field. Oikos 110, 344–352.

Herrnkind, W.F., Childress, M.J., Lavalli, K.L., 2001. Cooperative defence and other benefits among exposed spiny lobsters: inferences from group size and behaviour. Mar. Freshwater Res. 52, 1113–1124.

Herzing, D.L., Johnson, C.M., 1997. Interspecific interactions between Atlantic spotted dolphins (*Stenella frontalis*) and bottlenose dolphins (*Tursiops truncatus*) in the Bahamas, 1985–1995. Aquat. Mamm. 23, 85–99.

Heymann, E.W., 1997. The relationship between body size and mixed-species troops of tamarins (*Saguinus* spp.). Folia Primatol. 68, 287–295.

Heymann, E.W., Buchanan-Smith, H.M., 2000. The behavioural ecology of mixed-species troops of callitricine primates. Biol. Rev. 75, 169–190.

Hicklin, P.W., 1987. The migration of shorebirds in the Bay of Fundy. Wilson Bull. 99, 540–570.

Higashi, M., Yamamura, Y., 1993. What determines animals group size? Insider-outsider conflict and its resolution. Am. Nat. 142, 553–563.

Hill, R.A., Dunbar, R.I.M., 1998. An evaluation of the roles of predation rate and predation risk as selective pressures on primate grouping behaviour. Behaviour 135, 411–430.

Hilton, G.M., Cresswell, W., Ruxton, G.D., 1999. Intra-flock variation in the speed of response on attack by an avian predator. Behav. Ecol. 10, 391–395.

Hindwood, K.A., 1937. The flocking of birds with particular reference to the association of small insectivorous birds. Emu 36, 254–261.

Hingee, M., Magrath, R.D., 2009. Flights of fear: a mechanical wing whistle sounds the alarm in a flocking bird. Proc. R. Soc. London Ser. B. 276, 4173–4179.

Hingston, R.W.G., 1920. A Naturalist in Himalaya. H. F. and G. Witherby, London.

Hirsch, B.T., 2011. Spatial position and feeding success in ring-tailed coatis. Behav. Ecol. Sociobiol. 65, 581–591.

Hirsch, B.T., Morrell, L.J., 2011. Measuring marginal predation in animal groups. Behav. Ecol. 22, 648–656.

Hirth, D.H., McCullogh, D.R., 1977. Evolution of alarm signals in ungulates with special reference to white-tailed deer. Am. Nat. 111, 31–42.

Hoare, D.J., Ruxton, G.D., Godin, J.G.J., Krause, J., 2000. The social organization of free-ranging fish shoals. Oikos 89, 546–554.

Hodge, M.A., Uetz, G.W., 1996. Foraging advantages of mixed-species association between solitary and colonial orb-weaving spiders. Oecologia 107, 578–587.

Höjesjö, J., Johnsson, J.I., Petersson, E., Jarvi, T., 1998. The importance of being familiar: individual recognition and social behavior in sea trout (*Salmo trutta*). Behav. Ecol. 9, 445–451.

Holbrook, S.J., Schmitt, R.J., 2002. Competition for shelter space causes density-dependent predation mortality in damselfishes. Ecology 83, 2855–2868.

Holenweg, A.K., Noe, R., Sohnbel, M., 1996. Waser's gas model applied to associations between red colobus and Diana monkeys in the Tai National Park, Ivory Coast. Folia Primatol. 67, 125–136.

Holland, J.H., 1975. Adaptation in Natural and Artificial Systems. University of Michigan Press, Ann Arbor.

Hölldobler, B., Wilson, E.O., 1990. The Ants. Harvard University Press, Cambridge.

Hoogland, J.L., 1979. The effect of colony size on individual alertness of prairie dogs (Sciuridae: *Cynomys* spp.). Anim. Behav. 27, 394–407.

Hoogland, J.L., 1981. The evolution of coloniality in white-tailed and black-tailed prairie dogs (Sciuridae: *Cynomys leucurus* and *C. ludovicianus*). Ecology 62, 252–272.

Hopewell, L., Rossiter, R., Blower, E., Leaver, L., Goto, K., 2005. Grazing and vigilance by Soay sheep on Lundy island: influence of group size, terrain and the distribution of vegetation. Behav. Proc. 70, 186–193.

Houston, A.I., Sumida, B., 1987. Learning rules, matching and frequency dependence. J. Theor. Biol. 126, 289–308.

Hugie, D.M., 2003. The waiting game: a "battle of waits" between predator and prey. Behav. Ecol. 14, 807–817.

Humphries, D.A., Driver, P.M., 1970. Protean defence by prey animals. Oecologia 5, 285–302.

Hunter, A.F., 2000. Gregariousness and repellent defences in the survival of phytophagous insects. Oikos 91, 213–224.

Hurd, C.R., 1996. Interspecific attraction to the mobbing calls of black-capped chickadees (*Parus atricapillus*). Behav. Ecol. Sociobiol. 38, 287–292.

Hutto, R.L., 1988. Foraging behavior patterns suggest a possible cost associated with participation in mixed-species bird flocks. Oikos 51, 79–83.

Hutto, R.L., 1994. The composition and social organization of mixed-species flocks in a tropical deciduous forest in western Mexico. Condor 96, 105–118.

Iglesias, T.L., McElreath, R., Patricelli, G.L., 2012. Western scrub-jay funerals: cacophonous aggregations in response to dead conspecifics. Anim. Behav. 84, 1103–1111.

Ims, R.A., 1990. On the adaptive value of reproductive synchrony as a predator-swamping strategy. Am. Nat. 136, 485–498.

Inglis, I.R., Lazarus, J., 1981. Vigilance and flock size in brent geese: the edge effect. Z. Tierpsychol. 57, 193–200.

Inman, A.J., Krebs, J.R., 1987. Predation and group living. Trends Ecol. Evol. 2, 31–32.

Ioannou, C.C., Morrell, L.J., Ruxton, G.D., Krause, J., 2009. The effect of prey density on predators: conspicuousness and attack success are sensitive to spatial scale. Am. Nat. 173, 499–506.

Ioannou, C.C., Bartumeus, F., Krause, J., Ruxton, G.D., 2011. Unified effects of aggregation reveal larger prey groups take longer to find. Proc. R. Soc. London Ser. B. 278, 2985–2990.

Ioannou, C.C., Guttal, V., Couzin, I.D., 2012. Predatory fish select for coordinated collective motion in virtual prey. Science 337, 1212–1215.

Isbell, L.A., 1991. Contest and scramble competition: patterns of female aggression and ranging behavior among primates. Behav. Ecol. 2, 143–155.

Isbell, L.A., 1994. Predation on primates: ecological patterns and evolutionary consequences. Evol. Anthropol. 3, 61–71.

Ishihara, M., 1987. Effect of mobbing toward predator by the damselfish *Pomacentrus coelestis* (Pisces: Pomacentridae). J. Ethol. 5, 43–52.

Iwakuma, T., Morimoto, N., 1984. An analysis of larval mortality and development in relation to group size in *Dictyoploca japonica* Butler (Lepidoptera: Saturniidae), with special reference to field populations. Res. Popul. Ecol. 26, 51–73.

Jaatinen, K., Öst, M., 2013. Brood size matching: a novel perspective on predator dilution. Am. Nat. 181, 171–181.

Jackson, A.L., Ruxton, G.D., 2006. Toward an individual-level understanding of vigilance: the role of social information. Behav. Ecol. 17, 532–538.

Jackson, A.L., Brown, S., Sherratt, T.N., Ruxton, G.D., 2005. The effects of group size, shape and composition on ease of detection of cryptic prey. Behaviour 142, 811–826.

Jackson, A.L., Beauchamp, G., Broom, M., Ruxton, G.D., 2006. Evolution of anti-predator traits in response to a flexible targeting strategy by predators. Proc. R. Soc. London Ser. B. 273, 1055–1062.

James, R., Bennett, P.C., Krause, J., 2004. Geometry for mutualistic and selfish herds: the limited domain of danger. J. Theor. Biol. 228, 107–113.

Janson, C.H., 1998. Experimental evidence for spatial memory in foraging wild capuchin monkeys, *Cebus apella*. Anim. Behav. 55, 1229–1243.

Janson, C.H., Goldsmith, M.L., 1995. Predicting group size in primates: foraging costs and predation risks. Behav. Ecol. 6, 326–336.

Jarman, P.J., 1974. The social organization of antelope in relation to their ecology. Behaviour 48, 216–267.

Jennings, T., Evans, S.M., 1980. Influence of position in the flock and flock size on vigilance in the starling, *Sturnus vulgaris*. Anim. Behav. 28, 634–635.

Jensen, K.H., Larsson, P., 2002. Predator evasion in Daphnia: the adaptive value of aggregation associated with attack abatement. Oecologia 132, 461–467.

Jeschke, J.M., Tollrian, R., 2007. Prey swarming: which predators become confused and why? Anim. Behav. 74, 387–393.

Johnstone, R.A., 2000. Models of reproductive skew: a review and synthesis. Ethology 106, 5–26.

Jones, K.A., Jackson, A.L., Ruxton, G.D., 2011. Prey jitters; protean behaviour in grouped prey. Behav. Ecol. 22, 831–836.

Jouventin, P., Weimerskirch, H., 1990. Satellite tracking of wandering albatrosses. Nature 343, 746–748.

Jovani, R., Blanco, G., 2000. Resemblance within flocks and individual differences in feather mite abundance on long-tailed tits, Aegithalos caudatus (L.). Ecoscience 7, 428–432.

Jovani, R., Mavor, R., 2011. Group size versus individual group size frequency distributions: a nontrivial distinction. Anim. Behav. 82, 1027–1036.

Jovani, R., Serrano, D., Ursua, E., Tella, J.L., 2008. Truncated power laws reveal a link between low-level behavioural processes and grouping patterns in a colonial bird. PLoS One 3, e1992.

Judd, T.M., Sherman, P.W., 1996. Naked mole-rats recruit colony mates to food sources. Anim. Behav. 52, 957–969.

Jullien, M., Clobert, J., 2000. The survival value of flocking in neotropical birds: reality or fiction? Ecology 81, 3416–3430.

Jędrzejewski, W., Jędrzejewska, B., Okarma, H., Ruprecht, A.L., 1992. Wolf predation and snow cover as mortality factors in the ungulate community of the Bialowieża National Park, Poland. Oecologia 90, 27–36.

Kaby, U., Lind, J., 2003. What limits predator detection in blue tits (Parus caeruleus): posture, task or orientation? Behav. Ecol. Sociobiol. 54, 534–538.

Kahlert, H., Fox, A.D., Ettrup, H., 1996. Nocturnal feeding in moulting greylag geese Anser anser—an anti-predator response. Ardea 84, 15–22.

Källander, H., 2008. Flock-fishing in the great crested grebe Podiceps cristatus. Ardea 96, 125–128.

Kamilar, J.M., Bribiescas, R.G., Bradley, B.J., 2010. Is group size related to longevity in mammals? Biol. Lett. 6, 736–739.

Kastberger, G., Schmelzer, E., Kranner, I., 2008. Social waves in giant honeybees repel hornets. PLoS One 3, e3141.

Katsnelson, E., Motro, U., Feldman, M.W., Lotem, A., 2008. Early experience affects producer-scrounger foraging tendencies in the house sparrow. Anim. Behav. 75, 1465–1472.

Katsnelson, E., Motro, U., Feldman, M.W., Lotem, A., 2011. Individual-learning ability predicts social-foraging strategy in house sparrows. Proc. R. Soc. London Ser. B. 278, 582–589.

Katz, M.W., Abramsky, Z., Kotler, B.P., Rosenzweig, M.L., Alteshtein, O., Vasserman, G., 2013. Optimal foraging of little egrets and their prey in a foraging game in a patchy environment. Am. Nat. 181, 381–395.

Kent, R., Holzman, R., Genin, A., 2006. Preliminary evidence on group-size dependent feeding success in the damselfish Dascyllus marginatus. Mar. Ecol. Prog. Ser. 323, 299–303.

Kenward, R.E., 1978. Hawks and doves: factors affecting success and selection in goshawk attacks on woodpigeons. J. Anim. Ecol. 47, 449–460.

King, G.F., 2011. Venoms as a platform for human drugs: translating toxins into therapeutics. Expert Opin. Biol. Ther. 11, 1469–1484.

King, A.J., Isaac, N.J.B., Cowlishaw, G., 2009. Ecological, social, and reproductive factors shape producer-scrounger dynamics in baboons. Behav. Ecol. 20, 1039–1049.

King, A.J., Wilson, A.M., Wilshin, S.D., Lowe, J., Haddadi, H., Hailes, S., Morton, A.J., 2012. Selfish-herd behaviour of sheep under threat. Curr. Biol. 22, R561–R562.

Koboroff, A., Kaplan, G., Rogers, L.J., 2008. Hemispheric specialization in Australian magpies (*Gymnorhina tibicen*) shown as eye preference during response to a predator. Brain Res. Bull. 76, 304–306.

Koops, M.A., Giraldeau, L.-A., 1996. Producer-scrounger foraging games in starlings: a test of risk-maximizing and risk-sensitive models. Anim. Behav. 51, 773–783.

Kotler, B.P., Brown, J., Mukherjee, S., Berger-Tal, O., Bouskila, A., 2010. Moonlight avoidance in gerbils reveals a sophisticated interplay among time allocation, vigilance and state-dependent foraging. Proc. R. Soc. London Ser. B. 277, 1469–1474.

Krakauer, D.C., 1995. Groups confuse predators by exploiting conceptual bottlenecks: a connectionist model of the confusion effect. Behav. Ecol. Sociobiol. 36, 421–429.

Kramer, D.L., 1985. Are colonies supraoptimal groups? Anim. Behav. 33, 1031–1032.

Krause, J., 1993. The effect of 'Schreckstoff' on shoaling behaviour of the minnow—a test of Hamilton's selfish herd theory. Anim. Behav. 45, 1019–1024.

Krause, J., 1994a. Differential fitness returns in relation to spatial position in groups. Biol. Rev. 69, 187–206.

Krause, J., 1994b. The influence of food competition and predation risk on size-assortative shoaling in juvenile chub (*Leuciscus cephalus*). Ethology 96, 105–116.

Krause, J., Godin, J.G.J., 1994. Shoal choice in the banded killifish (*Fundulus diaphanus*, Teleostei, Cyprinodontidae): effects of predation risk, fish size, species composition and size of shoals. Ethology 98, 128–136.

Krause, J., Godin, J.G.J., 1995. Predator preferences for attacking particular prey group sizes: consequences for predator hunting success and prey predation risk. Anim. Behav. 50, 465–473.

Krause, J., Godin, J.G.J., 1996a. Influence of prey foraging posture on flight behavior and predation risk: predators take advantage of unwary prey. Behav. Ecol. 7, 264–271.

Krause, J., Godin, J.G.J., 1996b. Influence of parasitism on shoal choice in the banded killifish (*Fundulus diaphanus*, Teleostei, Cyprinodontidae). Ethology 102, 40–49.

Krause, J., Tegeder, R.W., 1994. The mechanism of aggregation behaviour in fish shoals: individuals minimise approach time to neighbours. Anim. Behav. 48, 353–359.

Krause, J., Butlin, R.K., Peuhkuri, N., Pritchard, V.L., 2000a. The social organization of fish shoals: a test of the predictive power of laboratory experiments for the field. Biol. Rev. 75, 477–501.

Krause, J., Hoare, D.J., Croft, D., Lawrence, J., Ward, A., Ruxton, G.D., Godin, J.G.J., James, R., 2000b. Fish shoal composition: mechanisms and constraints. Proc. R. Soc. London Ser. B. 267, 2011–2017.

Krebs, J.R., 1973. Social learning and the adaptive significance of mixed-species flocks of chickadees. Can. J. Zool. 51, 1275–1288.

Krebs, J.R., 1974. Colonial nesting and social feeding as strategies for exploiting food resources in the great blue heron (*Ardea herodias*). Behaviour 51, 99–134.

Krebs, J.R., McRoberts, M.H., Cullen, J.M., 1972. Flocking and feeding in the great tit *Parus major*—an experimental study. Ibis 114, 507–530.

Kruuk, H., 1972. The Spotted Hyena. University of Chicago Press, Chicago.

Külling, D., Milinski, M., 1992. Size-dependent predation risk and partner quality in predator inspection of sticklebacks. Anim. Behav. 44, 949–955.

Kurvers, R.H.J.M., Prins, H.H.T., van Wieren, S.E., van Oers, K., Nolet, B.A., Ydenberg, R.C., 2010. The effect of personality on social foraging: shy barnacle geese scrounge more. Proc. R. Soc. London Ser. B. 277, 601–608.

Kushlan, J.A., 1978. Nonrigorous foraging by robbing egrets. Ecology 59, 649–653.

Křivan, V., 2007. The Lotka–Volterra predator–prey model with foraging-predation risk trade-offs. Am. Nat. 170, 771–782.

Lahti, D.C., Johnson, N.A., Ajie, B.C., Otto, S.P., Hendry, A.P., Blumstein, D.T., Coss, R.G., Donohue, K., Foster, S.A., 2009. Relaxed selection in the wild. Trends Ecol. Evol. 24, 487–496.

Lajeunesse, M.J., 2009. Meta-analysis and the comparative phylogenetic method. Am. Nat. 174, 369–381.

Laland, K.N., Boogert, N.J., 2008. Niche construction, co-evolution and biodiversity. Ecol. Econ. 69, 731–736.

Lamprecht, J., 1978. The relationship between food competition and foraging group size in some larger carnivores. Z. Tierpsychol. 46, 337–343.

Landeau, L., Terborgh, J., 1986. Oddity and the 'confusion effect' in predation. Anim. Behav. 34, 1372–1380.

Lanham, E.J., Bull, C.M., 2004. Enhanced vigilance in groups in *Egernia stokesii*, a lizard with stable social aggregations. J. Zool. 263, 95–99.

Lau, J., Ioannidis, J.P.A., Terrin, N., Schmid, C.H., Olkin, I., 2006. The case of the misleading funnel plot. Br. Med. J. 333, 597–600.

Lavalli, K.L., Herrnkind, W.F., 2009. Collective defense by spiny lobster (*Panulirus argus*) against triggerfish (*Balistes capriscus*): effects of number of attackers and defenders. N. Z. J. Mar. Freshwater Res. 43, 15–28.

Lavalli, K.L., Spanier, E., 2001. Does gregariousness function as an antipredator mechanism in the Mediterranean slipper lobster, *Scyllarides latus*? Mar. Freshwater Res. 52, 1133–1143.

Lawrence, E.S., 1985. Vigilance during 'easy' and 'difficult' foraging tasks. Anim. Behav. 33, 1373–1375.

Lazarus, J., 1979. The early warning function of flocking in birds: an experimental study with captive quelea. Anim. Behav. 27, 855–865.

LeBaron, G.S., Heppner, F.H., 1985. Food theft in the presence of abundant food in herring gulls. Condor 87, 430–431.

Lehmann, J., Dunbar, R.I.M., 2009. Network cohesion, group size and neocortex size in female-bonded old world primates. Philos. Trans. R. Soc. London Ser. B. 276, 4417–4422.

Lendrem, D.W., 1983. Predation risk and vigilance in the blue tit (*Parus caeruleus*). Behav. Ecol. Sociobiol. 14, 9–13.

Lendrem, D.W., 1984. Flocking, feeding and predation risk: absolute and instantaneous feeding rates. Anim. Behav. 32, 298–299.

Lendrem, D.W., Stretch, D., Metcalfe, N.B., Jones, P., 1986. Scanning for predators in the purple sandpiper: a time-dependent or time-independent process? Anim. Behav. 34, 1577–1578.

Lendvai, A.Z., Barta, Z., Liker, A., Bokony, V., 2004. The effect of energy reserves on social foraging: hungry sparrows scrounge more. Proc. R. Soc. London Ser. B. 271, 2467–2472.

Lendvai, A.Z., Liker, A., Barta, Z., 2006. The effects of energy reserves and dominance on the use of social-foraging strategies in the house sparrow. Anim. Behav. 72, 747–752.

Leuthold, W., 1977. African Ungulates: A Comparative Review of Their Ethology and Behavioral Ecology. Springer-Verlag, New York.

Li, Z., Jiang, Z.G., Beauchamp, G., 2010. Nonrandom mixing between groups of Przewalski's gazelle and Tibetan gazelle. J. Mammal. 91, 674–680.

Li, C., Jiang, Z., Li, L., Li, Z., Fang, H., Li, C., Beauchamp, G., 2012. Effects of reproductive status, social rank, sex and group size on vigilance patterns in Przewalski's gazelle. PLoS One 7, e32607.

Ligon, J.D., Burt, D.B., 2004. Evolutionary origins. In: Koenig, W.D., Dickinson, J.L. (Eds.), Ecology and Evolution of Cooperative Breeding in Birds, Cambridge University Press, Cambridge, pp. 5–34.

Liker, A., Barta, Z., 2002. The effects of dominance on social foraging tactic use in house sparrows. Behaviour 139, 1061–1076.

Lima, S.L., 1987a. Distance to cover, visual obstructions, and vigilance in house sparrows. Behaviour 102, 231–238.

Lima, S.L., 1987b. Vigilance while feeding and its relation to the risk of predation. J. Theor. Biol. 124, 303–316.

Lima, S.L., 1990. The influence of models on the interpretation of vigilance. In: Bekoff, M., Jamieson, D. (Eds.), Interpretation and Explanation in the Study of Animal Behavior, Explanation, Evolution and Adaptation, vol. 2. Westview Press, Boulder, pp. 246–267.

Lima, S.L., 1994a. Collective detection of predatory attack by birds in the absence of alarm signals. J. Avian Biol. 25, 319–326.

Lima, S.L., 1994b. On the personal benefits of anti-predatory vigilance. Anim. Behav. 48, 734–736.

Lima, S.L., 1995a. Back to the basics of anti-predatory vigilance: the group-size effect. Anim. Behav. 49, 11–20.

Lima, S.L., 1995b. Collective detection of predatory attack by social foragers: fraught with ambiguity? Anim. Behav. 50, 1097–1108.

Lima, S.L., 2002. Putting predators back into behavioral predator–prey interactions. Trends Ecol. Evol. 17, 70–75.

Lima, S.L., Bednekoff, P.A., 1999. Back to the basics of antipredatory vigilance: can nonvigilant animals detect attack? Anim. Behav. 58, 537–543.

Lima, S.L., Bednekoff, P.A., 2011. On the perception of targeting by predators during attacks on socially feeding birds. Anim. Behav. 82, 535–542.

Lima, S.L., Zollner, P.A., 1996. Anti-predatory vigilance and the limits to collective detection— visual and spatial separation between foragers. Behav. Ecol. Sociobiol. 38, 355–363.

Lima, S.L., Zollner, P.A., Bednekoff, P.A., 1999. Predation, scramble competition, and the vigilance group size effect in dark-eyed juncos (*Junco hyemalis*). Behav. Ecol. Sociobiol. 46, 110–116.

Lind, J., Cresswell, W., 2005. Determining the fitness consequences of antipredation behavior. Behav. Ecol. 16, 945–956.

Lindstedt, C., Mappes, J., Paivinen, J., Varama, M., 2006. Effects of group size and pine defence chemicals on Diprionid sawfly survival against ant predation. Oecologia 150, 519–526.

Lindstedt, C., Huttunen, H., Kakko, M., Mappes, J., 2010. Disentangling the evolution of weak warning signals: high detection risk and low production costs of chemical defences in gregarious pine sawfly larvae. Evol. Ecol. 25, 1029–1046.

Lindstrom, A., 1989. Finch flock size and risk of hawk predation at a migratory stopover site. Auk 106, 225–232.

Lindström, L., Alatalo, R.V., Lyytinen, A., Mappes, J., 2001. Strong antiapostatic selection against novel rare aposematic prey. Proc. Natl. Acad. Sci. 98, 9181–9184.

Lingle, S., 2001. Anti-predator strategies and grouping patterns in white-tailed deer and mule deer. Ethology 107, 295–314.

Lipsey, M.W., Wilson, D.B., 2001. Practical Meta-Analysis. Sage, Beverly Hills.

Lloyd, M., 1967. Mean crowding. J. Anim. Ecol. 36, 1–30.

Lott, D.F., 1991. Intraspecific Variation in the Social Systems of Wild Vertebrates. Cambridge University Press, Cambridge.

Loughry, W.J., 1992. Ontogeny of time allocation in black-tailed prairie dogs. Ethology 90, 206–224.

Low, C., 2008. Grouping increases visual detection risk by specialist parasitoids. Behav. Ecol. 19, 532–538.

Low, C., Scheffer, S.J., Lewis, M.L., Gates, M.W., 2012. The relationship between variable host grouping and functional responses among parasitoids of *Antispila nysaefoliella* (Lepidoptera: Heliozelidae). Mol. Ecol. 21, 5892–5904.

Lucas, E., Brodeur, J., 2001. A fox in sheep's clothing: furtive predators benefit from the communal defense of their prey. Ecology 82, 3246–3250.

Lührs, M.L., Dammhahn, M., Kappeler, P., 2013. Strength in numbers: males in a carnivore grow bigger when they associate and hunt cooperatively. Behav. Ecol. 24, 21–28.

Lukoschek, V., McCormick, M.I., 2002. A Review of Multi-Species Foraging Associations in Fishes and Their Ecological Significance. 9th International Coral Reef Symposium, Bali 467–474.

Lung, M.A., Childress, M.J., 2007. The influence of conspecifics and predation risk on the vigilance of elk (*Cervus elaphus*) in Yellowstone National Park. Behav. Ecol. 18, 12–20.

MacDonald, D.W., 1983. The ecology of carnivore social behaviour. Nature 301, 379–384.

Machado, G., Bonato, V., Oliveira, P., 2002. Alarm communication: a new function for the scent-gland secretion in harvestmen (Arachnida: Opiliones). Naturwissenschaften 89, 357–360.

MacLean, E.L., Matthews, L.J., Hare, B.A., Nunn, C.L., Anderson, R.C., Aureli, F., Brannon, E.M., Call, J., Drea, C.M., Emery, N.J., Haun, D.B.M., Herrmann, E., Jacobs, L.F., Platt, M.L., Rosati, A.G., Sandel, A.A., Schroepfer, K.K., Seed, A.M., Tan, J., van Schaik, C.P., Wobber, V., 2012. How does cognition evolve? Phylogenetic comparative psychology. Anim. Cogn. 15, 223–238.

Magurran, A.E., 1986. Predator inspection behaviour in minnow shoals: differences between populations and individuals. Behav. Ecol. Sociobiol. 19, 267–273.

Magurran, A.E., 1990. The adaptive significance of schooling as antipredator defence in fish. Ann. Zool. Fenn. 27, 51–66.

Magurran, A.E., Higham, A., 1988. Information transfer across fish shoals under predatory threat. Ethology 78, 153–158.

Magurran, A.E., Oulton, W.J., Pitcher, T.J., 1985. Vigilant behaviour and shoal size in minnows. Z. Tierpsychol. 67, 167–178.

Mangel, M., 1990. Resource divisibility, predation and group formation. Anim. Behav. 39, 1163–1172.

Manor, R., Saltz, D., 2003. Impact of human nuisance disturbance on vigilance and group size of a social ungulate. Ecol. Appl. 13, 1830–1834.

Mappes, J., Alatalo, R.V., 1997. Effects of novelty and gregariousness in survival of aposematic prey. Behav. Ecol. 8, 174–177.

Marras, S., Batty, R.S., Domenici, P., 2012. Information transfer and antipredator maneuvers in schooling herring. Adapt. Behav. 20, 44–56.

Marshall, H.H., Carter, A.J., Rowcliffe, J.M., Cowlishaw, G., 2012. Linking social foraging behaviour with individual time budgets and emergent group-level phenomena. Anim. Behav. 84, 1295–1305.

Martella, M.B., Renison, D., Navarro, J.L., 1995. Vigilance in the greater rhea: effect of vegetation height and group size. J. Field Ornithol. 66, 215–220.

Martín, J., Luque-Larena, J.J., Lopez, P., 2006. Collective detection in escape responses of temporary groups of Iberian green frogs. Behav. Ecol. 17, 222–226.

Martínez, A.E., Gomez, J.P., 2013. Are mixed-species bird flocks stable through two decades? Am. Nat. 181, E53–E59.

Marzluff, J.M., Heinrich, B., 1992. Foraging by common ravens in the presence and absence of territory holders: an experimental analysis of social foraging. Anim. Behav. 42, 755–770.

Marzluff, J.M., Heinrich, B., Marzluff, C.S., 1996. Raven roosts are mobile information centres. Anim. Behav. 51, 89–103.

Le Masurier, A.D., 1994. Costs and benefits of egg clustering in *Pieris brassicae*. J. Anim. Ecol. 63, 677–685.

Mather, J.A., 2010. Vigilance and antipredator responses of Caribbean reef squid. Mar. Freshwater Behav. Physiol. 43, 357–370.

Mathis, A., Chivers, D.P., 2003. Overriding the oddity effect in mixed-species aggregations: group choice by armored and nonarmored species. Behav. Ecol. 14, 334–339.

Mathis, A., Chivers, D.P., Smith, R.J.F., 1996. Cultural transmission of predator recognition in fishes: intraspecific and interspecific learning. Anim. Behav. 51, 185–201.

Mathot, K.J., Giraldeau, L.-A., 2010a. Family-related differences in social foraging tactic use in the zebra finch (*Taeniopygia guttata*). Behav. Ecol. Sociobiol. 64, 1805–1811.

Mathot, K.J., Giraldeau, L.-A., 2010b. Within-group relatedness can lead to higher levels of exploitation: a model and empirical test. Behav. Ecol. 21, 843–850.

Mathot, K.J., van den Hout, P.J., Piersma, T., Kempenaers, B., Reale, D., Dingemanse, N.J., 2011. Disentangling the roles of frequency- vs. state-dependence in generating individual differences in behavioural plasticity. Ecol. Lett. 14, 1254–1262.

Maynard Smith, J., 1974. The theory of games and the evolution of animal conflicts. J. Theor. Biol. 47, 209–221.

Maynard Smith, J., Harper, D., 2003. Animal Signals. Oxford University Press, Oxford.

McComb, K., Packer, C., Pusey, A.E., 1994. Roaring and numerical assessment in contests between groups of female lions, *Panthera leo*. Anim. Behav. 47, 379–384.

McCormack, J.E., Jablonski, P.G., Brown, J.L., 2007. Producer-scrounger roles and joining based on dominance in a free-living group of Mexican jays (*Aphelocoma ultramarina*). Behaviour 144, 967–982.

McCrate, A.T., Uetz, G.W., 2010. Kleptoparasites: a twofold cost of group living for the colonial spider, *Metepeira incrassata* (Araneae, Araneidae). Behav. Ecol. Sociobiol. 64, 389–399.

McNamara, J.M., Houston, A.I., 1992. Evolutionarily stable levels of vigilance as a function of group size. Anim. Behav. 43, 641–658.

McVean, A., Haddlesey, P., 1980. Vigilance schedules among house sparrows *Passer domesticus*. Ibis 122, 533–536.

Metcalfe, N.B., 1984. The effects of visibility on the vigilance of shorebirds: is visibility important? Anim. Behav. 32, 981–985.

Michelena, P., Deneubourg, J.L., 2011. How group size affects vigilance dynamics and time allocation patterns: the key role of imitation and tempo. PLoS One 6, e18631.

Milinski, M., 1977a. Do all members of a swarm suffer the same predation? Z. Tierpsychol. 45, 373–388.

Milinski, M., 1977b. Experiments on the selection by predators on spatial oddity of their prey. Z. Tierpsychol. 43, 311–325.

Milinski, M., 1987. Tit-for-tat in sticklebacks and the evolution of cooperation. Nature 325, 433–437.

Milinski, M., Lüthi, J.H., Eggler, R., Parker, G.A., 1997. Cooperation under predation risk: experiments on costs and benefits. Proc. R. Soc. London Ser. B. 264, 831–837.

Miller, R.C., 1922. The significance of the gregarious habit. Ecology 3, 122–126.

Minderman, J., Lind, J., Cresswell, W., 2006. Behaviourally mediated indirect effects: interference competition increases predation mortality in foraging redshanks. J. Anim. Ecol. 75, 713–723.

Mitchell, W.A., 2009. Multi-behavioral strategies in a predator–prey game: an evolutionary algorithm analysis. Oikos 118, 1073–1083.

Mitchell, W.A., Lima, S.L., 2002. Predator–prey shell games: large-scale movement and its implications for decision-making by prey. Oikos 99, 249–259.

Mock, D.W., Lamey, T.C., Thompson, D.B.A., 1988. Falsifiability and the information centre hypothesis. Ornis Scand. 19, 231–248.

Moland, E., Eagle, J.V., Jones, G.P., 2005. Ecology and evolution of mimicry in coral reef fishes. In: Gibson, R.N. (Ed.), Oceanography and Marine Biology: An Annual Review, Aberdeen University Press, Aberdeen, pp. 455–482.

Monus, F., Barta, Z., 2008. The effect of within-flock spatial position on the use of social foraging tactics in free-living tree sparrows. Ethology 114, 215–222.

Moore, B.A., Doppler, M., Young, J.E., Fernández-Juricic, E., 2013. Interspecific differences in the visual system and scanning behavior of three forest passerines that form heterospecific flocks. J. Comp. Physiol. A. 199, 263–277.

Mooring, M.S., Hart, B.L., 1992. Animal grouping for protection from parasites: selfish herd and encounter-dilution effects. Behaviour 123, 173–193.

Morand-Ferron, J., Giraldeau, L.-A., 2010. Learning behaviorally stable solutions to producer-scrounger games. Behav. Ecol. 21, 343–348.

Morand-Ferron, J., Lefebvre, L., Reader, S.M., Sol, D., Elvin, S., 2004. Dunking behaviour in carib grackles. Anim. Behav. 68, 1267–1274.

Morand-Ferron, J., Giraldeau, L.-A., Lefebvre, L., 2007. Wild carib grackles play a producer scrounger game. Behav. Ecol. 18, 916–921.

Morand-Ferron, J., Varennes, E., Giraldeau, L.-A., 2011. Individual differences in plasticity and sampling when playing behavioural games. Proc. R. Soc. London Ser. B. 278, 1223–1230.

Morgan, M.J., Godin, J.G.J., 1985. Antipredator benefits of schooling behaviour in a cyprinodontid fish, the bandit killifish (*Fundulus diaphanus*). Z. Tierpsychol. 70, 236–246.

Morrell, L.J., Ruxton, G.D., James, R., 2011a. Spatial positioning in the selfish herd. Behav. Ecol. 22, 16–22.

Morrell, L.J., Ruxton, G.D., James, R., 2011b. The temporal selfish herd: predation risk while aggregations form. Proc. R. Soc. London Ser. B. 278, 605–612.

Morse, D.H., 1970. Ecological aspects of some mixed-species foraging flocks of birds. Ecol. Monogr. 40, 119–168.

Morton, T.L., Haefner, J.W., Nugala, V., Decino, R.D., Mendes, L., 1994. The selfish herd revisited: do simple movement rules reduce relative predation risk? J. Theor. Biol. 167, 73–79.

Mosser, A., Packer, C., 2009. Group territoriality and the benefits of sociality in the African lion, *Panthera leo*. Anim. Behav. 78, 359–370.

Mottley, K., Giraldeau, L.-A., 2000. Experimental evidence that group foragers can converge on predicted producer-scrounger equilibria. Anim. Behav. 60, 341–350.

Moynihan, M., 1962. The organization and probable evolution of some mixed species flocks of neotropical birds. Smithson. Misc. Collect. 143, 1–140.

Moynihan, M., 1968. Social mimicry: character convergence versus character displacement. Evolution 22, 315–331.

Mueller, H.C., 1975. Hawks select odd prey. Science 188, 953–954.

Mukherjee, S., Heithaus, M.R., 2013. Dangerous prey and daring predators: a review. Biol. Rev. 88, 550–563.

Munn, C.A., 1986. Birds that 'cry wolf'. Nature 319, 143–145.

Murie, A., 1944. The Wolves of Mount McKinley, Fauna of the National Parks of the United States. US National Park Service, Washington.

Møller, A.P., Jennions, M.D., 2002. How much variance can be explained by ecologists and evolutionary biologists? Oecologia 132, 492–500.

Møller, A.P., Thornhill, R., Gangestad, S.W., 2005. Direct and indirect tests for publication bias: asymmetry and sexual selection. Anim. Behav. 70, 497–506.

Nakagawa, S., Santos, E., 2012. Methodological issues and advances in biological meta-analysis. Evol. Ecol. 26, 1253–1274.

Nakayama, S., Masuda, R., Tanaka, M., 2007. Onsets of schooling behavior and social transmission in chub mackerel *Scomber japonicus*. Behav. Ecol. Sociobiol. 61, 1383–1390.

Neill, S.R.S.J., Cullen, J.M., 1974. Experiments on whether schooling by their prey affects the hunting behaviour of cephalopods and fish predators. J. Zool. 172, 549–569.

Neuhäuser, M., Kotzmann, J., Walier, M., Poulin, R., 2010. The comparison of mean crowding between two groups. J. Parasitol. 96, 477–481.

Niemitz, C., 2010. The evolution of the upright posture and gait—a review and a new synthesis. Naturwissenschaften 97, 241–263.

Norconk, M.A., 1990. Introductory remarks: ecological and behavioral correlates of polyspecific primate troops. Am. J. Primatol. 21, 81–85.

Norris, K.S., Dohl, T.P., 1980. The structure and function of cetacean schools. In: Herman, L.M. (Ed.), Cetacean Behavior: Mechanisms and Functions, Wiley, New York, pp. 211–261.

Odling-Smee, F.J., Laland, K.N., Feldman, M.W., 2003. Niche Construction. Princeton University Press, Princeton.

Ohguchi, O., 1981. Prey density and selection against oddity by three-spined sticklebacks. Z. Tierpsychol. 23, 1–79.

Okabe, A., Boots, B.N., Sugihara, K., Chiu, S.N., 1992. Spatial Tessellations: Concept and Applications of Voronoi Diagrams. John Wiley and Sons, New York.

Öst, M., Tierala, T., 2011. Synchronized vigilance while feeding in common eider brood-rearing coalitions. Behav. Ecol. 22, 378–384.

Owens, I.P.F., 2006. Where is behavioural ecology going? Trends Ecol. Evol. 21, 356–361.

Packer, C., Abrams, P., 1990. Should co-operative groups be more vigilant than selfish groups? J. Theor. Biol. 142, 341–357.

Packer, C., Caro, T.M., 1997. Foraging costs in social carnivores. Anim. Behav. 54, 1317–1318.

Packer, C., Scheel, D., Pusey, A.E., 1990. Why lions form groups: food is not enough. Am. Nat. 136, 1–19.

Pagel, M., 1998. Inferring evolutionary processes from phylogenies. Zool. Scr. 26, 331–348.

Pagel, M., Dawkins, M.S., 1997. Peck orders and group size in laying hens—futures contracts for non-aggression. Behav. Proc. 40, 13–25.

Palleroni, A., Miller, C.T., Hauser, M., Marler, P., 2005. Predation—prey plumage adaptation against falcon attack. Nature 434, 973–974.

Pappano, D.J., Snyder-Mackler, N., Bergman, T.J., Beehner, J.C., 2012. Social 'predators' within a multilevel primate society. Anim. Behav. 84, 653–658.

Parker, G.A., 1984. Evolutionarily stable strategies. In: Krebs, J.R., Davies, N.B. (Eds.), Behavioural Ecology, Blackwell Scientific Publications, Oxford, pp. 30–61.

Parker, G.A., Sutherland, W.J., 1986. Ideal free distributions when individuals differ in competitive ability: phenotype-limited ideal free models. Anim. Behav. 34, 1222–1242.

Pays, O., Goulard, M., Blomberg, S.P., Goldizen, A.W., Sirot, E., Jarman, P.J., 2009. The effect of social facilitation on vigilance in the eastern gray kangaroo, *Macropus giganteus*. Behav. Ecol. 20, 469–477.

Pays, O., Blomberg, S.P., Renaud, P.C., Favreau, F.R., Jarman, P.J., 2010. How unpredictable is the individual scanning process in socially foraging mammals? Behav. Ecol. Sociobiol. 64, 443–454.

Pays, O., Sirot, E., Fritz, H., 2012. Collective vigilance in the greater kudu: towards a better understanding of synchronization patterns. Ethology 118, 1–9.

Penney, H.D., Hassall, C., Skevington, J.H., Abbott, K.R., Sherratt, T.N., 2012. A comparative analysis of the evolution of imperfect mimicry. Nature 483, 461–464.

Pereira, A.G., Cruz, A., Lima, S.Q., Moita, M.A., 2012. Silence resulting from the cessation of movement signals danger. Curr. Biol. 22, R627–R628.

Peres, C.A., 1996. Food patch structure and plant resource partitioning in interspecific associations of Amazonian tamarins. Int. J. Primatol. 17, 695–723.

Pérez-Barberia, F.J., Shultz, S., Dunbar, R.I.M., 2007. Evidence for coevolution of sociality and relative brain size in three orders of mammals. Evolution 61, 2811–2821.

Périquet, S., Valeix, M., Loveridge, A.J., Madzikanda, H., Macdonald, D.W., Fritz, H., 2010. Individual vigilance of African herbivores while drinking: the role of immediate predation risk and context. Anim. Behav. 79, 665–671.

Périquet, S., Todd-Jones, L., Valeix, M., Stapelkamp, B., Elliot, N., Wijers, M., Pays, O., Fortin, D., Madzikanda, H., Fritz, H., Macdonald, D.W., Loveridge, A.J., 2012. Influence of immediate predation risk by lions on the vigilance of prey of different body size. Behav. Ecol. 23, 970–976.

Pescador-Rubio, A., 2009. Growth and survival of a tropical polyphagous caterpillar: effects of host and group size. Southwest. Entomol. 34, 75–84.

Pettifor, R.A., 1990. The effects of avian mobbing on a potential predator, the European kestrel, *Falco tinnunculus*. Anim. Behav. 39, 821–827.

Pitcher, T.J., Parrish, J.K., 1993. Function of shoaling behavior in teleosts. In: Pitcher, T.J. (Ed.), Behavior of Teleost Fishes, Chapman and Hall, New York, pp. 363–439.

Pitcher, T.J., Magurran, A.E., Winfield, I.J., 1982. Fish in larger shoals find food faster. Behav. Ecol. Sociobiol. 10, 149–151.

Pitcher, T.J., Green, D.A., Magurran, A.E., 1986. Dicing with death: predator inspection behaviour in minnow shoals. J. Fish Biol. 28, 439–448.

Pitman, R.L., Durban, J.W., 2012. Cooperative hunting behavior, prey selectivity and prey handling by pack ice killer whales (*Orcinus orca*), type B, in Antarctic Peninsula waters. Mar. Mammal Sci. 28, 16–36.

Pitman, R.L., Ballance, L.T., Mesnick, S.I., Chivers, S.J., 2001. Killer whale predation on sperm whales: observations and implications. Mar. Mammal Sci. 17, 494–507.

Poiani, A., 1991. Anti-predator behavior in the bell miner *Manorina melanophrys*. Emu 91, 164–171.

Pontzer, H., Kamilar, J.M., 2009. Great ranging associated with greater reproductive investment in mammals. Proc. Natl. Acad. Sci. 106, 192–196.

Popp, J.M., 1988. Scanning behaviour of finches in mixed-species groups. Condor 90, 510–512.

Porter, L.M., 2001. Benefits of polyspecific associations for the Goeldi's monkey (*Callimico goeldii*). Am. J. Primatol 54, 143–158.

Powell, G.V.N., 1974. Experimental analysis of the social value of flocking by starlings (*Sturnus vulgaris*) in relation to predation and foraging. Anim. Behav. 22, 501–505.

Powell, G.V.N., 1985. Sociobiology and the adaptive significance of interspecific flocks in the neotropics. Ornithol. Monogr. 36, 713–732.

Pöysä, H., 1987. Feeding-vigilance trade-off in the teal (*Anas crecca*): effects of feeding method and predation risk. Behaviour 103, 108–121.

Pozzi, L., Gamba, M., Giacoma, C., 2010. The use of artificial neural networks to classify primate vocalizations: a pilot study on black lemurs. Am. J. Primatol. 72, 337–348.

Prins, H.H.T., Ydenberg, R., Drent, R., 1980. The interaction of brent geese *Branta bernicla* and sea plantain *Plantago maritima* during spring staging: field observations and experiments. Acta Bot. Neerl. 29, 585–596.

Procaccini, A., Orlandi, A., Cavagna, A., Giardina, I., Zoratto, F., Santucci, D., Chiarotti, F., Hemelrijk, C.K., Alleva, E., Parisi, G., Carere, C., 2011. Propagating waves in starling, *Sturnus vulgaris*, flocks under predation. Anim. Behav. 82, 759–765.

Proctor, C.J., Broom, M., Ruxton, G.D., 2001. Modelling antipredator vigilance and flight response in group foragers when warning signals are ambiguous. J. Theor. Biol. 211, 409–417.

Pulliam, H.R., 1973. On the advantages of flocking. J. Theor. Biol. 38, 419–422.

Pulliam, H.R., Caraco, T., 1984. Living in groups: is there an optimal group size? In: Krebs, J.R., Davies, N.B. (Eds.), Behavioural Ecology, Blackwell Scientific Publications, Oxford, pp. 122–147.

Pulliam, H.R., Millikan, G.C., 1982. Social organization in the non-reproductive period. In: Farner, D.S., King, J.R., Parkes, K.C. (Eds.), Avian Biology, Academic Press, New York, pp. 169–197.

Pulliam, H.R., Pyke, G.H., Caraco, T., 1982. The scanning behavior of juncos: a game-theoretical approach. J. Theor. Biol. 95, 89–103.

Pycraft, W.P., 1910. A History of Birds. Methuen, London.

Quenette, P.Y., 1990. Functions of vigilance in mammals: a review. Acta Oecol. 11, 801–818.

Quinn, J.L., Cresswell, W., 2004. Predator hunting behaviour and prey vulnerability. J. Anim. Ecol. 73, 143–154.

Quinn, J.L., Cresswell, W., 2005. Escape response delays in wintering redshank, *Tringa totanus*, flocks: perceptual limits and economic decisions. Anim. Behav. 69, 1285–1292.

Quinn, J.L., Cresswell, W., 2006. Testing domains of danger in the selfish herd: sparrowhawks target widely spaced redshanks in flocks. Proc. R. Soc. London Ser. B. 273, 2521–2526.

Quinn, J.L., Ueta, M., 2008. Protective nesting associations in birds. Ibis 150, 146–167.

Quinn, D., Mott, R., Bollinger, E., Switzer, P., 2012. Size assortment in mixed-species groups of juvenile-phase striped parrotfish (*Scarus iserti*) in The Bahamas. Ichthyol. Res. 59, 212–215.

Radford, A.N., 2003. Territorial vocal rallying in the green woodhoopoe: influence of rival group size and composition. Anim. Behav. 66, 1035–1044.

Radford, A.N., Du Plessis, M.A., 2004. Territorial vocal rallying in the green woodhoopoe: factors affecting contest length and outcome. Anim. Behav. 68, 803–810.

Radford, A.N., Ridley, A.R., 2007. Individuals in foraging groups may use vocal cues when assessing their need for anti-predator vigilance. Biol. Lett. 3, 249–252.

Radford, A.N., Hollen, L.I., Bell, M.B.V., 2009. The higher the better: sentinel height influences foraging success in a social bird. Philos. Trans. R. Soc. London Ser. B. 276, 2437–2442.

Radford, A.N., Bell, M.B.V., Hollen, L.I., Ridley, A.R., 2011. Singing for your supper: sentinel calling by kleptoparasites can mitigate the cost to victims. Evolution 65, 900–906.

Randall, J.A., 2001. Evolution and function of drumming as communication in mammals. Am. Zool. 41, 1143–1156.

Randler, C., 2005a. Coots *Fulica atra* reduce their vigilance under increased competition. Behav. Proc. 68, 173–178.

Randler, C., 2005b. Vigilance during preening in coots *Fulica atra*. Ethology 111, 169–178.

Rands, S.A., 2010. Self-improvement for team-players: the effects of individual effort on aggregated group information. PLoS One 5, e11705.

Ranta, E., Rita, H., Lindstrom, K., 1993. Competition versus cooperation: success of individuals foraging alone and in groups. Am. Nat. 142, 42–58.

Ranta, E., Peuhkuri, N., Laurila, A., 1994. A theoretical exploration of antipredatory and foraging factors promoting phenotype-assorted fish schools. Ecoscience 1, 99–106.

Ranta, E., Peuhkuri, N., Laurila, A., Rita, H., Metcalfe, N.B., 1996. Producers, scroungers and foraging group structure. Anim. Behav. 51, 171–175.

Rasa, O.A.E., 1983. Dwarf mongoose and hornbill mutualism in the Taru Desert, Kenya. Behav. Ecol. Sociobiol. 12, 181–190.

Rasmussen, J.B., Downing, J.A., 1988. The spatial response of chironomid larvae to the predatory leech *Nephelopsis obscura*. Am. Nat. 131, 14–21.

Rattenborg, N.C., Lima, S.L., Amlaner, C.J., 1999. Half-awake to the risk of predation. Nature 397, 397–398.

Reader, T., Hochuli, D.F., 2003. Understanding gregariousness in a larval Lepidopteran: the roles of host plant, predation, and microclimate. Ecol. Entomol. 28, 729–737.

Reiczigel, J., Lang, Z., Rózsa, L., Tothmeresz, B., 2008. Measures of sociality: two different views of group size. Anim. Behav. 75, 715–721.

Reluga, T.C., Viscido, S., 2005. Simulated evolution of selfish herd behavior. J. Theor. Biol. 234, 213–225.

Repasky, R.P., 1996. Using vigilance behaviour to test whether predation promotes habitat partitioning. Ecology 77, 1880–1887.

Reynolds, A.M., Sword, G.A., Simpson, S.J., Reynolds, D.R., 2009. Predator percolation, insect outbreaks, and phase polyphenism. Curr. Biol. 19, 20–24.

Richner, H., Heeb, P., 1995. Is the information centre hypothesis a flop? Adv. Study Behav. 24, 1–45.

Richner, H., Heeb, P., 1997. Communal life: honest signaling and the recruitment center hypothesis. Behav. Ecol. 7, 115–118.

Ridgway, S., Carder, D., Finneran, J., Keogh, M., Kamolnick, T., Todd, M., Goldblatt, A., 2006. Dolphin continuous auditory vigilance for five days. J. Exp. Biol. 209, 3621–3628.

Ridley, A.R., Raihani, N.J., 2007. Facultative response to a kleptoparasite by the cooperatively breeding pied babbler. Behav. Ecol. 18, 324–330.

Rieucau, G., Giraldeau, L.-A., 2009a. Group size effect caused by food competition in nutmeg mannikins (*Lonchura punctulata*). Behav. Ecol. 20, 421–425.

Rieucau, G., Giraldeau, L.-A., 2009b. Video playback and social foraging: simulated companions produce the group size effect in nutmeg mannikins. Anim. Behav. 78, 961–966.

Rieucau, G., Morand-Ferron, J., Giraldeau, L.-A., 2010. Group size effect in nutmeg mannikin: between-individuals behavioral differences but same plasticity. Behav. Ecol. 21, 684–689.

Rieucau, G., Blanchard, P., Martin, J.G.A., Favreau, F.R., Goldizen, A.W., Pays, O., 2012. Investigating differences in vigilance tactic use within and between the sexes in eastern grey kangaroos. PLoS One 7, e44801.

Riipi, M., Alatalo, R.V., Lindstrom, L., Mappes, J., 2001. Multiple benefits of gregariousness cover detectability costs in aposematic aggregations. Nature 413, 512–514.

Ritchie, J., 1932. Systematic 'beating' by herons. Br. Birds 25, 228.

Ritz, D.A., 1994. Social aggregations in pelagic invertebrates. Adv. Mar. Biol. 30, 155–216.

Ritz, D.A., Hobday, A.J., Montgomery, J.C., Ward, A.J.W., 2011. Social aggregation in the pelagic zone with special reference to fish and invertebrates. Adv. Mar. Biol. 60, 161–227.

Roberts, G., 1995. A real-time response of vigilance behaviour to changes in group size. Anim. Behav. 50, 1371–1374.

Roberts, G., 1996. Why individual vigilance declines as group size increases. Anim. Behav. 51, 1077–1086.

Robertson, D.R., Sweatman, H.P.A., Fletcher, E.A., Clelland, M.G., 1976. Schooling as a mechanism for circumventing the territoriality of competitors. Ecology 57, 1208–1220.

Robinette, R.L., Ha, J.C., 2001. Social and ecological factors influencing vigilance by northwestern crows, *Corvus caurinus*. Anim. Behav. 62, 447–452.

Robinson, S.K., 1985. Coloniality in the yellow-rumped cacique as a defense against nest predators. Auk 102, 506–519.

Rode, N.O., Lievens, E.J.P., Flaven, E., Segard, A., Jabbour-Zahab, R., Sanchez, M.I., Lenormand, T., 2013. Why join groups? Lessons from parasite-manipulated Artemia. Ecol. Lett. 16, 493–501.

Rodgers, G., Kimbell, H., Morrell, L.J., 2013. Mixed-phenotype grouping: the interaction between oddity and crypsis. Oecologia 172, 59–68.

Rodman, P.S., 1981. Inclusive fitness and group size, with a reconsideration of group sizes of lions and wolves. Am. Nat. 118, 275–283.

Rodriguez-Girones, M.A., Vasquez, R.A., 2002. Evolutionary stability of vigilance coordination among social foragers. Proc. R. Soc. London Ser. B. 269, 1803–1810.

Rohwer, S.A., 1975. The social significance of avian winter plumage variability. Evolution 29, 593–610.

Rohwer, S., Butcher, G.S., 1988. Winter versus summer explanations of delayed plumage maturation in temperate birds. Am. Nat. 131, 556–572.

Rohwer, S., Ewald, P.W., 1981. The cost of dominance and the advantage of subordination in a badge-signalling system. Evolution 35, 441–454.

Romey, W.L., Walston, A.R., Watt, P.J., 2008. Do 3-D predators attack the margins of 2-D selfish herds? Behav. Ecol. 19, 74–78.

Ross, E.J., Deeming, D.C., 1998. Feeding and vigilance behaviour of breeding ostriches (*Struthio camelus*) in a farming environment in Britain. Br. Poult. Sci. 39, 173–177.

Roth, T.C., Lima, S.L., 2007. Use of prey hotspots by an avian predator: purposeful unpredictability? Am. Nat. 169, 264–273.

Roth, T.C., Cox, J.G., Lima, S.L., 2008. The use and transfer of information about predation risk in flocks of wintering finches. Ethology 114, 1218–1226.

Rowcliffe, J.M., Watkinson, A.R., Sutherland, W.J., Vickery, J.A., 1995. Cyclic winter grazing patterns in brent geese and the regrowth of salt-marsh grass. Funct. Ecol. 9, 931–941.

Ruxton, G.D., 1995. Foraging in flocks—non-spatial models may neglect important costs. Ecol. Model. 82, 277–285.

Ruxton, G.D., Beauchamp, G., 2008. The application of genetic algorithms in behavioural ecology, illustrated with a model of anti-predator vigilance. J. Theor. Biol. 250, 435–448.

Ruxton, G.D., Roberts, G., 1999. Are vigilance sequences a consequence of intrinsic chaos or external changes? Anim. Behav. 57, 493–495.

Ruxton, G.D., Hall, S.J., Gurney, W.S.C., 1995. Attraction toward feeding conspecifics when food patches are exhaustible. Am. Nat. 145, 653–660.

Ruxton, G.D., Sherratt, T.N., Speed, M.P., 2004. Avoiding Attack: The Evolutionary Ecology of Crypsis, Warning Signals and Mimicry. Oxford University Press, Oxford.

Ruxton, G.D., Jackson, A.L., Tosh, C.R., 2007. Confusion of predators does not rely on specialist coordinated behavior. Behav. Ecol. 18, 590–596.

Ryan, P.G., Edwards, L., Pichegru, L., 2012. African penguins Spheniscus demersus, bait balls and the Allee effect. Ardea 100, 89–94.

Ryer, C.H., Olla, B.L., 1992. Social mechanisms facilitating exploitation of spatially variable ephemeral food patches in a pelagic marine fish. Anim. Behav. 44, 69–74.

Rypstra, A.L., 1989. Foraging success of solitary and aggregated spiders: insights into flock formation. Anim. Behav. 37, 274–281.

Safranyk, L., Wilson, B., 2007. The Mountain Pine Beetle: A Synthesis of Biology, Management and Impacts on Lodgepole Pine. Canadian Forest Service, Victoria.

Saino, N., Fasola, M., Waiyaki, E., 1995. Do white pelicans *Pelecanus onocrotalus* benefit from foraging in flocks using synchronous feeding? Ibis 137, 227–230.

Sansom, A., Cresswell, W., Minderman, J., Lind, J., 2008. Vigilance benefits and competition costs in groups: do individual redshanks gain an overall foraging benefit? Anim. Behav. 75, 1869–1875.

Sazima, C., Jordano, P., Guimaraes, P.R., Dos Reis, S.F., Sazima, I., 2012. Cleaning associations between birds and herbivorous mammals in Brazil: structure and complexity. Auk 129, 36–43.

Scannell, J., Roberts, G., Lazarus, J., 2001. Prey scan at random to evade observant predators. Proc. R. Soc. London Ser. B. 268, 541–547.

Schaller, G.B., 1972. The Serengeti Lion. University of Chicago Press, Chicago.

Scheel, D., 1993. Watching for lions in the grass: the usefulness of scanning and its effects during hunts. Anim. Behav. 46, 695–704.

Scheel, D., Packer, C., 1991. Group hunting behaviour of lions: a search for cooperation. Anim. Behav. 41, 697–709.

Schlupp, I., Marler, C., Ryan, M.J., 1994. Benefit to male sailfin mollies of mating with heterospecific females. Science 263, 373–374.

Schmidt, P.A., Mech, L.D., 1997. Wolf pack size and food acquisition. Am. Nat. 150, 513–517.

Schmitt, R.J., Strand, S.W., 1982. Cooperative foraging by yellowtail, *Seriola lalandi* (Carangidae) on two species of fish prey. Copeia, 714–717.

Schoener, T.W., 1971. Theory of feeding strategies. Ann. Rev. Ecol. Syst. 2, 369–404.

Schuelke, O., 2001. Social anti-predator behaviour in a nocturnal lemur. Folia Primatol. 72, 332–334.

Seeley, T.D., 1985. Honeybee Ecology: A Study of Adaptation in Social Life. Princeton University Press, Princeton.

Selman, J., Goss-Custard, J.D., 1988. Interference between foraging redshanks *Tringa totanus*. Anim. Behav. 36, 1542–1544.

Selva, N., Drzejewska, B., Drzejewski, W., Wajrak, A., 2005. Factors affecting carcass use by a guild of scavengers in European temperate woodland. Can. J. Zool. 83, 1590–1601.

Semeniuk, C.A.D., Dill, L.M., 2005. Cost/benefit analysis of group and solitary resting in the cowtail stingray, *Pastinachus sephen*. Behav. Ecol. 16, 417–426.

Semeniuk, C.A.D., Dill, L.M., 2006. Anti-predator benefits of mixed-species groups of cowtail stingrays (*Pastinachus sephen*) and whiprays (*Himantura uarnak*) at rest. Ethology 112, 33–43.

Senar, J.C., 2006. Color displays as intrasexual signals of aggression and dominance. In: Hill, G.E., McGraw, K.J. (Eds.), Bird Coloration, Harvard University Press, Cambridge, pp. 87–136.

Senar, J.C., Camerino, M., 1998. Status signalling and the ability to recognize dominants: an experiment with siskins (*Carduelis spinus*). Proc. R. Soc. London Ser. B. 265, 1515–1520.

Seppä, T., Laurila, A., Peuhkuri, N., Piironen, J., Lower, N., 2001. Early familiarity has fitness consequences for Arctic char (*Salvelinus alpinus*) juveniles. Can. J. Fish. Aquat. Sci. 58, 1380–1385.

Seppänen, J.T., Forsman, J.T., Mönkönnen, M., Thomson, R.L., 2007. Social information use is a process across time, space and ecology, reaching heterospecifics. Ecology 88, 1622–1633.

Shattuck, M.R., Williams, S.A., 2010. Arboreality has allowed for the evolution of increased longevity in mammals. Proc. Natl. Acad. Sci. 107, 4635–4639.

Shaw, J.J., Tregenza, T., Parker, G.A., Harvey, I.F., 1995. Evolutionarily stable foraging speeds in feeding scrambles: a model and an experimental test. Proc. R. Soc. London Ser. B. 260, 273–277.

Sherman, P.W., 1977. Nepotism and the evolution of alarm calls. Science 197, 1246–1253.

Shi, J.B., Beauchamp, G., Dunbar, R.I.M., 2010. Group-size effect on vigilance and foraging in a predator-free population of feral goats (*Capra hircus*) on the Isle of Rum, NW Scotland. Ethology 116, 329–337.

Shima, J.S., 2001. Regulation of local populations of a coral reef fish via joint effects of density- and number-dependent mortality. Oecologia 126, 58–65.

Shultz, S., Dunbar, R.I.M., 2005. Both social and ecological factors predict ungulate brain size. Proc. R. Soc. London Ser. B. 273, 207–215.

Shultz, S., Dunbar, R.I.M., 2007. The evolution of the social brain: anthropoid primates contrast with other vertebrates. Proc. R. Soc. London Ser. B. 274, 2429–2436.

Sibly, R.M., 1983. Optimal group size is unstable. Anim. Behav. 31, 947–948.

Sih, A., 1987. Prey refuges and predator–prey stability. Theor. Popul. Biol. 31, 1–12.

Sih, A., 1998. Game theory and predator–prey response races. In: Dugatkin, L.A., Reeve, H.K. (Eds.), Game Theory and Animal Behavior, Oxford University Press, Oxford.

Sih, A., 2005. Predator–prey space use as an emergent outcome of a behavioral response race. In: Barbosa, P., Castellanos, I. (Eds.), Ecology of Predator–Prey Relationships, Oxford University Press, Oxford, pp. 241–255.

Silk, J.B., 2007. Social components of fitness in primate groups. Science 317, 1347–1351.

Sillen-Tullberg, B., 1988. Evolution of gregariousness in aposematic butterfly larvae: a phylogenetic analysis. Evolution 42, 293–305.

Similä, T., Ugarte, F., 1993. Surface and underwater observations of cooperatively feeding killer whales in northern Norway. Can. J. Zool. 71, 1494–1499.

Simons, A.M., 2004. Many wrongs: the advantage of group navigation. Trends Ecol. Evol. 19, 453–455.

Sirot, E., 2006. Social information, antipredatory vigilance and flight in bird flocks. Anim. Behav. 72, 373–382.

Sirot, E., 2012. Negotiation may lead selfish individuals to cooperate: the example of the collective vigilance game. Proc. R. Soc. London Ser. B. 279, 2862–2867.

Sirot, E., Pays, O., 2011. On the dynamics of predation risk perception for a vigilant forager. J. Theor. Biol. 276, 1–7.

Sirot, E., Touzalin, F., 2009. Coordination and synchronization of vigilance in groups of prey: the role of collective detection and predators' preference for stragglers. Am. Nat. 173, 47–59.

Skelhorn, J., Ruxton, G.D., 2006. Avian predators attack aposematic prey more forcefully when they are part of an aggregation. Biol. Lett. 2, 488–490.

Slaa, E.J., Wassenberg, J., Biesmeijer, J.C., 2003. The use of field-based social information in eusocial foragers: local enhancement among nestmates and heterospecifics in stingless bees. Ecol. Entomol. 28, 369–379.

Slotow, R., Coumi, N., 2000. Vigilance in bronze mannikin groups: the contributions of predation risk and intra-group competition. Behaviour 137, 565–578.

Smith, R.D., Ruxton, G.D., Cresswell, W., 2002. Do kleptoparasites reduce their own foraging effort in order to detect kleptoparasitic opportunities? An empirical test of a key assumption of kleptoparasitic models. Oikos 97, 205–212.

Smythe, N., 1970. On the existence of pursuit invitation signals in mammals. Am. Nat. 104, 491–494.

Sol, D., Garcia, N., Iwaniuk, A.N., Davis, K., Meade, A., Boyle, W.A., Szekely, T., 2010. Evolutionary divergence in brain size between migratory and resident birds. PLoS One 5, e9617.

Sonerud, G.A., Smedshaug, C.A., Brathen, O., 2001. Ignorant hooded crows follow knowledgeable roost-mates to food: support for the information centre hypothesis. Proc. R. Soc. London Ser. B. 268, 827–831.

Sorato, E., Gullett, P.R., Griffith, S.C., Russell, A.F., 2012. Effects of predation risk on foraging behaviour and group size: adaptations in a social cooperative species. Anim. Behav. 84, 823–834.

Spieler, M., Linsenmair, K.E., 1999. Aggregation behaviour of *Bufo maculatus* as an antipredator mechanism. Ethology 105, 665–686.

Sprague, A.J., Hamilton, D.J., Diamond, A.W., 2008. Site safety and food affect movements of semipalmated sandpipers (*Calidris pusilla*) migrating through the upper Bay of Fundy. Avian Conserv. Ecol. 3, 4.

Sridhar, H., Beauchamp, G., Shanker, K., 2009. Why do birds participate in mixed-species foraging flocks? A large-scale synthesis. Anim. Behav. 78, 337–347.

Sridhar, H., Srinivasan, U., Askins, R.A., Canales-Delgadillo, J.C., Chen, C.C., Ewert, D.N., Gale, G.A., Goodale, E., Gram, W.K., Hart, P.J., Hobson, K.A., Hutto, R.L., Kotagama, S.W., Knowlton, J.L., Lee, T.M., Munn, C.A., Nimnuan, S., Nizam, B.Z., Péron, G., Robin, V.V., Rodewald, A.D., Rodewald, P.G., Thomson, R.L., Trivedi, P., Wilgenburg, S.L.V., Shanker, K., 2012. Positive relationships between association strength and phenotypic similarity character-ize the assembly of mixed-species bird flocks worldwide. Am. Nat. 180, 777–790.

Srinivasan, U., Raza, R.H., Quader, S., 2010. The nuclear question: rethinking species importance in multi-species animal groups. J. Anim. Ecol. 79, 948–954.

Stahl, J., Tolsma, P.H., Loonen, M.J.J.E., Drent, R.H., 2001. Subordinates explore but dominants profit: resource competition in high Arctic barnacle goose flocks. Anim. Behav. 61, 257–264.

Stamp, N.E., 1981. Effect of group size on parasitism in a natural population of the Baltimore checkerspot, *Euphydryas phaeton*. Oecologia 49, 201–206.

Stander, P.E., 1992. Cooperative hunting in lions: the role of the individual. Behav. Ecol. Sociobiol. 29, 445–454.

Stanford, C.B., 1998. Chimpanzee and Red Colobus: The Ecology of Predator and Prey. Harvard University Press, Cambridge.

Stang, A.T., McRae, S.B., 2009. Why some rails have white tails: the evolution of white undertail plumage and anti-predator signaling. Evol. Ecol. 23, 943–961.

Stankowich, T., 2003. Marginal predation methodologies and the importance of predator prefer-ences. Anim. Behav. 66, 589–599.

Stankowich, T., Blumstein, D.T., 2005. Fear in animals: a meta-analysis and review of fear assess-ment. Proc. R. Soc. London Ser. B. 272, 2627–2634.

Stankowich, T., Coss, R., 2007. The re-emergence of felid camouflage with the decay of predator recognition in deer under relaxed selection. Proc. R. Soc. London Ser. B. 274, 175–182.

Stanley, C.R., Dunbar, R.I.M., 2013. Consistent social structure and optimal clique size revealed by social network analysis of feral goats, *Capra hircus*. Anim. Behav. 85, 771–779.

Steenbeek, R., Piek, R.C., van Buul, M., van Hooff, J.A.R.A.M., 1999. Vigilance in wild Thomas' langurs (*Presbytis thomasi*): the importance of infanticide risk. Behav. Ecol. Sociobiol. 45, 137–150.

Stenberg, M., Persson, A., 2005. The effects of spatial food distribution and group size on foraging behaviour in a benthic fish. Behav. Proc. 70, 41–50.

Stensland, E., Angerbjorn, A., Berggren, P., 2003. Mixed species groups in mammals. Mammal Rev. 33, 205–223.

Stephens, D.W., Krebs, J.R., 1986. Foraging Theory. Princeton University Press, Princeton.

Stillman, R.A., Goss-Custard, J.D., Caldow, R.W.G., 1997. Modelling interference from basic for-aging behaviour. J. Anim. Ecol. 66, 692–703.

Stillman, R.A., Goss-Custard, J.D., Alexander, M.J., 2000. Predator search pattern and the strength of interference through prey depression. Behav. Ecol. 11, 597–605.

Stringham, S.F., 2012. Salmon fishing by bears and the dawn of cooperative predation. J. Comp. Psychol. 126, 329–338.

Struhsaker, T.T., 1981. Polyspecific associations among tropical rain-forest primates. Z. Tierpsychol. 57, 268–304.

Studd, M., Montgomerie, R.D., Robertson, R.J., 1983. Group size and predator surveillance in foraging house sparrows (Passer domesticus). Can. J. Zool. 61, 226–231.

Sullivan, K.A., 1985. Vigilance patterns in downy woodpeckers. Anim. Behav. 33, 328–330.

Sullivan, K.A., 1988. Ontogeny of time budgets in yellow-eyed juncos: adaptation to ecological constraints. Ecology 69, 118–124.

Sumida, B., Houston, A.I., McNamara, J.M., Hamilton, W.D., 1990. Genetic algorithms and evolution. J. Theor. Biol. 147, 59–84.

Sumpter, D.J.T., 2010. Collective Animal Behavior. Princeton University Press, Princeton.

Sumpter, D.J.T., Krause, J., James, R., Couzin, I.D., Ward, A.J.W., 2008. Consensus decision making by fish. Curr. Biol. 18, 1773–1777.

Suter, R.B., Forrest, T.G., 1994. Vigilance in the interpretation of spectral analyses. Anim. Behav. 48, 223–225.

Sutherland, W.J., 1996. From Individual Behaviour to Population Ecology. Oxford University Press, Oxford.

Suzuki, T.N., 2012. Long-distance calling by the willow tit, Poecile montanus, facilitates formation of mixed-species foraging flocks. Ethology 118, 10–16.

Swynnerton, C.F.M., 1915. Mixed bird-parties. Ibis 67, 346–357.

Szulkin, M., Dawidowicz, P., Dodson, S.I., 2006. Behavioural uniformity as a response to cues of predation risk. Anim. Behav. 71, 1013–1019.

Tamura, N., 1989. Snake-directed mobbing by the Formosan squirrel Callosciurus erythraeus thaiwanensis. Behav. Ecol. Sociobiol. 24, 175–180.

Tan, K., Wang, Z., Li, H., Yang, S., Hu, Z., Kastberger, G., Oldroyd, B.P., 2012. An 'I see you' prey–predator signal between the Asian honeybee, Apis cerana, and the hornet, Vespa velutina. Anim. Behav. 83, 879–882.

Tania, N., Vanderlei, B., Heath, J.l.P., Edelstein-Keshet, L., 2012. Role of social interactions in dynamic patterns of resource patches and forager aggregation. Proc. Natl. Acad. Sci. 109, 11228–11233.

Taylor, R.J., 1984. Predation. Chapman & Hall, New York.

Teichroeb, J.A., Sicotte, P., 2012. Cost-free vigilance during feeding in folivorous primates? Examining the effect of predation risk, scramble competition, and infanticide threat on vigilance in ursine colobus monkeys (Colobus vellerosus). Behav. Ecol. Sociobiol. 66, 453–466.

Terborgh, J.W., 1983. Five New World Primates. Princeton University Press, Princeton.

Terborgh, J., 1990. Mixed flocks and polyspecific associations: costs and benefits of mixed groups to birds and monkeys. Am. J. Primatol. 21, 87–100.

Theodorakis, C.W., 1989. Size segregation and the effects of oddity on predation risk in minnow schools. Anim. Behav. 38, 496–502.

Thiollay, J.M., 1999. Frequency of mixed species flocking in tropical forest birds and correlates of predation risk: an intertropical comparison. J. Avian Biol. 30, 282–294.

Thiollay, J.M., Jullien, M., 1998. Flocking behaviour of foraging birds in a neotropical rain forest and the antipredator defence hypothesis. Ibis 140, 382–394.

Thompson, D.B.A., Barnard, C.J., 1983. Anti-predator responses in mixed species flocks of lapwings, golden plovers and gulls. Anim. Behav. 31, 585–593.

Thompson, D.B.A., Thompson, M.L.P., 1985. Early warning and mixed species association: the 'Plover's Page' revisited. Ibis 127, 559–562.

Thorpe, W.H., 1956. Learning and Instinct in Animals. Hazell, Watson and Viney Ltd, London.

Tickell, W.L.N., 2003. White plumage. Waterbirds 26, 1–12.

Tisdale, V., Fernández-Juricic, E., 2009. Vigilance and predator detection vary between avian species with different visual acuity and coverage. Behav. Ecol. 20, 936–945.

Tórrez, L., Robles, N., González, A., Crofoot, M.C., 2012. Risky business? Lethal attack by a jaguar sheds light on the costs of predator mobbing for capuchins (*Cebus capucinus*). Int. J. Primatol. 33, 440–446.

Tosh, C.R., Ruxton, G.D., 2007. The need for stochastic replication of ecological neural networks. Philos. Trans. R. Soc. London Ser. B. 362, 455–460.

Tosh, C.R., Ruxton, G.D., 2010. Modelling Perception with Artificial Neural Networks. Cambridge University Press, Cambridge.

Tosh, C.R., Jackson, A.L., Ruxton, G.D., 2006. The confusion effect in predatory neural networks. Am. Nat. 167, E52–E65.

Tosh, C., Jackson, A., Ruxton, G., 2007. Individuals from different-looking animal species may group together to confuse shared predators: simulations with artificial neural networks. Proc. R. Soc. London Ser. B. 274, 827–832.

Toth, Z., Bokony, V., Lendvai, A.Z., Szabo, K., Penzes, Z., Liker, A., 2009. Effects of relatedness on social-foraging tactic use in house sparrows. Anim. Behav. 77, 337–342.

Treherne, J.E., Foster, W.A., 1981. Group transmission of predator avoidance behaviour in a marine insect: the trafalgar effect. Anim. Behav. 29, 911–917.

Treves, A., 1997. Vigilance and use of micro-habitat in solitary rainforest mammals. Mammalia 61, 511–525.

Treves, A., 1998. The influence of group size and neighbors on vigilance in two species of arboreal monkeys. Behaviour 135, 1–29.

Treves, A., 1999. Within-group vigilance in red colobus and redtail monkeys. Am. J. Primatol. 48, 113–126.

Treves, A., 2000. Theory and method in studies of vigilance and aggregation. Anim. Behav. 60, 711–722.

Tripp, K.J., Collazo, J.A., 1997. Non-breeding territoriality of semipalmated sandpipers. Wilson Bull. 109, 630–642.

Tristram, H.B., 1859. On the ornithology of northern Africa. Ibis 1.

Turchin, P., Kareiva, P., 1989. Aggregation in *Aphis varians*: an effective strategy for reducing predation risk. Ecology 70, 1008–1016.

Turner, G.F., Pitcher, T.J., 1986. Attack abatement: a model for group protection by combined avoidance and dilution. Am. Nat. 128, 228–240.

Uetz, G.W., 1989. The "ricochet effect" and prey capture in colonial spiders. Oecologia 81, 154–159.

Uetz, G.W., 1992. Foraging strategies of spiders. Trends Ecol. Evol. 7, 155–159.

Uetz, G.W., Hieber, C.S., 1994. Group size and predation risk in colonial web-building spiders: analysis of attack abatement mechanisms. Behav. Ecol. 5, 326–333.

Uetz, G.W., Boyle, J., Hieber, C.S., Wilcox, R.S., 2002. Antipredator benefits of group living in colonial web-building spiders: the 'early warning' effect. Anim. Behav. 63, 445–452.

Ulbrich, K., Henschel, J.R., 1999. Intraspecific competition in a social spider. Ecol. Model. 115, 243–251.

Unglaub, B., Ruch, J., Herberstein, M.E., Schneider, J.M., 2013. Hunted hunters? Effect of group size on predation risk and growth in the Australian subsocial crab spider *Diaea ergandros*. Behav. Ecol. Sociobiol. 67, 785–794.

Vahl, W.K., van der Meer, J., Weissing, F.J., van Dullemen, D., Piersma, T., 2005. The mechanisms of interference competition: two experiments on foraging waders. Behav. Ecol. 16, 845–855.

Valone, T.J., 1989. Group foraging, public information, and patch estimation. Oikos 56, 357–363.

van der Veen, I.T., 2002. Seeing is believing: information about predators influences yellowhammer behavior. Behav. Ecol. Sociobiol. 51, 466–471.

van Eerden, M.R., Voslamber, B., 1995. Mass fishing by cormorants *Phalacrocorax carbo sinensis* at Lake Ijsselmeer, The Netherlands: a recent and successful adaptation to a turbid environment. Ardea 83, 199–212.

van Schaik, C.P., 1983. Why are diurnal primates living in groups? Behaviour 87, 120–144.

van Schaik, C.P., van Hooff, J., 1983. On the ultimate causes of primate social systems. Behaviour 85, 91–117.

van Schaik, C.P., van Noordwijk, M.A., 1986. The evolutionary effect of the absence of felids on the social organization of the Simeulue monkey (*Macaca fascicularis* fusca). Folia Primatol. 44, 138–147.

VanderWaal, K.L., Mosser, A., Packer, C., 2009. Optimal group size, dispersal decisions and post-dispersal relationships in female African lions. Anim. Behav. 77, 949–954.

Vaughn, R., Würsig, B., Packard, J., 2010. Dolphin prey herding: prey ball mobility relative to dolphin group and prey ball sizes, multispecies associates, and feeding duration. Mar. Mammal Sci. 26, 213–225.

Vehrencamp, S.L., 1983. Optimal degree of skew in cooperative societies. Amer. Zool. 23, 327–335.

Vickery, W.L., Giraldeau, L.-A., Templeton, J.J., Kramer, D.L., Chapman, C.A., 1991. Producers, scroungers and group foraging. Am. Nat. 137, 847–863.

Videan, E.N., McGrew, W.C., 2002. Bipedality in chimpanzee (*Pan troglodytes*) and bonobo (*Pan paniscus*): testing hypotheses on the evolution of bipedalism. Am. J. Phys. Anthropol. 118, 184–190.

Vine, I., 1971. Risk of visual detection and pursuit by a predator and the selective advantage of flocking behaviour. J. Theor. Biol. 30, 405–422.

Vine, I., 1973. Detection of prey flocks by predators. J. Theor. Biol. 40, 207–210.

Viscido, S.V., 2003. The case for the selfish herd hypothesis. Comm. Theor. Biol. 8, 665–684.

Viscido, S.V., Wethey, D.S., 2002. Quantitative analysis of fiddler crab flock movement: evidence for 'selfish herd' behaviour. Anim. Behav. 63, 735–741.

Viscido, S.V., Miller, M., Wethey, D.S., 2001. The response of a selfish herd to an attack from outside the group perimeter. J. Theor. Biol. 208, 315–328.

Viscido, S.V., Miller, M., Wethey, D.S., 2002. The dilemma of the selfish herd: the search for a realistic movement rule. J. Theor. Biol. 217, 183–194.

von Frisch, K., 1967. The Dance Language and Orientation of Bees. Harvard University Press, Cambridge.

Vucetich, J.A., Peterson, R.O., Waite, T.A., 2004. Raven scavenging favours group foraging in wolves. Anim. Behav. 67, 1117–1126.

Vulinec, K., Miller, M.C., 1989. Aggregation and predator avoidance in whirligig beetles (Coleoptera: Gyrinidae). J. N. Y. Entomol. Soc. 97, 438–447.

Wallace, D.J., Greenberg, D.S., Sawinski, J., Rulla, S., Notaro, G., Kerr, J.N.D., 2013. Rats maintain an overhead binocular field at the expense of constant fusion. Nature 498, 65–69.

Wang, Z., Li, Z.Q., Beauchamp, G., Jiang, Z.G., 2011. Flock size and human disturbance affect vigilance of endangered red-crowned cranes (*Grus japonensis*). Biol. Conserv. 144, 101–105.

Ward, P.I., 1985. Why birds in flocks do not co-ordinate their vigilance periods. J. Theor. Biol. 114, 383–385.

Ward, A.J.W., Hart, P.J., 2003. The effects of kin and familiarity on interactions between fish. Fish Fish. 4, 348–358.

Ward, A.J.W., Axford, S., Krause, J., 2002. Mixed-species shoaling in fish: the sensory mechanisms and costs of shoal choice. Behav. Ecol. Sociobiol. 52, 182–187.

Ward, A.J.W., Herbert-Read, J.E., Sumpter, D.J.T., Krause, J., 2011. Fast and accurate decisions through collective vigilance in fish shoals. Proc. Natl. Acad. Sci. 108, 2312–2315.

Ward, P., Zahavi, A., 1973. The importance of certain assemblages of birds as information-centres. Ibis 115, 517–534.

Waser, P.M., 1984. 'Chance' and mixed-species associations. Behav. Ecol. Sociobiol. 15, 197–202.

Watkins, J.L., Buchholz, F., Priddle, J., Morris, D.J., Ricletts, C., 1992. Variation in reproductive status of Antarctic krill swarms: evidence for a size-related sorting mechanism? Mar. Ecol. Prog. Ser. 82, 163–174.

Watson, M., Aebischer, N.J., Cresswell, W., 2007. Vigilance and fitness in grey partridges *Perdix perdix*: the effects of group size and foraging-vigilance trade-offs on predation mortality. J. Anim. Ecol. 76, 211–221.

Watt, D.J., Mock, D.W., 1987. A selfish herd of martins. Auk 104, 342–343.

Watt, P.J., Nottingham, S.F., Young, S., 1997. Toad tadpole aggregation behaviour: evidence for a predator avoidance function. Anim. Behav. 54, 865–872.

Watts, D.P., 1998. Long-term habitat use by mountain gorilla (*Gorilla gorilla* beringei). II. Reuse of foraging areas in relation to resource abundance, quality and depletion. Int. J. Primatol. 19, 681–702.

Wcislo, W.T., Danforth, B.N., 1997. Secondarily solitary: the evolutionary loss of social behavior. Trends Ecol. Evol. 12, 468–474.

Weatherhead, P.J., 1983. Two principal strategies in avian communal roosts. Am. Nat. 121, 237–243.

Webb, P.W., 1982. Avoidance responses of fathead minnow to strikes by four teleost predators. J. Comp. Physiol. A 147, 371–378.

Wey, T., Blumstein, D.T., Shen, W., Jordan, F., 2008. Social network analysis of animal behaviour: a promising tool for the study of sociality. Anim. Behav. 75, 333–344.

White, J.W., Warner, R.R., 2007. Behavioral and energetic costs of group membership in a coral reef fish. Oecologia 154, 423–433.

Whitehouse, M.E.A., Lubin, Y., 1999. Competitive foraging in the social spider *Stegodyphus dumicola*. Anim. Behav. 58, 677–688.

Whitesides, G.H., 1989. Interspecific associations of Diana monkeys, *Cercopithecus diana*, in Sierra Leone, West Africa, biological significance or chance? Anim. Behav. 37, 760–776.

Wiley, R.H., 1991. Both high- and low-ranking white-throated sparrows find novel locations of food. Auk 108, 8–15.

Williams, G.C., 1966. Adaptation and Natural Selection. Princeton University Press, Princeton.

Williams, C.K., Lutz, R.S., Applegate, R.D., 2003. Optimal group size and northern bobwhite coveys. Anim. Behav. 66, 377–387.

Willis, E.O., 1963. Is the zone-tailed hawk a mimic of the turkey vulture? Condor 65, 313–317.

Willis, E.O., 1966. Competitive exclusion and birds at fruiting trees in western Colombia. Auk 83, 479–480.

Willis, E.O., 1972. Do birds flock in Hawaii, a land without predators? Calif. Birds 3, 1–9.

Willis, E.O., 1976. Similarity of a tyrant-flycatcher and a silky-flycatcher: not all character convergence is competitive mimicry. Condor 74, 553.

Wilson, D.S., 1974. Prey capture and competition in the ant lion. Biotropica 6, 187–193.

Wilson, D.S., Clark, A.B., Coleman, K., Dearstyne, T., 1994. Shyness and boldness in humans and other animals. Trends Ecol. Evol. 9, 442–446.

Wilson, E.O., 1975. Sociobiology: The Modern Synthesis. Harvard University Press, Cambridge.

Wilson, W.G., Richards, S.A., 2000. Consuming and grouping: resource-mediated animal aggregation. Ecol. Lett. 3, 175–180.

Wise, M.J., Kieffer, D.L., Abrahamson, W.G., 2006. Costs and benefits of gregarious feeding in the meadow spittlebug, *Philaenus spumarius*. Ecol. Entomol. 31, 548–555.

Wittenberger, J.F., Hunt, G.L.J., 1985. The adaptive significance of coloniality in birds. In: Farner, D.S., King, J.R., Parkes, R.C. (Eds.), Avian Biology, vol. 8. Academic Press, New York, pp. 1–78.

Wolf, N.G., 1985. Odd fish abandon mixed-species groups when threatened. Behav. Ecol. Sociobiol. 17, 47–52.

Wolf, N., Mangel, M., 2007. Strategy, compromise, and cheating in predator–prey games. Evol. Ecol. Res. 9, 1293–1304.

Wolters, S., Zuberbuhler, K., 2003. Mixed-species associations of Diana and Campbell's monkeys: the costs and benefits of a forest phenomenon. Behaviour 140, 371–385.

Wood, A.J., Ackland, G.J., 2007. Evolving the selfish herd: emergence of distinct aggregating strategies in an individual-based model. Proc. R. Soc. London Ser. B. 274, 1637–1642.

Wood, J.B., Pennoyer, K.E., Derby, C.D., 2008. Ink is a conspecific alarm cue in the Caribbean reef squid, *Sepioteuthis sepioidea*. J. Exp. Mar. Biol. Ecol. 367, 11–16.

Wrangham, R.W., 1980. An ecological model of female-bonded primate groups. Behaviour 75, 262–300.

Wrangham, R.W., Gittleman, J.L., Chapman, C.A., 1993. Constraints on group size in primates and carnivores: population density and day-range as assays of exploitation competition. Behav. Ecol. Sociobiol. 32, 199–209.

Wrona, F.J., Dixon, R.W.J., 1991. Group size and predation risk: a field analysis of encounter and dilution effects. Am. Nat. 137, 186–201.

Wu, G.M., Giraldeau, L.-A., 2005. Risky decisions: a test of risk sensitivity in socially foraging flocks of *Lonchura punctulata*. Behav. Ecol. 16, 8–14.

Wynne-Edwards, V.C., 1962. Animal Dispersion in Relation to Social Behaviour. Oliver and Boyd, Edinburgh.

Yaber, M.C., Herrera, E.A., 1994. Vigilance, group size and social status in capybaras. Anim. Behav. 48, 1301–1307.

Ydenberg, R.C., Dill, L.M., 1986. The economics of fleeing from predators. Adv. Stud. Behav. 16, 229–249.

Yip, E.C., Powers, K.S., Aviles, L., 2008. Cooperative capture of large prey solves scaling challenge faced by spider societies. Proc. Natl. Acad. Sci. 105, 11818–11822.

Zajonc, R., 1965. Social facilitation. Science 149, 269–274.

Zemel, A., Lubin, Y., 1995. Inter-group competition and stable group sizes. Anim. Behav. 50, 485–488.

Zheng, W., Beauchamp, G., Jiang, X., Li, Z., Yang, Q., 2013. Determinants of vigilance in a reintroduced population of Père David's deer. Curr. Zool. 59, 265–270.

Index

Note: Page numbers followed by "f" denote figures; "t" tables; "b" boxes.

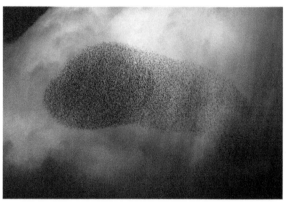

FIGURE 3.10 A murmuration of European starlings. Waves of turning in tight flocks of starlings propagate as waves, which are thought to deter predation from birds of prey. *Photo credit: Muffin.*

FIGURE 4.1 A meerkat on the lookout. *Photo credit: Kevin Ryder.*

FIGURE 8.1 Mixed-species groups in nature: examples of mixed-species groups composed of two species: (a) A cattle egret and the cow that provides flushed insects (*photo credit: Sarah and Iain*) and (b) Thomson's and Grant's gazelles (the larger of the two gazelle species) (*photo credit: NH53*). Which one represents a mixed-species group?

Model: Semipalmated
sandpiper

  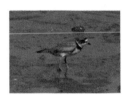

| Putative mimic | Ancestry control | Habitat control |
| Western sandpiper | Little stint | Semipalmated plover |

Score: 1   1   3

FIGURE 8.6 Mimicry and mixed-species flocking: a case of putative mimicry in a mixed-species flock that is probably due to shared ancestry. The top picture represents the model species. The bottom row of pictures contains, from left to right, the putative mimic, the sister species of the putative mimic, and a species that participates in the mixed-species flock with the model and putative mimic but to a lesser extent. Scores represent the match judged by a panel of blinded raters, with 1 being a close match and 4 a poor match. *Adapted from Beauchamp and Goodale (2011). Photo credits: Semipalmated sandpiper (Guy Beauchamp); Western sandpiper (Winnu); Semipalmated plover (Dendroicacerulea); Little stint (Omarrun).*